BRAIN MECHANISMS *and* PSYCHOTROPIC DRUGS

Pharmacology and Toxicology: Basic and Clinical Aspects

Mannfred A. Hollinger, Series Editor
University of California, Davis

Forthcoming Titles

BRAIN MECHANISMS *and* PSYCHOTROPIC DRUGS

Edited by
Andrius Baskys
Gary Remington

CRC Press
Boca Raton   New York   London   Tokyo

Library of Congress Cataloging-in-Publication Data

Brain mechanisms and psychotropic drugs / edited by Andrius Baskys and
    Gary Remington.
            p.   cm. -- (Pharmacology and toxicology)
        Includes bibliographical references and index.
        ISBN 0-8493-8386-2 (alk. paper)
        1. Psychotropic drugs--Physiological effect.
    2. Neuropsychopharmacology.   3. Neurotransmitters.   4. Brain--Effect
    of drugs on.   I. Baskys, Andrius.   II. Remington, Gary.
    III. Series: Pharmacology & toxicology (Boca Raton, Fla.)
        [DNLM: 1. Psychotropic Drugs--pharmacology.   2. Brain--drug
    effects.   3. Neurotransmitters--physiology.   QV 77.2 B814 1996]
    RM315.B657   1996
    615′.788--dc20
    DNLM/DLC
    for Library of Congress   95-47545
                                                                                                                CIP

ABSTRACT

This book provides a potential bridge over the existing gap between the experimental neurosciences and treatment strategies used in the clinical practice of psychopharmacology. Written by recognized experts in their fields, it integrates clinical psychopharmacology with basic neuroscience and offers the latest in treatment approaches for major psychiatric disorders. The text is divided into three major sections. The first two sections focus on the basic neurosciences, while the third section moves the reader into topics relevant to clinical practice. Topics covered in the first two sections include such fundamental concepts as ion channels, synapses, second messenger mechanisms, and the aging brain. The second section includes chapters on serotonin, dopamine, acetylcholine, GABA, glutamate, and the neuropeptides. The final section is clinically oriented and includes chapters on major psychotropic drug classes: antidepressants, neuroleptics, mood stabilizers, benzodiazepines, and cognition enhancing drugs. This book is for all those who are involved in clinical use of psychotropic medications including medical students, interns and residents, general practitioners, practicing psychiatrists, and researchers.

INTRODUCTION

Modern psychopharmacology dates back to the 1950s, a decade which witnessed the clinical development of neuroleptics, antidepressants, mood stabilizers, and anxiolytics. Striking in this process has been the role of serendipity, as the psychotropic actions of particular compounds were frequently identified only coincidentally. This is true, for example, with lithium, the neuroleptics, and monoamine oxidase inhibitors, and it is therefore not so surprising that the mechanisms by which these different compounds exerted their psychotropic effects were, at least initially, poorly understood. Indeed, elucidating the mechanisms of action for these drugs became the means by which our theories of pathophysiology regarding particular psychiatric disorders grew. In many cases, it was only in later years that the reverse occurred, whereby theories of pathophysiology offered the framework for developing new compounds.

Such theories have become increasingly complex in the face of technological advances and expanding knowledge, and the time when the biochemical underpinnings of major mental disorders were explained by one, or even several, neurotransmitters is long past. Moreover, changes at the basic science level now occur with increasing rapidity, making it difficult to integrate advances in knowledge with the ongoing practice of clinical psychopharmacology. Information from major textbooks on neurochemistry and psychopharmacology often demands a special training in these fields, hampering clinicians in their efforts to remain updated. Conversely, textbooks which focus on the treatment of mental disorders frequently fail to provide sufficiently detailed coverage of neuronal or molecular mechanisms accounting for drug actions.

This book was conceived as a potential bridge over the existing gap between experimental neurosciences and the clinical practice of psychopharmacology. The text is divided into three major sections: Part I. Basic Physiology; Part II. Neurotransmitters and Neuromodulators; and, Part III. Psychotropic Drugs. The first two sections focus on the basic sciences, while the third section moves the reader into topics relevant to clinical practice. In the first section, Chapters 1 to 3 introduce the reader to a number of fundamental concepts of neurophysiology: ion channels, synapses, second messenger mechanisms, and the aging brain. The second section, with its focus on neurotransmitters and neuromodulators, includes chapters reviewing serotonin and dopamine, acetylcholine, GABA, glutamate, and the neuropeptides. The final section, Psychotropic Drugs, is clinically oriented, with chapters specific to the major classes of psychotropic drugs: antidepressants, neuroleptics, mood stabilizers, and the benzodiazepines. In addition, a chapter has been devoted to the burgeoning field of nootropic and cognition enhancing drugs, which have received increased attention worldwide during recent years. While each of these chapters is clinically focused, there has been an effort to elucidate the mechanisms of action for each class of medications based on current knowledge provided through research in basic sciences and reviewed in the earlier sections.

For those involved in the practice of clinical psychopharmacology, it is essential that at least a fundamental understanding of how these drugs work be maintained if they are to be used optimally in the clinical setting. This is of more than heuristic value, as it can improve treatment strategies and is particularly relevant to issues of polypharmacy and potential drug interactions. To this end, the present book is for all those involved in the clinical use of psychotropic medications, including medical students, interns and residents, general practitioners, practicing psychiatrists, and researchers in the field of clinical psychopharmacology.

Andrius Baskys, M.D., Ph.D.
Gary Remington, M.D., Ph.D., F.R.C.P.(C).

THE EDITORS

Andrius Baskys, M.D., Ph.D., is an Associate Investigator with the Medical Research Council of Canada Group on Nerve Cells and Synapses at the Department of Physiology, University of Toronto.

Dr. Baskys graduated in 1975 from the Medical School of Vilnius University in Lithuania and received his Ph.D. degree from Moscow University in 1980. He did postdoctoral work at the University of Toronto and was appointed as an Assistant Professor at the Department of Psychology, University of Toronto. He worked as a visiting scientist at the School of Pharmacy, University of London, and the Department of Psychiatry, University of California, San Francisco. He has recently completed a Psychiatry residency program at the University of Toronto.

Dr. Baskys is a member of the American Society of Neuroscience, the Canadian Physiological Society, the International Brain Research Organisation, and the American Association for the Advancement of Science.

He has been a recipient of the Ontario Mental Health Foundation Fellowship in Alzheimer's Disease, the Ontario Alzheimer's Society Award, and awards from private industry. He has published over 30 research papers and 2 books, and has made over 40 presentations at national and international meetings. His current major research interests include mechanisms of glutamate-mediated synaptic transmission and psychopharmacology of schizophrenia and Alzheimer's disease.

Gary Remington, M.D., Ph.D., F.R.C.P.(C), Schizophrenia Research Program, Clarke Institute of Psychiatry, Toronto, Ontario, is Associate Professor, Department of Psychiatry, University of Toronto.

Dr. Remington completed his M.D. at McMaster University, Hamilton, Ontario, as well as a year of training in Internal Medicine at the University of Western Ontario, London, Ontario, before obtaining his fellowship in Psychiatry at the University of Toronto. His Ph.D. focused on the influence of catecholaminergic maturation on activity levels during neurodevelopment and the modulatory role of cholinergic mechanisms.

Dr. Remington is a member of the American College of Psychiatrists (Laughlin Fellow), Canadian College of Neuropsychopharmacology, Canadian Medical Association, Canadian Psychiatric Association, College of Physicians and Surgeons of Ontario, Ontario Psychatric Association, Royal College of Physicians and Surgeons of Canada, and Society for Neuroscience.

In 1984, Dr. Remington was awarded a Laughlin Fellowship by the American College of Psychiatrists. He has been appointed to a World Health Organization Task Force on neuroleptics, and is presently a member of the Psychopharmacology Panel of the Canadian Alliance for Research on Schizophrenia (CAROS).

Dr. Remington's research interests include the psychopharmacology of schizophrenia, with particular interest in novel neuroleptics and the evaluation of clinical efficacy and side effects using positron emission tomography (PET) to establish their mechanisms of action. He is currently a reviewer for 7 international journals, has published over 50 articles, and has presented approximately 125 invited lectures related to the pharmacotherapy of schizophrenia.

CONTRIBUTORS

Andrius Baskys, M.D., Ph.D.
Department of Physiology
University of Toronto
Toronto, Ontario, Canada

Guy Chouinard, M.D., M.Sc., F.R.C.P.(C)
Fernand-Sequin Research Center
Louis-H. Lafontaine Hospital
and
Department of Psychiatry
University of Montreal
Montreal, Quebec, Canada

M. Frances Davies, Ph.D.
Molecular Research Institute
Palo Alto, California

Gary M. Hasey, M.D., F.R.C.P.(C), M.Sc.
Department of Psychiatry
University of Toronto
Toronto, Ontario, Canada

Craig Hudson, M.D., F.R.C.P.(C)
Clarke Institute of Psychiatry
University of Toronto
Toronto, Ontario, Canada

Gwen O. Ivy, Ph.D.
Departments of Psychology and
  Anatomy and Cell Biology
University of Toronto
Scarborough, Ontario, Canada

Russell T. Joffe, M.D.
Department of Psychiatry
McMaster University
Hamilton, Ontario, Canada

Rifaat Kamil, M.D., C.M., F.R.C.P.(C)
Department of Psychiatry
and
Clarke Institute of Psychiatry
University of Toronto
Toronto, Ontario, Canada

Anna Lin, B.Sc.
Clarke Institute of Psychiatry
University of Toronto
Toronto, Ontario, Canada

John F. MacDonald, Ph.D.
Department of Physiology
University of Toronto
Toronto, Ontario, Canada

John Nelson, Ph.D.
Fernand-Seguin Research Center
Louis-H. Lafontaine Hospital
and
Department of Psychiatry
University of Montreal
Montreal, Quebec, Canada

Gary Remington, M.D., Ph.D., F.R.C.P.(C)
Clarke Institute of Psychiatry
University of Toronto
Toronto, Ontario, Canada

Heidi H. Swanson, Ph.D., D.Sc.
Department of Psychobiology
Division of Ethopharmacology
University of Seville
Seville, Spain

Manfred Windisch, Ph.D.
Department of Central Nervous System Pharmacology
EBEWE Pharmaceuticals
Unterach/Attersee, Austria

J. Martin Wojtowicz, Ph.D.
Department of Physiology
University of Toronto
Toronto, Ontario, Canada

L. Trevor Young, M.D., Ph.D.
Department of Psychiatry
McMaster University
Hamilton, Ontario, Canada

TABLE OF CONTENTS

Part I

BASIC PHYSIOLOGY

Chapter **1**

MEMBRANES, SYNAPSES, AND ION CHANNELS

J. Martin Wojtowicz

CONTENTS

0-8493-8386-0/96/$0.00+$.50

I. STRUCTURE AND FUNCTION OF NEURONS

It is estimated that billions of neurons are present in the mammalian brain. Together with an even larger number of glia cells, the neurons form a unique network of connections responsible for our actions and thoughts.

A reductionist approach to the understanding of the brain is to examine and understand its functional units, neurons, and glia. The glia cells play significant roles during development, maintenance of structure, and homeostasis within the nervous system. We will only consider neuronal cells in the topics described in this chapter. For information on glial functions see Reference 1.

A typical neuron is made up of a bushy dendritic tree, a cell body, and an axon. The cell body contains a nucleus and is the site where genetic information is stored and transcribed as need arises. Dendrites are the main sites of synaptic contacts. The shapes of dendrites vary among different neuronal types, presumably to provide for their different functional needs.

Examination of one neuronal type, the granule cell in the dentate gyrus of the hippocampus, can give a good idea of how dendrites receive and integrate synaptic input (Figure 1). It is estimated that approximately 5,000 to 10,000 synapses make contact with each granule cell. The majority of these synapses are excitatory, each providing a small increment of depolarizing influence on the resting membrane potential of the cell. By far the greatest number of these excitatory synapses is formed by axonal terminals of the perforant pathway, projecting into the dentate gyrus from the adjacent part of the cerebral cortex. It has been shown by anatomical and physiological methods that two major parts of the perforant pathway carry information conveyed from different places within the brain and target specific parts of the dendritic trees of granule neurons.[2] Hence, the location of synapses on dendrites is important.

The functional characteristics of individual synapses are also important. Basically, synapses can be either excitatory or inhibitory. Usually, the excitatory synapses by far outnumber the inhibitory synapses on neurons, and, as a rule, the excitatory ones are located on the whole extent of the dendritic branches, including the distant dendrites far from the cell body. Their individual contributions can add together, resulting in summated excitatory currents which travel toward the cell body (Figure 1). The inhibitory synapses are fewer in number but are strategically located near the cell body. At this location, the inhibitory synapses can block excitation by effectively shunting the excitatory currents and preventing them from reaching the cell body. At this particular neuron the dendritic currents are mainly passive. Action potentials arise at the axon hillock upon depolarization above a preset threshold for initiation of the event (Figure 1B). The balance of excitation and inhibition determines if and how many nerve impulses (action potentials) will be fired by the cell. The net effect of the integrated synaptic input is sent along the axon as a train of action potentials toward a target cell.

II. BASIC STRUCTURE OF IONIC CHANNELS

Understanding of various facets of neuronal function is based on our knowledge of ionic channels. The channels are large proteins that form ion-permeant pores in lipid membranes. Specialized, transmitter-sensitive receptor channels are present at the synaptic junctions. Other channels are located on dendrites, on cell bodies, and on axons. The action potentials are usually fired at the initial segment of the axon

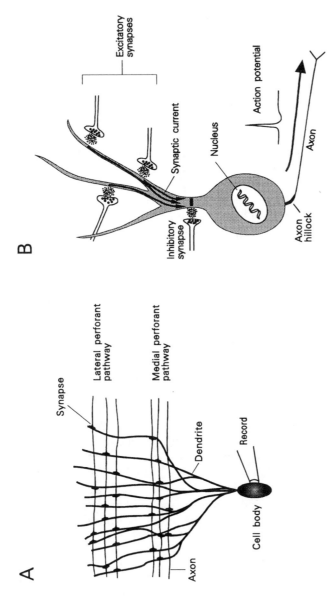

Figure 1

A. A single neuron, represented here by a dentate granule cell, receives thousands of synaptic contacts imparting either excitatory or inhibitory influence from many sources. In dentate gyrus (a part of the hippocampal formation), most of the synapses belong to lateral and medial perforant pathways. The two pathways innervate separate portions of the dendritic tree and carry different sensory signals. Information from many synaptic inputs is integrated in the cell body. The cell body but not the dendrites is easily accessible to electrophysiological recordings carried out by means of small glass pipettes attached to the membrane. **B.** Three excitatory synapses are shown to puff a small cloud of transmitter substance and produce excitatory currents (arrows), which flow toward the cell body. The current inside the nerve cells is carried primarily by K^+ ions. An inhibitory synapse releases another transmitter substance that shunts, as need arises, the excitatory current away from cell body. The horizontal bar at the level of the inhibitory synapse illustrates the inhibitory effect. Action potentials are the signals carrying information from the cell body toward a target cell.

(axon hillock) and rely on voltage-dependent sodium channels. Thanks to recent developments in molecular biology, these sodium channels have been cloned and studied in detail.[3] Expression of the channel proteins in artificial systems, such as frog oocytes, allows the investigator to alter the primary structure by changing the amino acid residues and to study the function of individual channels. These studies have led to partial understanding of the relationship between the molecular structure and the functional properties of the voltage-dependent sodium channels.

The secondary structure of the channel is partly understood, and certain functional features can be deduced from the properties of amino acids found in different parts of the protein molecule.[3] For example, the sodium voltage-gated channel structure indicates that certain regions have hydrophobic properties and consequently are likely to be found within the membrane. Other segments of the molecule have hydrophilic properties and are likely located on either side of the membrane. Four regions containing a large proportion of hydrophobic amino acids are thought to constitute the intramembranous domains labeled I to IV in Figure 2. Thus, the channel molecule is thought to be woven through the membrane. Consequently, other parts of the channel ought to be found either inside the cell or in the extracellular fluid.

The three-dimensional arrangement of the intramembranous domains of the channel in a functional, ion-conducting unit is only partly understood from X-ray and electron microscopy studies. Structural analysis indicates that the four domains come together to form a pore which allows sodium ions to pass in a regulated manner. The diameter of this pore, as well as the electrical charges of amino acids lining the pore, determines selectivity of the channel. It has been proposed that the walls of the pore are formed by segments S5 and S6 and that the negatively charged amino acids near segment S6 attract the positively charged sodium ions toward the pore (see Figure 2). Next to the pore are the positively charged segments S4, which act as voltage sensors. In response to a change in electrical potential across the membrane, the positively charged amino acids in the S4 segments are thought to move, thereby altering the arrangement of the adjacent amino acids forming the pore and increasing the diameter enough to allow ionic flux.

Certain amino acids, located on the extracellular surface of the channel, bind selectively to toxins such as saxitoxin (ScTx in Figure 2A). Toxins often block channels by impeding their opening or closing. A large number of toxins, selective for a variety of channels, are very much in demand by neuroscientists, since the toxins provide selective pharmacological tools to study various types of channels. Amino acids located on the cytoplasmic side of the channel can often be phosphorylated by intracellular kinases. Some possible phosphorylation sites are indicated by the letter P on the intracellular loop between the domains I and II in Figure 2A. Such phosphorylation is known to alter the function of the channels either by changing their gating properties or by increasing their sensitivity to neurotransmitters.[4,5]

III. PATCH CLAMPING AND MOLECULAR CLONING ARE TWO KEY METHODS EMPLOYED IN STUDIES OF IONIC CHANNELS

Knowledge of the DNA sequences coding for individual channel proteins makes it possible to modify the protein structure by point mutations. Expression of such modified proteins in frog oocytes makes them accessible to electrical measurements by means of patch clamping. Patch clamping is a method of measuring small electrical currents flowing through single channel openings[6] (Figure 3). It has been found

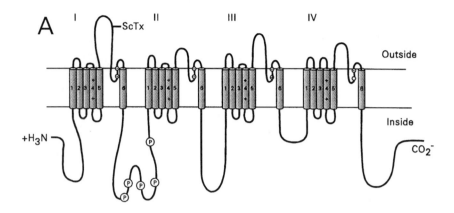

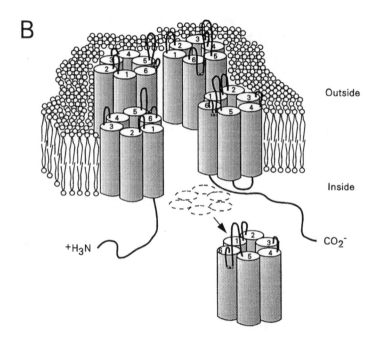

Figure 2
A. Model for the transmembrane arrangement of the voltage-activated sodium channel protein. Only the main (alpha) subunit is shown for simplicity. The channel is composed of a long chain of amino acids interconnected by peptide bonds. Amino acids perform specific functions in the channel, depending on their position. The cylinders, for example, represent assemblies of amino acids located within the cellular membrane and responsible for the formation of the conductive pore. (Modified from W.A. Catterall, 1992, *Physiol. Rev.*, 72: S15–S48. With permission.) **B.** A model for the possible three-dimensional arrangement of the four domains in the membrane. It is proposed that four homologous domains make up a channel. One of the domains is shown separately outside the membrane.

that even a single amino acid substitution or deletion can change the behavior of channels significantly. Such minor modifications of the molecular structure result in channels opening for longer or shorter periods of time or carrying larger or smaller currents. Also, the selectivity of the channels for certain ions can be altered by substitutions of either negatively or positively charged amino acids located at the

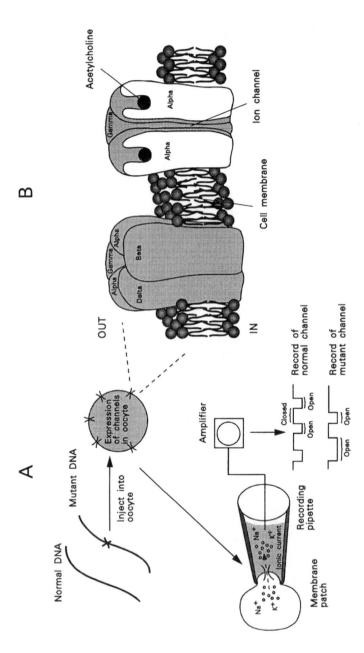

Figure 3

A. Cloned DNA (or RNA) coding for various ionic channels can be injected into frog eggs (oocytes). The cellular metabolic machinery produces the desired proteins and inserts them in the membrane as functional channels. A glass pipette can be attached to the membrane to form a tight bond with the membrane around its rim. In the isolated patch of the membrane, even a small current can be recorded and amplified. A typical patch-clamp recording will exhibit steplike signals representing small currents carried by ions such as Na+ and K+. Each type of channel has a characteristic amplitude and duration of the current. **B.** An expanded view of two channels shows five subunits (two alpha, beta, gamma, and delta). Each subunit is composed of a large number (hundreds) of amino acids. The amino acid composition varies among the subunits and determines their respective functions. Two alpha subunits, for example, form parts of the ionic channels and, in addition, form the receptor "pockets" where a transmitter substance (acetylcholine in this case) binds and causes channel opening.

mouth of the channel. As a result of such studies, the functional properties of channels and receptors are gradually becoming understood at the molecular level.

The studies have led to many conceptual breakthroughs. We now realize a number of channels are typically made up of three to five subunits. For example, the acetylcholine-sensitive channel is composed of five subunits (Figure 3B). The mix of subunits in the acetylcholine channel changes with age of an animal and is often different among different cell types. This finding explains why there are functional and pharmacological differences among the acetylcholine channels. In addition, the relationship between the receptor and the channel has been clarified. For example, in the acetylcholine channel the receptor sites for acetylcholine occupy only a small part of the alpha subunits (of which there are two on each channel).

The explosion of information about the molecular structure of ionic channels revealed many channel subtypes with unknown functions, the existence of which was not even suspected a few years ago. For example, until the early 1980s only two basic subtypes of the acetylcholine receptors were known. They were recognized on the basis of binding to selective agonists and antagonists, such as muscarine, nicotine, curare, atropine, etc. Similarly, there were three known types of glutamate receptors distinguished on the basis of three agonists (kainate, NMDA, and quisqualate). We now realize that there are more than 20 subtypes of glutamate receptors and many subtypes of acetylcholine receptors.[7,8] Similar numbers of receptor subtypes have been identified for all known transmitter substances. Research into the functional consequences of the diverse structural properties of receptors poses a major challenge to neuroscientists.

IV. REASONS FOR CHANNEL DIVERSITY

In each nerve cell there is a need for strict control of the properties and the localization of the channels. For example, the glutamate-sensitive channels need to be localized at synaptic junctions where glutamate is released from axonal terminals. Adjacent inhibitory synapses require other types of channels sensitive to inhibitory transmitters, such as γ-aminobutyric acid (GABA). Sodium-permeable, voltage-sensitive channels need to be expressed in parts of the cell where action potentials are to be fired, primarily in the cell body and the axon.

Identification and characterization of ionic channels have made it possible to map their location on neurons and correlate the location with a particular function. A key technique used in identification and localization of ionic channels is immunostaining. Antibodies raised against ionic channel proteins can be labeled with a fluorescent dye or with an electron-dense marker, which can be identified under a fluorescent or electron microscope. The labeled antibodies can be used as probes to map the channels of interest on different parts of neurons. An elegant example was presented by Black et al.,[9] who identified and labeled a certain type of voltage-dependent sodium channel on nodes of Ranvier of myelinated axons. These channels are distinct from another type of sodium channel found on the initial axonal segments in neurons.[10] The functional differences between the channels in the initial segment and in the nodes remain to be determined.

Another example of precise channel localization comes from the work of Robitaille et al.,[11] who found that calcium channels and acetylcholine channels are located in close apposition to each other, on two adjacent cells. This apposition of the channels is crucial for operation of synapses (Figure 4).

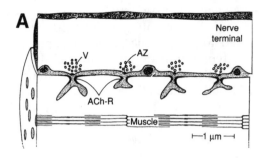

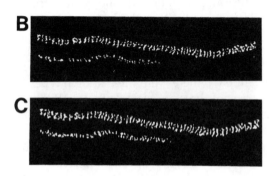

Figure 4
A. Cross section through a muscle cell and an adjacent nerve terminal illustrates synaptic release sites (active zones, AZ) located directly above the postsynaptic acetylcholine receptors (ACh-R) in the frog neuromuscular junction. **B.** When viewed from above, each active zone can be identified by a row of calcium channels. The calcium channels were labeled with omega-conotoxin conjugated with a color dye (Texas Red). **C.** α-Bungarotoxin outlines bands of acetylcholine receptor channels on the muscle that correspond to the bands of the calcium channels on the presynaptic terminal. (From R. Robitaille et al., 1990, *Neuron*, 5: 773–779. With permission.)

V. DO PROPERTIES OF SINGLE CHANNELS PREDICT OR EXPLAIN THE MACROSCOPIC NEUROPHYSIOLOGICAL PHENOMENA?

Recordings of single channel activities from patches (1 to 2 μm²) of excitable membranes reveal the presence of sodium and potassium channels. When challenged by sudden, depolarizing stimuli, sodium channels open in 1 to 2 ms but inactivate shortly thereafter (Figure 5A). This maneuver repeated a number of times reveals a pattern, which is clearly evident after summing all events to form an average response. The resulting ensemble average current is similar to the macroscopic sodium current that can be recorded in many cells capable of firing action potentials.[12] Originally, this type of sodium current was observed by Hodgkin and Huxley[13] in the squid giant axon. From their pioneering work Hodgkin and Huxley inferred the existence of voltage-dependent sodium channels. Moreover, they predicted how the interplay of sodium and potassium channels gives rise to action potentials. Their hypothesis required that potassium channels open more slowly than sodium channels but stay open for a longer time. These predictions have been proven correct by patch-clamp recordings of single channel activity (Figure 5B). Hence, in the case of the action potential, the basic features of the macroscopic phenomenon can be explained by properties of single channels. With further progress it should be possible to fully understand the detailed operations of these and other channels in terms of

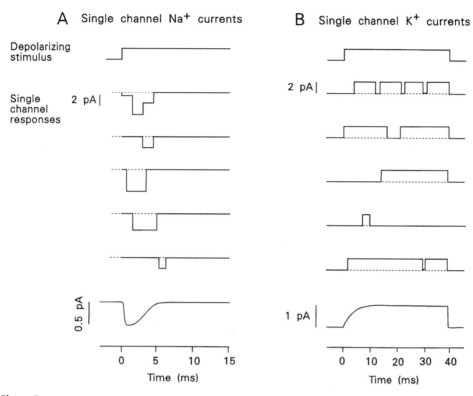

Figure 5
A. Depolarizing stimulus applied to a patch of membrane evokes channel openings recorded as steps of current. Five representative responses show typical patterns of channel openings. The smooth curve at the bottom was obtained by averaging several hundred consecutive responses. Note that channels close after about 5 ms but the pulse is maintained. **B.** Similar depolarizing pulse opens potassium-selective channels in a membrane patch, where sodium channels had been blocked pharmacologicaly. Typically, potassium channels open with a slight delay but remain open for the duration of the stimulus. These channels tend to open for longer periods of time. Sodium and potassium channel openings are shown in the opposite directions to illustrate that the currents flowing through sodium channels are inward (from outside to inside the cell) and those of the potassium channels are outward.

their molecular structure and to carry this understanding toward full explanation of cellular functions.

Detailed understanding of the molecular structure and function of sodium channels has led to the understanding of certain neural disorders. Hyperkalemic periodic paralysis is a muscle disorder characterized by sporadic muscle weakness and associated with an elevated serum potassium level.[14] Electrophysiological recordings from muscle fibers in patients with this disease show persistent depolarization of muscle fibers and abnormal sodium channel openings. The genetic cause of this disease has been traced to the alpha subunit of the sodium channel gene located on human chromosome 17. As is often the case with mutations, the cause of the disease is not a missing part of the gene but the presence of multiple gene copies with a few erroneous substitutions of amino acids. In the case of the periodic paralysis, valine is substituted for methionine near the intracellular end of transmembranous segment IV and methionine is substituted for threonine in segment II S5. Evidently, even such subtle mutations can cause malfunctioning of the channel. As a result, the channels seem to open for abnormally long times and cause an imbalance of ions in the muscle cells. This, in turn, produces the paralysis.[15]

VI. SYNAPTIC TRANSMISSION

Synaptic transmission can be best understood from examination of various ionic channels participating in the process. The sequence of events during synaptic transmission is shown in Figure 6. It begins with a depolarizing wave in the axon which opens voltage-dependent calcium channels in the presynaptic terminal. The calcium channels are often concentrated in small patches corresponding to the active sites of transmitter release (refer to Figure 4). Apparently, it is crucial that calcium concentration rises to relatively high levels near the release sites, just at the inner surface of the cellular membrane. The calculated concentration of calcium ions is estimated to be as high as 100 μM, which represents a dramatic increase over the resting levels of free calcium inside the terminals of around 0.1 μM. The emerging view of the release process is that a large but brief calcium signal triggers a chain of biochemical reactions which propel synaptic vesicles toward the membrane and facilitate the release of a transmitter substance into the synaptic cleft by exocytosis[17] (Figure 6). Subsequently, the vesicular membrane is reclaimed by the axonal terminal and recycled to package more of the neurotransmitter substance.

Studies of easily accessible neuromuscular synapses in crustaceans have shown that synapses possess a variable number of sites where the vesicle fusion can occur and that the efficacy or strength of individual synapses depends on the number and the location of such "active" sites. It appears that synapses with two, three, or more active sites have a sufficient "fire power" for the release of greater amounts of transmitter substance than are released by synapses with a single active zone. Conversely, synapses with one or no active zones are essentially ineffective but can be recruited as a result of repeated stimulation of the presynaptic terminal.[16]

The diffusion of the transmitter across the synaptic cleft towards the postsynaptic membrane occurs in a fraction of a millisecond. Upon reaching the postsynaptic surface, where the receptors are located, the transmitter molecules bind to the receptor sites and trigger an appropriate postsynaptic response. Naturally, the receptors are matched to the transmitter substance normally released at the synapse. However, often two types of transmitter substances can be co-released and activate different receptors.

Basically, transmitters can have two effects at synapses: one is to open or close ionic channels, and the other is to trigger a change in the rate of a chemical reaction inside the postsynaptic cell via a second messenger cascade. Conventional transmitters, such as glutamate, GABA, acetylcholine, and glycine, usually act as fast transmitters triggering the opening or closing of a channel and a subsequent change in electrical potential of the receiving cell. This, in turn, can lead to more excitation or inhibition. Peptide transmitters are often co-localized with the conventional transmitters listed above but act more slowly and often stimulate enzymatic reactions.

The peptide transmitters form the largest group of transmitter substances in the brain. At least 40 peptide transmitters have been identified, and the list is still growing.[18] The mode of release of peptides is slightly different from that of the conventional transmitters. Peptides are localized in large, dense core vesicles which appear to require more prolonged and widespread diffusion of calcium into the terminals for their release.[19] The postsynaptic receptors for peptides are of the metabotropic type. Their postsynaptic actions are mediated by membrane-bound G proteins and not directly by ionic channels.[7] The stimulation pattern of the presynaptic cell often determines which transmitter type (conventional or peptide) will be released and what the effect on the receiving cell will be. For example, it has been shown that the mossy fiber terminals which use glutamate for rapid synaptic

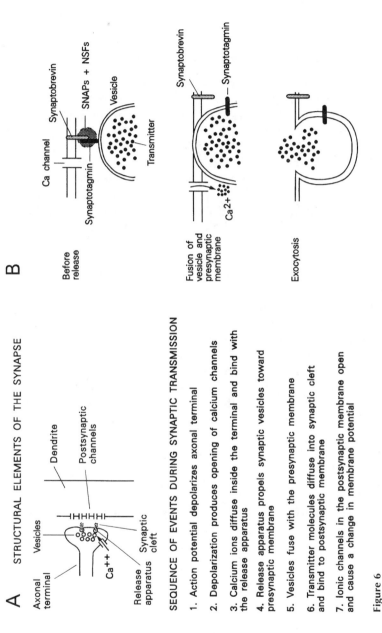

A STRUCTURAL ELEMENTS OF THE SYNAPSE

B

SEQUENCE OF EVENTS DURING SYNAPTIC TRANSMISSION

1. Action potential depolarizes axonal terminal

2. Depolarization produces opening of calcium channels

3. Calcium ions diffuse inside the terminal and bind with the release apparatus

4. Release apparatus propels synaptic vesicles toward presynaptic membrane

5. Vesicles fuse with the presynaptic membrane

6. Transmitter molecules diffuse into synaptic cleft and bind to postsynaptic membrane

7. Ionic channels in the postsynaptic membrane open and cause a change in membrane potential

Figure 6

A. Basic elements of the synapse. B. Details of the exocytotic release process. Synaptobrevin, synaptotagmin, SNAPs, and NSFs are some of the proteins known to participate in the release process. In this scheme vesicles are held near the membrane by an anchor composed of several synaptic proteins. Influx of calcium dissolves the anchor and causes fusion of vesicles with the cytoplasmic membrane. (Modified from T. Sollner and J.E. Rothman, 1994, *TINS*, 17: 344-348. With permission.)

transmission in the hippocampus can also release the peptide endorphin when stimulated with bursts of action potentials. However, unlike glutamate, the endorphin has a slow and inhibitory effect.[20]

This traditional division between fast and slow transmitters is becoming blurred because of the discovery of a great number of metabotropic receptors for the fast-acting transmitters. These receptors appear to function in a manner similar to that of the peptide receptors described above. (Refer also to Chapter 7 in this volume.)

VII. QUANTAL SYNAPTIC TRANSMISSION

The concept of quantal synaptic transmission is central in the understanding of the nervous system. Synapses are the primary interneuronal connections which serve not only as transmitting devices but also as signal amplifiers or attenuators, coincidence detectors, or storage devices. As described above, one of the essential components of each synapse is a cluster of vesicles filled with a transmitter substance. It is estimated that 5,000 to 10,000 molecules are contained in each packet.[1] Each vesicle is capable of sending a packet of transmitter substance toward the receiving cell.

Soon after the basic concept of the synapse with its essential structural elements was discovered, physiologists developed techniques suitable for recording of the electrical activity at synapses. del Castillo and Katz[21] were the first to record and characterize such activity at synapses of the neuromuscular junction of the frog. A surprising feature of these signals was their "noisy" appearance, characterized by sporadic and variable amplitudes. According to Sir Bernard Katz, this unexpected type of activity was at first dismissed as some sort of artifact and neglected. Only a year later the discovery was appreciated and confirmed, eventually leading to the Nobel Prize for Katz. Since then, this random synaptic noise has been observed at many synapses and is still under considerable scrutiny in many laboratories around the world. In all cases these miniature synaptic responses appear to represent a spontaneous spillage of the transmitter from synaptic vesicles. In all synapses studied to date, the quantal synaptic events vary in size. It is still a matter of debate as to what this variance represents. Among the most likely factors are the uneven filling of synaptic vesicles and a variable number of postsynaptic receptors at synapses. The variable quantal size adds further noise to synaptic transmission.

During normal synaptic transmission, evoked by action potentials in presynaptic nerve terminals, a number of the miniature potentials are released almost simultaneously. Hence, the miniature synaptic potentials are the quantal units of the evoked synaptic potentials. It follows that the size of evoked responses can be regulated by either decreasing or increasing the number of participating quanta, which is commonly measured as the quantal content, m, of synaptic responses, where m is given by the formula:

$$m = \frac{\text{average evoked response}}{\text{average miniature response}} \qquad (1)$$

Depending on the strength of the synapse, m can vary from a "weak" $m \leq 0.1$ to a "strong" $m \geq 10$.

In summary, the synaptic transmission can be seen as a transient increase in probability of release of quantal units. If probability is near 1 and if the number of participating quanta is large, say over 20, the size of evoked potentials becomes invariant during repeated stimulation. Such a process makes for a very secure synaptic transmission. However, more often the synaptic responses fluctuate in size and in some cases fail (Figure 7). This is intriguing because it indicates that transmission of information among individual neurons is probabilistic and noisy. The reasons for this type of transmission in the nervous system remain unknown. Nevertheless, a low probability of release at many synapses suggests that there is a reserve pool of connections which are relatively underused. There is evidence that these weak synapses can be strengthened by brief episodes of activity, providing a possible cellular mechanism for learning.

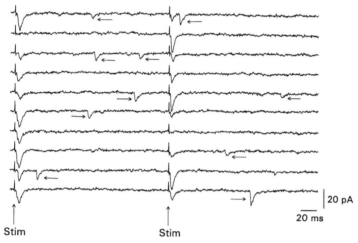

Figure 7
Evoked and spontaneous synaptic currents recorded in a dentate granule neuron. Each horizontal trace shows synaptically evoked responses occurring shortly after the repeated stimuli (Stim arrows). These responses fluctuate in size because of variable quantal content. In the intervals between the stimulations, many spontaneous quantal responses occurred. Some are indicated by horizontal arrows.

VIII. POSTSYNAPTIC EFFECTS OF TRANSMITTERS ARE DETERMINED BY THE PROPERTIES OF POSTSYNAPTIC CHANNELS

Single ligand-activated channels are not readily accessible to studies by patch clamping, but sufficient indirect evidence has been accumulated to explain how the single channel currents give rise to complete synaptic currents. For example, fragments of muscular or neuronal membranes adjacent to synapses can be studied on the assumption that the properties of single channels in those areas closely resemble the synaptic channels in the synaptic cleft.

To simulate rapid activation of the channels in patched membranes, one can apply transmitters by ejecting them from small tubes.[22] The speed of onset and offset of the neurotransmitter applications is important because the receptors respond differently to brief or long applications (Figure 8). Moreover, a single substance such as glutamate can produce both short and long synaptic currents. A typical example of a short-lasting current is the AMPA receptor–mediated current seen at many synapses in the brain. Channels gated by this receptor open for a few milliseconds immediately after the glutamate application. These kinetic properties of the channels determine rapid onset and brief duration of synaptic currents. The mechanisms responsible for rapid closure of the channels are called inactivation or desensitization. These mechanisms are common features of many types of channels.

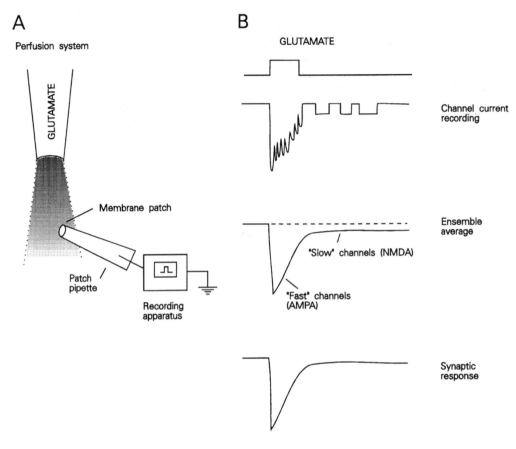

Figure 8
A. Experimental arrangement illustrates a perfusion apparatus ejecting glutamate onto a patch of membrane at the tip of the patch pipette. **B.** A brief pulse of glutamate produces rapid channel openings. The ensemble average of several such responses shows "fast" AMPA and "slow" NMDA currents in a single patch. **C.** Synaptic response recorded in an intact neuron resembles the patch responses. (Modified from Lester et al., 1990, *Nature*, 346: 565–567. With permission. ©1990 Macmillan Magazines Limited.)

Long-lasting synaptic responses are often the result of channels opening for long periods of time. Single channel recordings from membranes possessing NMDA-type glutamate channels show slow onset of the openings. However, these channels stay open for prolonged periods, even in response to brief glutamate applications.

Both fast and slow glutamate-sensitive channels can be present at the same synapses, presumably in close proximity. Thus, a synapse can have a "double personality," depending on the dominant channel type. As a rule, the fast glutamate channels are active at all membrane potentials, providing the capability of transmitting fast synaptic signals. The slower, NMDA channels open preferentially at membrane potentials above –40 mV. The voltage dependence, as well as the long duration of openings, gives these channels the ideal properties to serve as coincidence detectors in neurons.

IX. SYNAPTIC PLASTICITY

Certain brain structures can now be studied in detail, thanks to the development of the "brain slice" technology. Mammalian hippocampus, for example, can be cut into 500 μm thick slices with the preservation of some essential features of its synaptic connections. Basically, each slice can be viewed as a trisynaptic circuit beginning with the synapses of the perforant pathway in dentate gyrus. The second synaptic relay is formed by mossy fiber synapses terminating on dendrites of CA3 area. The third component of the circuit is located in CA1 area, where axons of the CA3 pyramidal neurons make synapses on dendrites of the CA1 pyramidal neurons (Figure 9A).

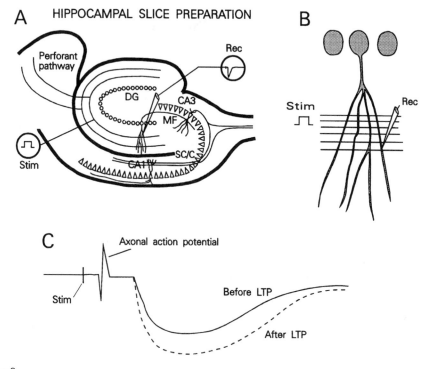

Figure 9
A. Schematic illustration of the hippocampal slice from the rat brain shows three main neuronal fields, DG (dentate gyrus), CA3, and CA1. The three regions are connected in a circuit that begins in DG and ends in CA1. A major afferent pathway into the hippocampus enters through DG as the perforant pathway. **B.** Horizontal lines represent axons of the perforant pathway traversing the dendrites of the dentate granule neurons. Synaptic responses can be evoked by stimulation of the axons (Stim) and recorded extracellularly as field potentials in the vicinity of the dendrites (Rec). **C.** Superimposed traces of field potentials recorded before and after induction of LTP. Note increased amplitude of the synaptic response without change in the axonal action potentials.

Each of the synaptic relays can be scrutinized in detail with the use of most modern electrophysiological techniques, including patch clamping and ion imaging with fluorescent probes.

One of the intriguing features of the hippocampal synapses is the phenomenon of long-term potentiation (LTP). Discovered in the early 1970s by Bliss and Lomo,[23] LTP represents one of the many types of synaptic plasticity. LTP can be induced in any of the three relays of the hippocampal circuit, for example, in the dentate gyrus. The induction is typically produced by a relatively strong and rapid sequence of action potentials in the presynaptic axons. For example, four 100 Hz bursts, each lasting 500 ms, will reliably produce LTP at synapses in dentate gyrus (Figure 9B).

LTP is a type of synaptic amplification, whereby the signals processed by the hippocampus are magnified. This magnification can last several hours in the slice preparation and much longer *in vivo*. The physiological role of LTP in animal behavior is still not certain, but its possible contribution to certain types of learning and memory has been proposed by a number of investigators.[24]

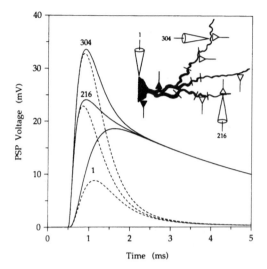

Figure 10
Inset shows a part of a neuronal structure that was reconstructed in the computer as an electrical circuit made of several compartments. The borders between the compartments are indicated by the vertical lines crossing the dendrites. Five excitatory (open triangles) and two inhibitory (closed triangles) synapses have been incorporated into the model. Voltage changes produced by activation of all five excitatory synapses were recorded at points 1, 216, and 304 as solid lines in the graph. Note strong attenuation of the amplitude of the responses as it spreads along dendrites. The dashed lines show responses at the same points during activation of the inhibitory synapses together with the excitatory ones. Note that the durations, but not the peak amplitudes, are altered in the dendrites, whereas the cell body response is drastically reduced in amplitude. (From C. Koch and I. Segev, 1989, *Methods in Neuronal Modelling*, MIT Press. With permission.)

X. SYNAPTIC INTERACTIONS ON DENDRITES

In the central nervous system, particularly in structures such as the cerebral cortex, the cerebellum, and the hippocampus, neurons possess long dendrites studded with thousands of synaptic spines. The cerebellar Purkinje neurons possess on the order of 100,000 spines each, while the hippocampal neurons possess 5,000 to 10,000 spines each. Every spine is usually contacted by one synapse. Hence, a large part of the synaptic array feeds information to the cell via dendritic spines. Understanding the significance of such an arrangement and of possible interactions among synapses on dendrites is a challenge for future investigations. Recordings obtained so far show a wealth of ionic channels such as calcium, sodium, and potassium voltage-dependent channels on dendrites. Together with the synaptic ligand-gated channels responsive to such transmitters as glutamate or GABA, they may serve to modulate (reduce or enhance) synaptic signals and make them more or less "visible" to the cell soma and to other synaptic contacts.

The complexity of dendritic morphology and the overwhelming numbers of synapses ($10^9/mm^3$ of cortical tissue) present on neurons prompted computer simulations of synaptic interactions on dendrites. The structure of neurons can be reconstructed in a computer and gradually furnished with a complement of ionic channels to simulate real cells. Computer simulations can at times test theoretical predictions that are difficult to address with the presently available experimental techniques. For example, in a simulation in Figure 10 the inhibitory synapses were shown not only to reduce the peak amplitude of the excitatory responses seen in the cell body but also to greatly diminish the duration of the excitatory potentials in the dendrites. The latter result cannot be seen by direct measurements because dendrites are difficult to impale with recording microelectrodes. In other cases, the results of computer simulations often lead to testable predictions which can be addressed in electrophysiological experiments.[25] It is expected that complementary results of computer modeling and of experiments on living tissue will lead to gradual refinement of both approaches. The results of these combined approaches show promise for further understanding of single cells, cell assemblies, and ultimately of the whole brain structure.

REFERENCES

1. Kuffler, S.W., Nicholls, J.G., and Martin, A.R., *From Neuron to Brain*, 2nd ed., Sinauer Associates, Sunderland, MA, 1984.
2. Witter, M.P., Connectivity of the rat hippocampus, in *The Hippocampus. New Vistas*, Chan-Palay, V. and Kohler, C., Eds., Alan R. Liss, New York, 1989, 53.
3. Catterall, W.A., Cellular and molecular biology of voltage-gated sodium channels, *Physiol. Rev.*, 72, S15, 1992.
4. Blackstone, C., Murphy, T.H., Moss, S.J., Baraban, J.M., and Huganir, R.L., Cyclic AMP and synaptic activity-dependent phosphorylation of AMPA-preferring glutamate receptors, *J. Neurosci.*, 14, 7585, 1994.
5. Wang, L.-Y., Salter, M.W., and MacDonald, J.F., Regulation of kainate receptors by cAMP-dependent protein kinase and phosphatases, *Science*, 253, 1132, 1991.
6. Neher, E. and Sakmann, B., The patch clamp technique, *Sci. Am.*, 1, 44, 1992.
7. Nicoll, R.A., Malenka, R.C., and Kauer, J.A., Functional comparison of neurotransmitter receptor subtypes in mammalian central nervous system, *Physiol. Rev.*, 70, 513, 1990.
8. Hollmann, M. and Heinemann, S., Cloned glutamate receptors, *Annu. Rev. Neurosci.*, 17, 31, 1994.
9. Black, J.S., Friedman, B., Waxmann, S.G., Elmer, L.W., and Angelides, K.J., Immuno-ultrastructural localization of sodium channels at nodes of Ranvier and perinodal astrocytes in rat optic nerve, *Proc. R. Soc. London*, 238, 39, 1989.
10. Wollner, D.A. and Catterall, W.A., Localization of sodium channels in axon hillocks and initial segments of retinal ganglion cells, *Proc. Natl. Acad. Sci. U.S.A.*, 83, 8424, 1986.
11. Robitaille, R., Adler, E.M., and Charlton, M.P., Strategic location of calcium channels at transmitter release sites of frog neuromuscular synapses, *Neuron*, 5, 773, 1990.
12. Hille, B., *Ionic Channels of Excitable Membranes*, Sinauer Associates, Sunderland, MA., 1992.
13. Hodgkin, A.L. and Huxley, A.F., Current carried by sodium and potassium ions through the membrane of the giant axon of loligo, *J. Physiol. (London)*, 116, 449, 1952.
14. Rudel, R. and Ricker, K., The primary periodic paralysis, *TINS*, 8, 467, 1985.
15. Cannon, S.C., Brown, R.H., and Corey, D.P., A sodium channel defect in hyperkalemic periodic paralysis: potassium-induced failure of inactivation, *Neuron*, 6, 619, 1991.
16. Atwood, H.L., Cooper, R.L., and Wojtowicz, J.M., Nonuniformity and plasticity of quantal release at crustacean motor nerve terminals, in *Neurotransmitter Release*, Stjärne, L., Greengard, P., and Grillner, S., Eds., Raven Press, New York, 1994, 363.
17. Sollner, T. and Rothman, J.E., Neurotransmission: harnessing fusion machinery at the synapse, *TINS*, 17, 344, 1994.
18. Hall, Z.W., *An Introduction to Molecular Neurobiology*, Sinauer Associates, Sunderland, MA., 1992.
19. Verhage, M., Differential release of amino acids, neuropeptides, and catecholamines from isolated nerve terminals, *Neuron*, 6, 517, 1991.
20. Weisskopf, M.G., Zalutsky, R.A., and Nicoll, R.A., The opioid peptide dynorphin mediates heterosynaptic depression of hippocampal mossy fiber synapses and modulates long term potentiation, *Nature (London)*, 362, 423, 1993.
21. del Castillo, J. and Katz, B., Quantal components of the end-plate potential, *J. Physiol. (London)*, 124, 560, 1954.

22. Lester, R.A.J., Clements, J.D., Westbrook, G.L., and Jahr, C.E., Channel kinetics determine the time course of NMDA receptor mediated synaptic currents, *Nature (London)*, 346, 565, 1990.
23. Bliss, T.V.P. and Lomo, T., Long-lasting potentiation of synaptic transmission in the dentate area of the anaesthetized rabbit following stimulation of the perforant path, *J. Physiol. (London)*, 232, 331, 1973.
24. Baudry, M. and Davis, J.L., *Long Term Potentiation. A Debate of Current Issues*, MIT Press, Cambridge, MA, 1991.
25. Koch, C. and Segev, I., *Methods in Neuronal Modelling*, MIT Press, Cambridge, MA, 1989.

Chapter 2

ALTERED SIGNAL TRANSDUCTION IN PSYCHIATRIC ILLNESS

Craig Hudson and Anna Lin

CONTENTS

0-8493-8386-0/96/$0.00+$.50

I. INTRODUCTION

Signal transduction can be likened to the process by which a specific radio signal is beamed out of a radio station, combines with other broad frequency broadcast bands for transmission, and is then selected and translated into a recognizable signal by an appropriately tuned radio receiver. This anaolgy illustrates the need not only to look at the individual components, but also the need to understand the interaction among the individual elements in the communication system. Naturally, signal transduction in the cell is somewhat more complex than radio transmission, in that individual signals can interact with each other to create an entirely new signal. The interaction between cellular communication and development of neuronal circuits was first suggested by physiological psychologists almost 50 years ago.[1] The Hebbian notion that various signals can be integrated to establish a new signal was revolutionary for its day but not pursued outside the realm of psychology until decades later. With the advent of recombinant DNA methods, some of the key elements of neuronal signal transduction have now been well characterized and their specific roles in signal transduction defined. Consequently, it is now possible to use this knowledge of signal transduction to partially understand the pathophysiology of some major mental illnesses, which may lead to the development of new pharmacotherapy. For the sake of clarity, the chapter is divided into four parts: an overview of receptor signal transduction; signal transduction mechanisms implicated in the pathophysiology and treatment of affective disorders; signal transduction mechanisms implicated in the pathophysiology and treatment of schizophrenia; and a summary.

II. OVERVIEW

In the simplest case, the synaptic signal begins when a presynaptic electrical signal leads to the release of a neurotransmitter into the synaptic cleft. Activation of neuronal receptors by their respective neurotransmitters (first messengers) activates the production of intracellular second messengers and ion fluxes, through one of several signal transduction processes. In the most direct condition, the primary messenger binds to a receptor that incorporates an ion channel within its own structure. With the binding of a neurotransmitter to this sort of receptor, the ion channel undergoes a conformational change, which, in turn, alters the ion flux through the channel. The so-called ligand-gated ion channel receptor allows for rapid changes in ion fluxes and can be used as a model to understand the more complex interactions when several neurotransmitters interact to produce differing effects in the signal transduction cascade. To understand these more complex interactions, however, it is important to understand the various ways in which a receptor couples with the intracellular messenger systems.

In the majority of cases, coupling of the receptor to an intracellular effector enzyme or ion channel occurs through a family of intermediary guanine nucleotide proteins (G proteins) located in the plasma membrane. The various G proteins may be both inhibitory and excitatory, allowing for the inhibition or activation of effector enzymes. Schematic representations of G protein–coupled and uncoupled signal transduction are found in Figure 1. This overview section will discuss each component of that signal in turn.

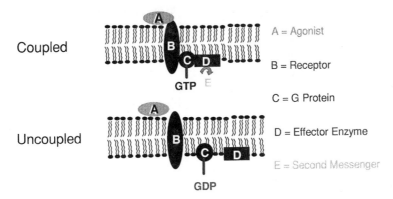

Figure 1
Transmembrane signal transduction.

A. Primary Messengers

Just as a radio signal contains many frequency bands, the primary messengers in signal transduction (neurotransmitters) are varied and many. In fact, there are so many primary messengers that there is still debate over which primary messengers are neurotransmitters. Most investigators would agree that a neurotransmitter should meet the following criteria: a substance that is synthesized by a neuron and released in response to electric impulses; and a substance that, after release, acts on other neurons to alter their electrical properties.[2]

Although the vast majority of neurotransmitters are amino acids, the monoamine-based neurotransmitters appear to be the most important with respect to treating major psychiatric illness. The monoamine neurotransmitters can be divided into two broad categories: catecholamines (dopamine, norepinephrine, and epinephrine) and indolamines (serotonin and melatonin). The more prevalent amino acid–based neurotransmitters can be divided into inhibitory, gamma aminobutyric acid (GABA) and glycine, and excitatory, glutamate.

Other primary messengers include hormones and neuropeptides. These broad groups of primary messengers likely play a major role in central nervous system (CNS) function, but because of the scope of this chapter, the interested reader is referred elsewhere.[2]

B. Receptors

Just as radios have antennae of different shapes tuned to unique wave characteristics, neurons have various receptors embedded in the cell membrane so that primary messengers can translate their signal into one recognized by intracellular enzymes. The notion that primary messengers exert their influence over cellular signal transduction has been pursued since the beginning of the 20th century. Receptors are proteins that straddle the neuronal membrane, containing both an extracellular and an intracellular component. There is general agreement that, by definition, a receptor must not only bind specific molecules but must also lead to a change in the signal transduction pathway. Those proteins that bind ligand but have no measurable impact on signal transduction are referred to as acceptors. The receptors themselves may be divided into two broad classes: ligand-gated ion channel receptors, described above, which incorporate an ion channel within the receptor structure;

and the much more prevalent G protein–linked receptors. Recent evidence indicates that there can be considerable individual variation in receptor structure even within the same receptor subtype.

Until recent developments in molecular biology, most receptor subtypes were defined and characterized over the years by the study of response and antagonism of responses to various selective agonists and antagonists. This procedure, which was performed on post-mortem CNS tissue, was time-consuming and difficult to interpret since a blunted response may arise from true competition for a particular receptor or from a number of other steps in the signal transduction cascade. Despite the difficulties inherent in this methodology, a number of receptor and receptor subtypes have been identified. Over the past 20 years, receptors were identified by their binding affinity for various radioactive-labeled ligands. More recently, with the advent of recombinant DNA and cloning technology, the number of receptors and receptor subtypes has expanded rapidly and will likely continue to do so. The discussion of receptor types in this chapter is not exhaustive, but rather focuses on those receptor subtypes which have a putative role in psychiatric illness.

C. G Proteins

The vast majority of receptors (>90%) require an intermediary protein in order to couple to intracellular enzymes. The coupling proteins gain their name from the specific nucleotide (guanine) that activates them and are referred to as G proteins. After the neurotransmitter binds to a receptor, the G proteins allow the receptor to "couple" with an intracellular enzyme to produce second messengers. In the resting state, G proteins are bound by the nucleotide guanosine diphosphate (GDP) and have no contact with receptors. After the neurotransmitter binds to a receptor, this binding causes the G protein to exchange GDP for guanosine triphosphate (GTP) which activates the G protein. Depending on the type, the GTP-bound G protein then either activates or inhibits an effector enzyme. The activated G protein is, then, analogous to a switch on the radio which allows signals carried by the neurotransmitter (first messenger) to be translated into intracellular messengers (second messengers). After a brief time, a catalytic enzyme on the G protein converts the GTP back to GDP, thus inactivating itself and turning off the signal. The situation is thus analogous to an alarm clock radio, which is activated when the alarm goes off; then, after a given amount of time, if the alarm is not turned off by external means, it automatically shuts itself off.

Although the radio switch analogy works reasonably well, the cell chemistry is much more complex, based in part on the complexity of the structure of G proteins. Structurally, G proteins are composed of three subunits, which have been cloned: alpha, beta, and gamma. The largest of the subunits, the alpha subunit, has the greatest variation. In fact, each member of the G protein superfamily has a unique alpha subunit. The exact role of the beta and gamma subunits is yet to be determined, but it is likely the beta–gamma dimer also impacts on effector enzymes in a direct or indirect manner (i.e., inhibition by binding free alpha subunits). It is important to note that the multiplicity of the alpha subunit, and to a lesser degree the beta and gamma subunits, provides for the coupling of a wide variety of receptors to the same or different second messenger systems. In this way, various types of receptors have the ability to regulate each other, allowing for greater signal divergence, convergence, or filtering than could be achieved solely on the basis of receptor diversity. Consequently, G protein–linked signal transduction allows for a variety of signals to be relayed and/or filtered simultaneously.

Neurotransmitter stimulation of the effector enzyme adenylyl cyclase (AC) is a classic model of G protein–modulated second messenger–forming processes. Briefly, agonist stimulation of a receptor results in the association of the receptor with the inactive G protein. As described above, this activation results in the replacement of GDP with GTP. The binding of GTP results in a conformational change in G protein which in turn stimulates AC, leading to increased levels of cyclic adenosine monophosphate (cAMP). An inhibitory process activates an inhibitory G protein, leading to decreased production of cAMP. Increases in cAMP levels activate cAMP-dependent protein kinase which modulates the function of a broad range of membrane receptors, intracellular enzymes, membrane ion channels, and transcription regulatory factors. Signal termination is brought about by the hydrolysis of GTP to GDP and a return of the G protein to its inactive form. Good reviews of G protein function are readily available.[3]

D. Second Messengers

The regulation of cAMP production was the first to be characterized, but a variety of other second messenger systems are also regulated by G protein–coupled receptors. The hydrolysis of inositol phospholipids (PPI), specifically phosphatidyinositol 4,5-bisphosphate (PIP$_2$), by phospholipase C (PLC) yields two second messengers, inositol 1,4,5-triphosphate (IP$_3$), and diacylglycerol (DAG). Regulation of PLC is thought to be mediated by a separate group of G proteins collectively known as Gp.[4] IP$_3$ stimulates the release of intracellular calcium from stores in the endoplasmic reticulum. DAG stimulates protein kinase C (PKC) which phosphorylates, and thereby regulates, a variety of receptors and ion channels. The complete characterization of PPI, IP$_3$, and DAG is not nearly complete, but the regulation of intracellular calcium implies a critical role for these phospholipid-derived second messengers in neuronal function.[4]

Metabolism of other membrane phospholipids, including phoshatidylcholine, phosphatidylethanolamine, phosphatidylserine, and phosphatidylinositol, also leads to a distinct group of phospholipid-derived second messengers. Phospholipase A2 (PLA$_2$) cleaves the sn-2-acyl bond of phospholipids, producing equimolar amounts of lysophospholipids and free fatty acids.[5] The specific G protein isoform involved in the activation of PLA$_2$ is unknown. Jelsma and Axelrod[6] have suggested that the beta-gamma dimer may activate the enzyme, but this method of activation must be viewed as speculative. Release of long carbon chain fatty acids leads to the formation of eicosanoids (the *eico* root denoting the number of carbon molecules, 20, in the chain), including arachidonic acid, prostaglandins, and leukotrienes.

Receptor G protein coupling is also involved in the regulation of ion channels, including K$^+$ and Ca^{2+} channels.[7] The exact mechanism of the regulation of ion channels by G proteins is complex, and the interested reader is referred elsewhere.[3] Adding a further layer of complexity, ion channels may be regulated by protein kinases which are activated by other second messenger systems.

As was seen at the level of receptors or G proteins, it would be too simplistic to view signal transduction as many autonomous systems propagating signals independent of each other. There is considerable capacity for integration and modification by communication between second messenger systems as well. This ability for cross regulation between systems is referred to as cross-talk. Defining and characterizing cross-talk mechanisms between second messengers will likely prove much more difficult than characterizing individual signal transduction systems. Nevertheless, considerable effort is likely to be paid to cross regulation between systems in terms of defining the etiopathology and, more importantly, developing treatment to complex psychiatric disorders which may well affect more than one signal transduction system at a time. The best characterized of the cross-talk mechanisms is that of the interaction between cAMP and PPI second messenger systems in which elevated cAMP leads to increased cAMP-dependent kinase activity which, in turn, leads to reduced agonist-stimulated PPI hydrolysis. It is important to note that cross-talk mechanisms underlie the processes by which receptors become more and less sensitive. In a process known as heterologous desensitization, a receptor in one particular signal pathway may be "down regulated," or made less sensitive, by the actions of another signal transduction pathway.[8] The processes involved in down regulation of receptors exceed the scope of this chapter but include changes in receptor sensitivity through the phosphorylation of a receptor, G protein, or effector enzyme.

III. ABNORMAL SIGNAL TRANSDUCTION IN SCHIZOPHRENIA

A. Neurotransmitters

The notion that schizophrenia resulted from an excess of dopamine acting at the D_2 receptor was first proposed over 20 years ago.[9] Although appealing because of its elegance and simplicity, the dopamine hypothesis fails to explain some important clinical phenomena such as the significant delay between dopamine receptor blockade by neuroleptics and resolution of psychotic symptoms. In search of an alternative hypothesis, other investigators have proposed schizophrenia results from a deficit of glutamate. The glutamate deficiency hypothesis is supported by the similarity between the psychosis induced by glutamate receptor blockade PCP and the symptoms of schizophrenia. Still others have attempted to integrate the dopamine and glutamate theories of schizophrenia, suggesting that schizophrenia may result from an excess of dopamine in neurons originating in the midbrain and a deficiency of glutamate originating in the cerebral cortex.[10] The field of neurotransmitter research is in constant evolution, so it is likely that these current theories of neurotransmitter function in schizophrenia will continue to evolve as well. Already, the role of glutamate in schizophrenia must be reevaluated in terms of the recently discovered putative neurotransmitter, nitric oxide, since glutamate appears to control its release.[11]

B. Receptors

The potency of typical neuroleptics demonstrates a strong negative correlation to D_2 affinity and therapeutic potency (with an increasing D_2 affinity, the amount of neuroleptic required for a therapeutic effect decreases).[12] Although clinically important, in terms of neuroleptic development to date, there is no evidence of genetic linkage of the D_2 receptor, essentially ruling out the D_2 receptor as an etiological cause of schizophrenia. More recently, there is evidence that the D_4 receptor is elevated as much as six times in post-mortem brain samples from patients with schizophrenia as compared with neuroleptic-treated Alzheimer brains. Again, this finding takes on clinical importance because the atypical neuroleptic, clozapine, demonstrates particularly strong affinity to the D_4 receptor. But, the genetic evidence linking the D_4 receptor to schizophrenia is negative.

Just as the affinity of typical neuroleptics to the D_2 receptors implicated dopamine receptors in the pathophysiology of schizophrenia, the efficacy of risperidone in the treatment of schizophrenia implicates serotonin (5-HT). Risperidone demonstrates strong affinity to both D_2 as well as $5\text{-}HT_2$ receptors, implicating both sets of receptors in the treatment of schizophrenia. Interestingly, it appears that it is the combination blockade of $D_2/5\text{-}HT_2$ that may be important, for risperidone demonstrates improved clinical efficacy over haloperidol (high D_2, weak $5\text{-}HT_2$ antagonist) and ritanserin (weak D_2, high $5\text{-}HT_2$ antagonist). This increased clinical efficacy with combination receptor blockade raises speculation about the interaction of receptors in the signal transduction system. In one of the few studies to date, Seeman et al.[13] demonstrate an absence of the normal communication between D_1 and D_2 receptors in post-mortem schizophrenic brain. Undoubtedly, more such studies will follow, and it underlines the importance of analyzing signal transduction systems which allow the cross -talk between receptors such as G proteins.

C. G Proteins

The fact that neuroleptics which act on more than one receptor subtype can be more effective implies that the switching or regulation of interreceptor communication may be in dysregulation in schizophrenia. Consequently, G protein structure and function has been studied in schizophrenia. A few groups using various methodologies have reported changes in the G protein concentrations in post-mortem schizophrenic brain samples, but the results are inconclusive as of yet.[4] It should be noted that small changes in G protein abundance can have substantial impact on the efficiency/inefficiency of receptor coupling. For example, inactivation of inhibitory G_i with pertussis toxin causes marked increase in dopaminergic transmission.[17] Moreover, there is preliminary evidence to suggest that the altered D_1/D_2 communication may have a partial role in altered G protein function.[13] The G protein story in schizophrenia is an evolving one, and the absence of a genetic linkage should not deter further reading in this area as it is likely that development of future neuroleptics may be directed at modifying G protein function.

D. Second Messengers

Given the complexity of working with post-mortem brain tissue, several investigators have turned to analogous peripheral cellular models to characterize abnormal second messengers in schizophrenia. Research on leukocytes has given conflicting results. Pandey et al.[15] reported reduced isoproterenol-stimulated cAMP production in a mixed leukocyte preparation from patients with schizophrenia when compared with controls. More recent attempts to stimulate cAMP activity in peripheral leukocyte preparations from patients with schizophrenia using the agonist PGE_1 failed to find a difference between patients. Similarly, other studies utilizing PGE_1, isoproterenol, and histamine have failed to demonstrate differences in cAMP production between the patients with schizophrenia and normal controls. This absence of findings draws into question the validity of peripheral leukocytes as a model to assess altered second messenger production in schizophrenia.

The first studies using platelets as a peripheral model of CNS second messenger production were also conducted by Pandey and co-workers[16] in a small number of acutely psychotic schizophrenic patients. They found that PGE_1-stimulated cAMP formation was enhanced in schizophrenic patients with an acute exacerbation of symptoms compared with chronic schizophrenic patients and healthy controls. Subsequent studies with larger numbers of subjects, however, found PGE$_1$-stimulated cAMP production to be reduced in both acutely psychotic and chronic schizophrenic patients.[17,18] Moreover, direct activation of AC with forskolin was unchanged in schizophrenic patients, suggesting that reduced PGE$_1$ activation involves a regulatory component of the signal transduction system upstream from AC.[17,19]

Interestingly, the decreased cAMP response was significantly negatively correlated with symptoms in schizophrenia but not in depression. This finding of reduced cAMP production in one peripheral cell type, but not another, also highlights the need for caution in extrapolating data derived from studies of peripheral cells to CNS function. Taken together, the results of studies of beta adrenergic and PGE$_1$-activated signal transduction in leukocytes and platelets have yielded conflicting and, for the most part, negative findings regarding the occurrence of signal transduction disturbances manifest in these peripheral cells of patients with schizophrenia. This contrasts to the agreement among investigators regarding the reduced leukocyte beta adrenergic receptor (BAR) response in patients with affective disorders.

IV. ABNORMAL SIGNAL TRANSDUCTION IN AFFECTIVE DISORDERS

A. Neurotransmitters

As with neuroleptic development, serendipity played a significant role in the discovery of neuroleptics. Reserpine, which depletes catecholamines, was not only the first effective antihypertensive but was also noted to induce severe depression in some patients treated with this drug. Recognizing the potential importance of catecholamines in depression, Schildkraut[20] and others proposed the catecholamine hypothesis of depression. More recent research indicates that other noncatecholamine-based neurotransmitters such as serotonin are at least equally important in understanding the pathophysiology and in the development of novel treatments for affective disorders.[2]

B. Receptors

Unlike schizophrenia, no specific receptor has been associated with a general theory of affective disorders. Down regulation or "turning off" the BARs appears to be universally associated with treatment of depression. First reported by Susler[21] and others, the down regulation of BARs appears to occur in chronic treatment of depression whether the antidepressants affect catecholamines or serotonin. Even successful electroconvulsive therapy is associated with down-regulated BARs. It is hoped that further elucidation of this phenomenon will indicate a potential therapeutic avenue rather than just the existence of an interesting epiphenomenon.

C. G Proteins

The most direct line of evidence supporting altered signal transduction in bipolar affective disorder was a marked elevation of the stimulatory G protein, G$_s$, in the frontal and occipital cortex of post-mortem brain from patients with an established, lifetime diagnosis of bipolar affective disorder compared with nonpsychiatric controls who had no evidence of neuropathology at the time of post-mortem examination of their brains.[22] These elevations in G_s were measured with well-characterized antibodies which were specific to the alpha subunit of G_s. The measured immunoreactivity was not likely to be related to ante-mortem lithium treatment, post-mortem delay, or age and sex differences. No differences were evident in other G protein subtypes between bipolar affective disorders and controls. Intuitively, it would seem that elevation in G_s would enhance the production of certain second messengers. Continuing with the timed switch analogy, increased G_s is similar to leaving the switch on for a longer period of time. In support of this notion, agonist-induced modulation of G protein subunit abundance in several cell lines is associated with altered receptor-effector responsivity.[23] While the mechanisms involved in this regulation of G protein abundance are still to be elucidated, the capacity for regulation of receptor-effector responsivity through changes in G protein abundance appears to be another important process regulating signal transduction. Although the G_s finding in bipolar disorder is striking, the evidence for altered G protein function in other forms of affective disorder is not as compelling.

D. Second Messengers

As in schizophrenia, there is conflicting research on peripheral blood cells, but for the most part the research indicates a reduction in cAMP production in response to various agonists. Given the down regulation of the BAR receptor, it might be anticipated that beta adrenergic–stimulated cAMP production would be suppressed, but there is considerable evidence that other agonists such as PGE_1 also result in decreased cAMP production in affective disorders.[4]

Intracellular calcium levels are of particular interest, since many enzymes are calcium dependent. In a study of bipolar patients in various states of clinical status, Carman et al.[24] found increased intracellular levels of calcium as patients with bipolar disorder became manic. It may be that these increased intracellular levels of calcium result from increased production of IP_3. As described above, IP_3 releases calcium from the smooth endoplasmic reticulum. Adding support to this notion of increased phospholipid turnover leading to increased intracellular calcium levels is the fact that lithium blocks inositol recycling by inhibiting inositol monophosphatase.[25] Unfortunately, the data on calcium channel blockers in the treatment of bipolar disorder are not compelling.[4]

V. SUMMARY

The Hebbian notion that altered communication between cells may lead to altered cellular function appears to have come of age in the work of signal transduction in psychiatric illness. Even with the advent of recombinant DNA and imaging technology, it is still very difficult, however, to analyze the components of signal transduction in schizophrenia and affective disorder. One is forced to choose from peripheral cellular models or post-mortem brain samples, both with their own limitations. These limitations notwithstanding, results of basic science studies provide a substantial and compelling body of evidence implicating alterations of signal transduction, possibly involving G proteins, in the pathophysiology of affective disorder. Delineating the various levels of altered signal transduction in affective disorders raises the possibility of defining multiple loci for the action of existing psychotropic medications, which may aid in the development of new psychotropic agents.

To date, the evidence for altered signal transduction in schizophrenia is inconsistent, likely reflecting the heterogeneity of the illness. Although the inconsistencies may be discouraging, some authors[4] have suggested that it may be possible to subtype schizophrenia by defining homogeneous patterns of altered signal transduction within subsets of schizophrenic patients. Moreover, elucidating mechanisms of altered signal transduction beyond the receptor may allow for the development of novel neuroleptics which are directed at G protein or effector enzyme function.

REFERENCES

1. Hebb, D. O., *The Organization of Behaviour*, John Wiley, New York, 1949.
2. Hyman, S. E. and Nestler, E. N., *The Molecular Foundations of Psychiatry*, American Psychiatric Press, Washington, D.C., 1993.
3. Gilman, A., G proteins: transducers of receptor-generated signals, *Annu. Rev. Biochem.*, 56, 615-649, 1987.
4. Hudson, C., Young, T., Li, P., and Warsh, J., CNS signal transduction in the pathophysiology and pharmacotherapy of affective disorders and schizophrenia, *Synapse*, 13, 278-293, 1993.
5. Chang, J., Musser, J., and McGregor, H., Phospholipase A2: function and pharmacological regulation, *Biochem. Pharmacol.*, 36, 2429-2436, 1987.
6. Jelsma, C. and Axelrod, J., Stimulation of phospholipase A2 activity in bovine rod outer segments by the beta-gamma subunits of transducin and its inhibition by the alpha subunit, *Proc. Natl. Acad. Sci. U.S.A.*, 84, 3623-3627, 1987.
7. Birnbaumer, L., Abramiwitz, J., and Brown, A., Receptor-effector coupling by G proteins, *Biochem. Biophys. Acta*, 1031, 163-224, 1990.
8. Stadel, J. M. and Lefkowitz, R., Beta-adrenergic receptors, in *The Beta-Adrenergic Receptor*, Perkins, J. P., Ed., Humana Press, Clifton, NJ, 1991.
9. Snyder, S. H., The dopamine hypothesis of schizophrenia: focus on the dopamine receptor, *Am. J. Psychiatry*, 133, 197-202, 1972.
10. Carlsson, M. and Carlsson, A., Interactions between glutaminergic and monoaminergic systems within the basal ganglia — implications for schizophrenia and Parkinson's disease, *Trends Neurosci.*, 13, 272-276, 1990.
11. Bredt, D. and Snyder, S., Nitric oxide: a novel neuronal messenger, *Neuron*, 8, 8-11, 1992.
12. Seeman, P., Lee, T., Chau-Wong, M., and Wong, K., Antipsychotic drug doses and neuroleptic/dopamine receptors, *Nature (London)*, 261, 717-719, 1976.
13. Seeman, P., Niznik, H., Guan, H., Booth, G., and Ulpian, C., Link between D_1 and D_2 dopamine receptors is reduced in schizophrenia and Huntington's diseased brain, *Proc. Natl. Acad. Sci. U.S.A.*, 86, 10156-10160, 1989.
14. Innis, R. and Aghjnian, G., Pertussis toxin blocks autoreceptor-mediated inhibition of dopaminergic neurons in rat substantia nigra, *Brain Res.*, 411, 139-143, 1987.
15. Pandey, G. N., Dysken, M. W., Garver, D. L., and Davis, J., Changes in the lymphocyte beta-adrenergic receptor function in affective illness, *Am. J. Psychiatry*, 136, 675-678, 1979.
16. Pandey, G. N., Garver, D. L., Tamminga, C., Ericksen, S., Ali, S., and Davis, J., Postsynaptic supersensitivity in schizophrenia, *Am. J. Psychiatry*, 134, 518-522, 1977.
17. Kafka, M. S., van Kammen, D., and Bunney, W., Reduced cAMP production in blood platelets from schizophrenic patients, *Am. J. Psychiatry*, 136, 685-687, 1979.
18. Rotrosen, J., Miller, A., Mandio, D., Traficante, L. J., and Gershon, S., Phospholipid and prostaglandin hypothesis of schizophrenia, in *Psychopharmacology. The Third Generation of Progress*, Meltzer, H. Y., Ed., Raven Press, New York, 1978, 759-764.
19. Garver, G. L., Johnson, C., and Kanter, D. R., Schizophrenia and reduced cAMP production: evidence for the role of receptor linked events, *Life Sci.*, 31, 1987-1992, 1982.
20. Schildkraut, J. J., The catecholamine hypothesis of affective disorders: a review of supporting evidence, *Am. J. Psychiatry*, 122, 509-522, 1965.
21. Susler, F., New perspectives on the molecular pharmacology of affective disorders, *Eur. Arch. Psychiatry Neurol. Sci.*, 238, 231-239, 1989.

22. Young, L. T., Li, P., Kish, S. J., Siu, K. P., and Warsh, J. J., Postmortem cerebral cortex Gs alpha-subunit levels are elevated in bipolar affective disorder, *Brain Res.*, 553, 323-326, 1991.
23. Milligan, G. and Green, A., Agonist control of G protein levels, *Trends Pharmacol. Sci.*, 12, 207-209, 1991.
24. Carman, J. S., Post, R. M., Runkle, D. C., Bunney, W. E., and Wyatt, R. J., Increased serum calcium and phosphorus with the switch into manic or psychotic states, *Br. J. Psychiatry*, 135, 55-61, 1979.
25. Hallacher, L. M. and Sherman, W. R., The effects of lithium ion and other agents on the activity of myso-insitol-1-phosphatase from bovine brain, *J. Biol. Chem.*, 255, 10896-10901, 1980.

Chapter **3**

THE AGING NERVOUS SYSTEM

Gwen O. Ivy

CONTENTS

0-8493-8386-0/96/$0.00+$.50
© 1996 by CRC Press, Inc.

I. INTRODUCTION

The chapter explores changes in our sensory systems, which affect our perceptions, as well as changes in our central nervous system (CNS), which affect all aspects of our behavior as well as our general health status. The vast majority of changes that occur during the normal aging process are not reversible and often lead to a cascade of events that are deleterious to an individual's ability to function. However, it is important to note that aging does not affect all of us in the same ways or at the same rate, because of both genetic and environmental factors. Indeed, many individuals over 80 years of age retain excellent sensory, motor, and cognitive abilities.

II. AGE-RELATED CHANGES IN SENSORY SYSTEMS

The loss of one's contact to the outside world through one's sensory systems would be a frightening ordeal if it happened all at once. During aging, however, the loss of our sensory capacities happens generally quite slowly — so that we mentally and/or physically compensate and, therefore, tend not to notice our losses. If we do notice, we generally seek corrective aids (prostheses) when possible. However, none of us wants to admit that we are "growing old," that is, that the aging process can actually affect us, and so most of us tend to procrastinate in realizing any sensory incapacitations as they very stealthily creep up on us. This causes changes in our perceptions of the world and thus the way we interact with it.

A. Vision

Vision and audition play major roles in spatial orientation, localization, and social communication. An age-related decline of these functions may contribute significantly to deterioration of cognitive abilities in the elderly. Changes in visual acuity for the aged may pose many difficulties, ranging from decreased mobility to dependency on others to loss of self-confidence and depression. From the practical point of view, evaluation of cognitive functions must therefore include evaluation of the sensory pathology as well. Visual acuity declines linearly between age 20 and 50 and then exponentially from age 60 to 80[1] with age effects being significantly greater for dynamic as opposed to static acuity.[2] Both the cornea and lens are responsible for focusing a clear image on the retina, and both undergo alterations in composition, refraction, and power leading to distorted images to the retina. At about age 60, the cornea can show the formation of a senile ring known as the *arcus senilis,* an opaque border between the cornea and sclerotic coat thought to result in the decrease in peripheral vision and visual field.[3] The basal surface of the epithelium becomes wavy, irregular, and gets thicker, and the epithelial cell cytoplasm exhibits brownish, finely granular ferric substances bound within lysosomes.[4]

After age 40, the lens undergoes a thickening process characterized by a laying down of cortical fibers, compaction of central nuclear fibers, and a decrease in elasticity of the lens capsule. This, along with a yellowing of the lens, leads to a decreased ability of the lens to focus as a result of reduced malleability.[5] This loss for near vision is universal, progressive, and known as presbyopia. As a result of the above, there is a decrease in the amount of light reaching the retina, increased glare as a result of increased lens opacity, and a consequent need for greater illumination.[5] Changes in the lens also account for altered color perception for the short wavelengths of blue and green, especially after age 60. The longer wavelengths of red appear to remain relatively stable. Depth perception (binocular stereopsis) remains relatively constant between ages 25 and 45, then declines at an accelerating rate until age 75 as a result of peripheral changes.[6] The retina undergoes thickening and breaks in the Bruch's membrane, leading to decreased blood circulation, clustering of pigment in the pigment epithelium, a decrease in the number of photoreceptors, and glial proliferation in the macula. Retinal ganglion cells decrease in number, while their dendrites enlarge and become packed with lipofuscin (age pigment, see below), which is followed by ultimate optic nerve loss.[1] Finally, the visual cortex shows a significant reduction in the number of neurons (50% decline in 80-year-olds) leading to an altered recognition of form, color, movement, and depth hierarchy.[1]

B. Audition

It has been estimated that more than 50% of people over 60 years of age have some degree of hearing loss, and the incidence increases significantly with age. Irreversible hearing loss occurring with age is known as presbycusis and is characterized by diffuse, symmetrical loss of outer and inner hair cells of the organ of Corti in the lower basal turn of the cochlea and accompanying degeneration of cochlear nerve fibers. Presbycusis is associated with high-tone loss and reduced speech understanding. Further, individuals have problems comprehending both interrupted and speeded speech as early as the fifth decade.[8] By age 80, speech comprehension may be reduced by more than 25%.[7] Changes in hearing threshold occur continuously throughout adulthood[8] with progressive loss starting at an average age of 32 for men and 37 for women.[7] An approximately linear hearing loss has been noted for lower frequencies, while an exponential hearing loss is·seen at high frequencies in people between 30 and 80 years of age.[7]

Four types of presbycusis have generally been acknowledged. These are

1. *Sensory*, characterized by atrophy of hair cells, degeneration of the organ of Corti, and associated degeneration of spinal ganglion cells and dendrites in the basal turn of the cochlea. Regeneration of hair cells does not occur, and, consequently, this hair cell loss, along with loss of supporting cells, is thought to lead to degeneration of cochlear neuron dendrites. Affected persons show a sharp decrease in hearing for all frequencies, but especially high frequencies.

2. *Neural*, associated with primary degeneration of neurons of the auditory pathway and thought to parallel general neuron loss in the CNS. Reduced speech discrimination is seen with no impairment of pure tone discrimination.

3. *Metabolic or strial*, stria vascularis atrophy leading to uniform hearing loss across all frequencies. Little or no hair cell loss is observed. Some researchers believe the marked devascularization resulting from strial atrophy plays a central role in sensory cell loss.

4. *Mechanical*, atrophic changes in the basilar membrane. Loss of elasticity results in hearing loss, especially for high frequencies.

Age-related changes at a cellular level reveal degeneration, loss of nerve fibers, cell shrinkage, intraneuronal pigment accumulation, and dendritic mitochondrial alterations of the spinal ganglion cells within the organ of Corti. Extensive neuronal loss has been reported in the human superior temporal gyrus of the auditory cortex, and animal studies have revealed cortical thinning, axonal abnormalities, and increased glial proliferation.[9] Brody[10] cited neuronal loss of 50% from ages 20 to 75 years in the superior temporal gyrus. Scheibel and Scheibel[11] found progressive dendritic changes in layer III and V pyramidal cells of the superior temporal (the primary auditory area of the cerebral cortex) and middle frontal gyri. Changes begin with swelling in the cell body and base of the apical dendrite with ultimate loss of both basal and apical dendrites, leading to progressive neuronal loss and subsequent thinning of the cortex. Scheibel and Scheibel also noted age-related changes of increased lumpiness and dimensions of the human auditory cortex and attributed these changes to an abnormal accumulation of tubular material. Finally, other age-related changes noted to date are increased lipofuscin and neuronal loss in the superior olive, inferior colliculus, medial geniculate nucleus, and auditory cortex. A significant age-related decline may occur in neurotransmitters and their enzymes in the medial geniculate and auditory cortex.[7]

Hearing loss has social, psychological, and personal implications for individuals suffering from it. Personal safety may be compromised because of an inability to hear verbal warnings, approaching traffic, or imminent threats. Occupational status may be jeopardized if one's job relies on a high level of communication with others. Paranoia may arise from the feeling that others are whispering about you, and frustration, withdrawal, and isolation can result when only fragments of conversation are heard or when constant repetition is required. Although aids are available to those with hearing deficits, the aids are often inadequate when enhancing the higher frequencies.

C. Somatosensation

The somatosensory system is responsible for the five sensations: tactile, vibratory, hot, cold, and pain. It includes the kinesthetic apparatus and vestibular system as well. The main sensory receptors of this system are thought to be the Meissner and Pacinian corpuscles, which are located in the dermis of the skin and which provide information about the above sensations to the spinal cord and CNS. Other receptors include intraepidermal nerve endings, the superficial dermal nerve network, the hair follicle network, and mucocutaneous end-organs.[12]

The Meissner corpuscles are responsible for extreme tactile sensitivity in the fingertips and for low-vibration detection below 80 Hz. Their concentration is greatest in the glabrous skin of the fingertips, followed by the palmar skin, thenar eminence, and plantar surface of the big toe. With age, the corpuscles become enlarged, coiled, and irregular. They also elongate, decrease in number, loosen from the epidermis, and may detach totally. Meissner corpuscles show a decrease with age of up to 90%.[12] Pacinian corpuscles are found in the glabrous skin of the viscera around joints, muscle tendons, intraosseous membranes, and periosteum. Considerable numbers are lost with age along with such morphological changes as increases in size, because of layer upon layer of lamellae being laid down along the innervating axon, and the development of an irregular shape.[13] The role of other receptors, such as Kraus end-bulbs, ruffinian cylinders, and Merkel disks, is less well known, and age-related changes have not (to this author's knowledge) been studied.

It is generally agreed that tactile acuity decreases with age[13,14] and that a decrease in the number, rather than function, of Meissner corpuscles or Merkel disks may be the cause for impairment.[15] Elderly hands are found to be more sensitive than the feet, in keeping with the fact that the hands house greater numbers of Meissner corpuscles than do the feet. The aged also have a decreased ability to detect "double stimulation" which results in a "facial dominant response" upon simultaneous stroking of the cheek and hand. This is observed even in those elderly people with high cognitive levels, leading to the conclusion that the loss of tactile sensitivity is real and not a consequence of decreased cognitive ability.[15] Although some disagreement exists among researchers regarding the thinning of epidermis and dermis with age, decreased amounts of collagen and elastin and a general thinning of the epidermis have been acknowledged.[13] These mechanical changes relate to the depth of indentation produced by any force used to detect tactile acuity and may predict loss of sensitivity by way of age-related biological changes. Stevens[14] has proposed that a breakdown of peripheral receptors or their neural attachments may be caused by changes in mechanical properties of the skin or vascular changes rendering peripheral circulation sluggish. It is important to acknowledge that many elderly retain a high level of tactile acuity.[15] More research is needed to evaluate if loss of tactile sensitivity affects one's ability to locate, manipulate, and identify objects.

With regard to vibratory sensitivity, the elderly are significantly less sensitive to vibrations on the hands and feet. The loss starts at about the fifth decade, is more severe for the lower extremities (approximately 23 to 40%), and is relatively uniform at all frequencies.[13] Impaired sensitivity may be because of diminished receptor populations in the hands and toes and reduced circulation, especially in the legs and spinal cord. Also, a decrease in vibratory detection on the inside of the ankle has been implicated in increased postural sway in the aged.[16]

Research on pain and temperature sensitivity in the aged has yielded contradictory results. When electrical stimulation to the teeth and radiant heat on the skin are used to induce pain, no differences are found between young and aged subjects.[12] However, Harkins et al.[17] observed increased pain thresholds for middle-aged and elderly individuals exposed to contact heat of 43 to 45°C. Small differences observed in the responses of the aged to radiant heat may be because of changes in thermal dispersion properties of the skin. It is important to recognize that pain is more than a sensory experience, for there are cognitive as well as motivational aspects that will affect the subject's response. For example, anxious subjects report lower pain thresholds. Also, subjects' thresholds increase as they gain experience with the stimulus and deem it unharmful, leading to a higher criterion of what is classified as painful. On the other hand, because endogenous opiates are our natural analgesics, Hamm and Knisely[18] studied pain and aging in relation to these morphinelike compounds and concluded that pain perception increases with age, possibly by reducing the amount of opioid binding to a decreased number of receptors. Thus, in contrast to the clear loss of touch information, the change in the perception of pain is more complicated. Certain changes may lead to a loss in pain sensitivity whereas others may actually heighten perception of pain when the pain signal finally gets through. The ratio of these two types of change may well differ across individuals, preventing a straightforward generalization.

The aged seem to be susceptible to falls, gait disturbances, muscle weakness, slowed reaction time, and dizziness. Little is known about age effects on the kinesthetic receptors, though changes in metabolism and function of skeletal muscles exist with age. Finally, the incidence of falls and their possible relation to the issue of postural sway may be considered here. Postural sway has been correlated with age and appears to be associated with low aerobic and anaerobic capacity, increased percentage of body fat, and decreased muscular strength.[16] It may be possible to improve control over postural sway by way of physical training, but it is important to consider that an impairment of other senses, such as visual, vibratory, and vestibular senses, can also lead to increased postural sway.[16]

D. Olfaction and Gustation

The senses of smell and taste are known as the "chemical senses" with both involving psychological and physiological experiences while at the same time having separate receptors, nervous pathways, and brain regions. These two senses will be discussed in tandem because smell is responsible for most food flavors and a decrease in olfactory acuity subsequently results in a significant decline in food recognition.

Smell sensitivity has been estimated to be 10,000 times greater than taste sensitivity[19] and shows a general decline, loss of discrimination, and recognition with age. Total anosmia is considered rare and usually follows a head injury resulting in severance of olfactory nerves; however, declines in acuity are significant with age when compared with the fine discrimination that occurs in youth.[20] There is evidence that suggests the olfactory system is more susceptible to injury than is the taste system and that aging appears to slow recovery.[21] Although the use of olfaction over one's life span may retard its decline, olfactory acuity appears to peak at age 20 and decline thereafter, with a noticeable decrease in odor detection after age 70.[22] However, detection of mercaptans, an odorant used to enhance natural gas detection, declines at about age 50, posing a serious health and safety risk. Russel et al.[23] have found evidence supporting a selective, rather than a uniform, alteration of odor perception, with sweet-smelling items exhibiting the most vulnerable age-related change.

The olfactory receptors are specialized bipolar neurons located in the olfactory epithelium along with supporting cells and basal cells. Their numbers are vast, being estimated at about 10,000,000,[24] and they retain the capability of undergoing continuous renewal from undifferentiated basal cells at the base of the epithelium. This turnover is unique to the nasal epithelium, unlike other CNS neurons, which steadily die and are not replaced over one's life span.[20] For some researchers, this renewal process makes the olfactory receptors much more vulnerable to decrements in division potential and function with age, poor nutrition, drugs, and hormonal status.[25] Other investigators argue that this constant turnover allows for "a second chance" in this neuronal population. The olfactory neurons are often said to be the meeting place of the brain and the environment.[20] The dendrites of these neurons possess motile cilia covered with mucus to create a greater surface area exposed to the environment. The axons make up the olfactory nerve, and each axon travels without branching through holes in the cribriform plate to the olfactory bulb, where much of the information processing that leads to odor identification takes place.

Age-related declines in olfaction detection and recognition may be due to a number of physiological and cellular changes. First, there is a decrease in the yellowish brown pigment found in the olfactory epithelium.[3] Contrary to earlier speculation that this is lipofuscin, the pigment is now believed to be important for olfactory perception; its decline with age may reflect either decreased receptor cell numbers or a decrease in the amount of pigment per cell. Either decreased receptor numbers or the age-related thickening of the cribriform plate (and subsequent narrowing of holes through which the olfactory nerves pass) may lead to a decrease in the number of primary axons reaching the glomeruli.[25] The glomeruli demonstrate a moth-eaten appearance as these fibers degenerate and die. In the olfactory bulb, neuronal loss, decreased numbers of synapses, increased amyloid bodies, and hypertrophied astrocytes have been documented, suggesting that such degenerative changes may be secondary to the loss of receptors, which in turn may be secondary to external (environmental) factors.[26] A transsynaptic transfer of degenerative changes may well occur in the olfactory system, since a decrease in olfactory perception is often accompanied by more central age-related changes, such as neurofibrillary tangles and senile plaques in hippocampus and neurofibrillary tangles in the amygdala.[25]

Decreased olfactory acuity has also been cited in smokers[3] and those with cerebral vascular disturbances.[27] Age-related diseases such as Alzheimer's, Parkinson's, and Huntington's,[28] as well as epilepsy,[29] also impair olfaction, with olfactory deficits appearing to be among the first signs (for review see Reference 21). Evidence suggesting that pathological changes are because of an uptake of toxic substances, such as aluminum or viral involvement has not been conclusive. Olfactory deficits in patients with Alzheimer's disease are associated with neuritic plaques and neurofibrillary tangles throughout the olfactory-related brain structures, a decrease in mitral cells in the olfactory bulb, and plaques, neurofibrillary tangles, granulovacuolar degeneration, and cell loss in the anterior olfactory nuclei. Patients with Parkinson's disease exhibit a significant reduction in uptake of L-DOPA in the caudate and putamen, with olfactory deficits appearing to be unrelated to the severity of motor or cognitive symptoms and with no improvement with L-DOPA therapy.

A comparison of sex differences and olfactory acuity has revealed that female olfactory function is superior across the human life span and across cultural groups, with the differences possibly reflecting anatomical and physiological variations in nasal airways, olfactory neural pathways, or endocrine systems.[28] Both olfactory and taste acuity appear to decrease most in aging males, followed by the sick and those taking medications.[19]

Though some elderly retain olfactory acuity levels which approximate those of the young,[30] research clearly indicates that sensory decline in this area exists, with important implications regarding safety and independent living. Since olfaction plays a major role in taste, a subsequent loss of interest in food may arise, leading to nutritional deficits and digestive disorders. Decreased olfactory acuity can also lead to a decrease in personal hygiene, which may contribute to social isolation of the aged. Associated with a decrease in olfaction is a decline in taste acuity. The major taste receptors are the taste buds scattered throughout the oral cavity on the soft palate, pharynx, and larynx, and are very heavily concentrated on the tongue. Some discrepancy exists as to whether the number of specific taste buds declines with age and whether taste bud number is a criterion for acuity. In brief, a major assumption has existed among researchers that, since the number of taste receptors declines with age, there must be a resultant decline in taste perception. However, there has been, to date, no evidence to support this since it is not known how many taste buds are needed to provide adequate sensory input for perceptual function. Investigations have shown the importance of the referral of olfactory input to the taste system.[21] It has also been acknowledged that taste thresholds of detection and recognition decrease with age. However, Weiffenbach et al.[31] point out that taste complaints may exist in the absence of any gustatory sensory deficit and that memory distortions and social and emotional changes within the individual are possible causes.

Taste thresholds decline differentially for different substances. Studies conflict as to increases or decreases in taste thresholds for sour, bitter, salt, and sweet. When comparing taste acuity of institutionalized and noninstitutionalized elderly men, Spitzer[32] found detection for sour, salt, and bitter increased with age while sweet detection appeared not to be age related. Among medicated males, sour thresholds increased, and medicated hypertensives were found to have significantly higher salt thresholds. Variations in sweet detection among previous studies may be due to experimental shortcomings, poor oral hygiene, inadequate volume of tastant, and short tastant time. Schiffman[33] found elderly subjects preferred their food amplified with a few drops of commercial flavoring. She also found that amplification of food odors helped decrease complaints of food bitterness — in essence, enhancing the odor enhances the flavor. An interesting view posed by Chauhan[34] suggests that ethnic background may also have an effect on adaptation to tastes and smells. Taste is adversely affected by dentures, missing teeth, caries, periodontal disease, poor oral hygiene, and salivary changes. Good oral hygiene improved the ability for salt acuity by 68%.[35]

Saliva acts as a solvent, dissolving molecules into solution and consequently evoking taste responses. Salivary flow is adversely affected by anticholinergic medications and such systemic conditions as malignant lymphomas associated with Sjogren's syndrome, rheumatoid arthritis, diabetes, cirrhosis, parkinsonism, and depression.[36] Medications are also suspected of introducing tastable substances into the blood which can then alter taste thresholds.[37] In conclusion, whether a direct decline in taste acuity is associated with aging or is more a result of secondary factors as previously mentioned, a decreased taste acuity can impact on one's life in ways that range from oversalting food to nutritional deficits induced by a lack of interest in eating to danger of food poisoning because of an inability to detect contaminants.

III. GROSS CHANGES IN THE BRAIN

A. Neuron Loss

In humans, neuronal dropout appears to be a consistent feature of the aging brain, although some brain regions are more affected than others. This complex and conflicting literature has been extensively reviewed by Coleman and Flood,[38] who point out that numerous technical difficulties plague such quantitative studies. For example, neurons not only die but shrink in size, and the smallest ones soon become practically indistinguishable from glial cells. To complicate matters further, glial cells themselves proliferate, thus replenishing cell density (though not cell volume, since glial cells are smaller than most neurons). These changes are then superimposed upon a generally shrinking brain (due mainly to loss of neurons and their extensive dendritic arbors), with shrinkage affecting neocortex in particular. Nonetheless, when all factors are considered, the bulk of the evidence indicates a substantial decrease in neuron density, as well as in absolute numbers, in a variety of brain regions.

In neocortex, Brody[10,39] reported a 50 to 60% decrease in neuron density in superior temporal gyrus, a 50% decrease in superior frontal gyrus, a 20 to 30% decrease in precentral gyrus and area striata, and a 10 to 20% decrease in postcentral and inferior temporal gyri between young to middle adulthood and the ninth to tenth decades of life. Other authors find decreases in neuron density from 10 to 60% in different neocortical areas, thus corroborating the general finding that neuron numbers decrease from youth to old age and that some neocortical areas are more susceptible than others.

There is also substantial evidence that cells are lost with age in human hippocampus and subiculum, structures intimately involved in learning and memory processes. Although specific estimates of loss vary across studies, most hippocampal regions appear to undergo a moderate loss of approximately 20% with age. In aged nonhuman primates, there is a substantial reduction of both the depth of the CA1 pyramidal cell layer and the density of neurons in this layer (by about 40%), indicating a high degree of neuronal loss in this hippocampal region.[40] Data on neuronal density and volume in various regions of rat hippocampus similarly reflect neuronal loss with age.

Subcortical brain regions display variable neuron loss with age. For example, the olfactory bulbs and some hypothalamic nuclei show neuronal loss, while other hypothalamic nuclei display stable numbers of neurons.[41] Specifically, the supraoptic and paraventricular nuclei of the hypothalamus do not show neuronal loss with age in either humans or rodents although, at least in humans, the medial preoptic nucleus shows pronounced loss throughout life in both sexes. It is presently not known how neuron loss in this sexually dimorphic brain region might correlate with gonadal aging or with sexual behavior.

In the cerebellum, there is substantial evidence that at least 20 to 30% of Purkinje cells are lost with age in humans[42] and nonhuman primates.[43] Although this may lead to motor disturbances, such as uneven gait, any effects of such loss on cognitive processes are not obvious.

Despite conflicting reports on loss of cholinergic neurons in human nucleus basalis of Meynert during normal aging, there does seem to be real loss if the pre-adult period of life is compared with the senescent period. However, given that the death of some cholinergic neurons is probably compensated for by increased transmitter production by neighboring neurons, we might not expect this loss to result in cognitive deficits unless it proceeds to pathological levels such as are found in Alzheimer's disease.

Finally, there is an approximate 30 to 40% decline in numbers of pigmented neurons in both locus coeruleus[44,45] and in substantia nigra[46] of humans. Although the exact time of the neuron loss is not clear, it is apparent after 50 years of age. A severe loss of locus coeruleus neurons that project to neocortex has been related to attentional impairment in both Alzheimer's and Parkinson's diseases,[47,48] and a severe loss of substantia nigra neurons that project to the striatum has been linked to motor disturbances in Parkinson's disease.[49]

B. Brain Vascular System Changes

Blood vessels transport oxygen and glucose, as well as other nutrients, growth factors, and hormones, to virtually all of the cells of the body. Like other tissues, blood vessels show signs of aging which may impair their function and impact heavily on the tissues supplied. In the case of nonproliferative populations such as neurons, decreased oxygen (ischemia) leads to a frank loss of cells. Brain microvasculature is specialized both morphologically and with active transport mechanisms to maintain a blood–brain barrier, which it does at some energy cost. Age-related changes to brain circulatory systems may have the added consequence of permitting entry into the neural tissues of substances not normally found there, such as amyloid, to be discussed below. Erosion of the blood–brain barrier is of special concern because increased permeability permits the entry into the brain of blood-borne toxic or infectious substances.

In fact, a number of morphological changes have been noted in brain microvasculature during normal aging. Fang[50] has demonstrated a significant increase with age in the "winding" or "coursing" effects of cerebral blood vessels in normal human brain. For example, there is a pronounced decrease in vascularity in deeper laminae of the cortex in areas where there is also cell loss. Fang[50] proposes that this rearrangement of microvasculature is related to neuron dropout and concomitant glial infiltration of cortical regions, although neither the initial cause nor the effect of such changes is currently known. Certainly, such dramatic cell loss and vascular reorganization in the cortex would be expected to result in some type of functional or cognitive deficit, but the nature of this deficit would hinge critically on the cortical region(s) affected.

Some morphological changes with age may well reflect altered blood-brain barrier function. In aged monkeys, both thickened basement laminae of cortical capillaries and aberrant interendothelial tight junctions — anatomical substrates of main portions of the blood-brain barrier — have been observed.[51,52] A similar thickening of the basement lamina was observed in cortical capillaries of Alzheimer's disease and was thought to be involved in altered blood-brain barrier permeability.[53] The number of mitochondria per cerebral capillary profile was also shown to decline with increasing age in both rat[52] and monkey.[51] Such a decrease in energy-producing mitochondria, coupled with changes in interendothelial tight junctions and the overall thinning of the endothelial component of the capillary wall, suggests a decline in the work capability of these capillaries with serious perturbations in ionic homeostasis between blood and brain.[52] Once again, this puts the brain at risk from factors present in the blood that are normally blocked from access to the brain.

The above noted microscopic changes in cerebral vasculature with age may contribute to decreased cerebral blood flow. Cerebral blood flow and oxygen uptake in a normal elderly population (mean age 71 years) were found to be lower than the values for a young group (mean age 21 years), and cerebral oxygen consumption was 6 to 10% lower in the elderly.[54,55] The reduction in metabolic rate was found to be greater on the left side than on the right side,[56] suggesting implications for language in particular. In contrast to normal aged individuals, both cerebral blood flow and oxygen consumption are markedly reduced in patients with presenile and senile dementia; most important, this is in general proportion to intellectual deterioration.[56]

Specific subsystems of the clinical picture also correlate with both decreased cerebral blood flow and neuronal degeneration. For example, decreased blood flow occurs in the temporal region of demented patients who display primarily memory deficits, in occipitoparietotemporal regions of patients displaying agnosia and disorientation, and in both frontal and post-central-temporal regions of patients with severe mental deterioration.[57] The regions with lowest blood flow were also those with the most pronounced degenerative changes at autopsy,[58] although the direction of causality has not been established. Further, it should be emphasized that it is not known if the loss of neurons or the decreased blood flow is the primary pathogenetic factor in the various subtypes of dementia.

C. Brain Weight

An obvious place to begin any examination of gross changes in brain anatomy is with the overall weight of the brain. In humans, there is a progressive loss of brain tissue with advancing age. This is characterized, in particular, by a marked atrophy of neocortical gyri and a widening of sulci, with a corresponding secondary dilation of the ventricular system.[11,59] Total brain mass shrinks by approximately 5 to 10% per decade in the normal aged individual, leading to losses of 5% by age 70, 10% by age 80, and 20% by age 90.[60,61] The majority of the tissue loss in normal human aging is in the cerebral cortex. Focusing just on the cortex, Corsellis[62] found a 2% (female) to 3.5% (male) drop per decade from age 20. The atrophy is most marked over the frontal lobes, although parietal and temporal lobes suffer considerable losses as well. Several investigators have attempted to correlate declines in overall human brain mass with behavioral indexes, particularly in cases of dementia. There does not appear to be a clear correlation between these two factors in humans.[61-63] Terry and

Davies[64] did not find differences in overall brain weight or in more specific measures of cortical thickness in the frontal and superior temporal regions when they compared demented and age-matched controls. However, it should be noted that most studies attempting to correlate declines in brain mass with loss of cognitive abilities have not focused on the mass of individual structures as, for example, hippocampus. In sum, the existing work on gross brain mass would lead to the conclusion that, despite the clear decline of this mass with age in humans, there is no correspondingly clear behavioral consequence.

D. Glial Cell Changes

There are three types of glial cells in the CNS: astrocytes, oligodendrocytes, and microglia. These cells perform a wide variety of "support" functions in the nervous system; while these cells are absolutely integral to neuronal function, they do play more of a support and modulatory than an information transmission role. For example, astrocytes are known to function in the removal of degenerative debris in the nervous system, as well as in maintaining ion homeostasis and in metabolism of putative and acknowledged neurotransmitters. In response to practically any form of neural insult, astrocytes become hypertrophic, displaying a response that includes enlargement of the cell body and increases in both the number and length of astrocytic processes. In fact, these are the cells that form glial scars after brain or spinal cord injury. Oligodendrocytes provide the insulating myelin sheath for many CNS axons and are responsible for maintaining this sheath, as well as for control of the local ionic environment of the axon. The functions of microglia are not as well understood, but, like astrocytes, these cells are thought to play a role in phagocytosis, the cellular engulfment and digestion of neural debris following brain trauma.

Changes in glial cells with age have not been widely investigated, probably because of both past difficulties in distinguishing the different glial subtypes in the same region and our lack of understanding of specific glial functions. It is generally accepted that most glial cells retain the ability to proliferate throughout an individual's life span, but it is not known if all glial subtypes do or if they do to the same extent. Korr[65] has shown that several nonneuronal cell types can proliferate in brains of aged rodents. These proliferative cells include astrocytes, oligodendrocytes, cells of the subependymal layer (thought to contain undifferentiated glioblast cells), and endothelial cells, which line the vascular system. Extensive data on proliferation of two of these glial subtypes, astrocytes and microglia, have been reviewed by Finch and Morgan.[66] These findings, along with those showing decreased numbers of neurons with age, are consistent with the idea that more glial cells are created to handle the increased incidence of neural trauma with age. These data also imply that there should be an increase in the glia-to-neuron ratio with age.

IV. CELLULAR CHANGES

A. Lipofuscin

One of the most consistent age-related morphological changes to occur in cells is an approximately linear accumulation of lipofuscin, or age pigment. Although lipofuscin accumulates in virtually all organisms and all cell types, neurons accumulate larger amounts of the pigment because they generally do not divide during the postnatal period. The term lipofuscin derives from the Greek *lipo* (fat) and *fuscus* (dark or dusty) and was first given to the pigment because of its staining with lipidic dyes and its brownish color in unstained sections. It is now known, however, that a major portion of the pigment is proteinaceous rather than lipidic. Lipofuscin granules, as seen under the electron microscope, have several emblematic morphological properties. The granules are surrounded by a single limiting membrane of lysosomal origin and contain an electron-dense matrix which may be finely or roughly granular and may contain lamellar (membranous) bodies and/or vacuoles (previously containing lipid); under ultraviolet light, the lipofuscin pigment emits a characteristic yellowish autofluorescence.

It should be emphasized that lipofuscin is a normal component of the cell cytoplasm; it represents the normal process of cellular catabolism taking place in digestive organelles called lysosomes. Thus, even young, healthy neurons and glial cells contain some lipofuscin; the amount present represents the balance between anabolic and catabolic processes in the cell. The increased amount of cellular lipofuscin seen with age may be because of decreased efficiency of the lysosomal system. This, in turn, may be caused by decreased synthesis of enzymes or by decreased efficiency of enzymes because of free radical damage to either the enzymes themselves or to their substrates.[67,68]

The accumulation of lipofuscin is not uniform throughout the brain, but displays regional differences. In a nonhuman primate (*Macaca mulatta*), the rank order of lipofuscin accumulation from highest to lowest was found to be medulla, hippocampus, midbrain, pons, neocortex, and cerebellum, as determined by percentage of neurons in a region displaying aggregates of lipofuscin.[69] Further, within a given brain region, some cell types accumulate more lipofuscin than do others. For example, in neocortical association areas, medium and large pyramidal neurons of layers II, III, and V accumulate more pigment than do stellate or small pyramidal cells of layer IV, and the large cells of the brain stem acquire more pigment than do the small cells.

A specialized form of lipofuscin in neurons is neuromelanin. The ultrastructure and composition of neuromelanin are closely related to those of lipofuscin, with the exception that neuromelanin contains an extra electron-dense component that is believed to consist of catecholamine degradation products. Neuromelanin is a prominent component of lipofuscin in neurons of substantia nigra and locus coeruleus, in particular, as well as in nucleus paranigralis, nucleus subcoeruleus, the dorsal motor nucleus of the vagus, and a few other nuclei.[70,71] While neurons in substantia nigra and locus coeruleus are especially susceptible to death during normal aging and (particularly) in Parkinson's disease, pigment accumulation is not known to be responsible for the death of these neurons.

In sum, lipofuscin is a well-known marker of aging in virtually all types of neural cells. However, there are as yet no firm data linking the age-related accumulation of this pigment to any functional deficits at the cellular or organismic level.

B. Dendritic Changes

A number of researchers have examined dendritic structure in the aging rat and have reported reduced branching complexity of basal and oblique dendrites of layer V pyramidal cells, as well as a thinning of dendritic spines in the visual and auditory cortex (reviewed in Reference 72). In the rat and mouse occipital cortex, layer III pyramidal cells also show a decrease in dendritic arborization with age; this is especially evident in the distal apical dendritic branches of the oldest animals. A similar reduction has been observed in layer III pyramidal cells in the auditory cortex. However, examinations of layer II pyramidal cells in entorhinal cortex and of layer IV stellate cells in mouse somatosensory cortex revealed no dendritic changes over the life span. It is very difficult to do these studies well in humans because of increased (progressive) dendritic spine deterioration with increasing time post-mortem before proper tissue fixation is achieved.

In contrast to findings of dendritic decline with age in certain rat brain regions, in others an age-related increase in dendrites is actually found. Flood and Coleman[73] view these apparently discrepant age-related changes as a manifestation of the shifting balance between degenerative processes and compensatory proliferation of both synapses and dendritic branches. Thus, although some brain regions initially show compensatory dendritic proliferation to age-related neuronal loss, ultimately in old age, a regression is found.

C. Synaptic Changes

There appears to be an age-related decline in the number of synapses in the cerebral cortex (reviewed in Reference 72). Again, these studies have been done in the rat. This has been observed as decreased synaptic density in the molecular layers of both the occipital and the parietal cortex. In addition, the number of synapses per neuron has been estimated to decline following young adulthood and then to level off into old age.

In the hippocampal dentate gyrus of rat, decreases with age in synaptic density and in estimated number of synapses per neuron have been reported, although there are also reports of synaptic stability in this region. In the CA4 region of the rat hippocampus, however, no age-related changes in synaptic density were observed. Thus, as was found to be the case with dendritic branching, synaptic density and number vary according to the specific brain region examined, but generally appear to decline with age. The status of changes in shape and size of hippocampal synapses with age is not clear at present.

D. Decreased Ability to Cope with Brain Damage

During development, there is a proliferation of dendrites and synapses, and even after reaching a mature configuration, the dendritic tree and synaptic component of the neuron continue to be plastic. The capacity of the neuron to respond anatomically by producing new dendritic and synaptic material is assumed to underlie learning and memory processes. Therefore, an examination of these plastic capacities of the neuron is of particular significance.

Plasticity has been induced or examined in aging using such techniques as environmental enrichment, lesion-induced synaptogenesis, and naturally occurring reactions to age-associated synaptic loss. At least in rat (again, immediate fixation is necessary and thus precludes such studies in humans, who also vary in so many genetic and environmental ways which lab animals largely avoid), increases in dendrites and synapses can be induced through environmental stimulation or by the elimination of neighboring afferents. Studies examining environmental enrichment effects on general brain features have found changes during early adulthood that include increased occipital cortex thickness and decreased neuronal packing density, indicating more "neuropil" that contains dendrites, synaptic connections, glial cells, and vasculature. In later adulthood and aging, these enrichment effects appear to be reduced.

In addition to age-related alterations in enrichment-induced synaptogenesis, researchers have also examined the response of the aged brain to lesion-induced synaptogenesis, which has more bearing on events that may take place during the aging process. During aging, synaptic "sprouting" seems to decline. Sprouting may be defined as the new growth of axon collaterals in response to a nearby denervated dendrite (due to lesion, disease, or aging). This has been most clearly shown as a reduced sprouting response in the hippocampus following lesions of the hippocampal commissure. Briefly, McWilliams and Lynch (see Reference 74 for review) have shown that when the hippocampal commissure of rats is lesioned at various ages, the ability of the entorhinal cortical afferents to grow down into the proximal dendritic territory normally occupied by the (now nonexistent) commissural synapses decreases dramatically with age: at postnatal day 15, full recovery of synaptic number is achieved by 1 week, while at 2 years the synapses are not noticeably replaced. This indicates that the aged brain does not recover from damage as well as the young brain does.

Flood and Coleman[73] have suggested that while certain populations of neurons are undergoing atrophy during aging, others are undergoing a compensatory plastic proliferatory response. Neuronal connections are likely quite dynamic even during aging, with a continuous interplay between growth and regression processes. Initial cell loss appears to be compensated for by a growth of dendrites and/or synapses in adjacent, or connected, neurons. However, in addition to the increased cell loss during aging, there is a reduction in the capacity or speed of compensatory growth processes, leading to only a partial compensation or recovery. Thus, although retaining "plasticity," the brain cannot completely accommodate the changes caused by age.

E. Amyloid Plaques and Neurofibrillary Tangles

Alzheimer's disease (AD) is characterized by numerous β-amyloid plaques (neuritic plaques, NP) and neurofibrillary tangles (NFT) throughout the neocortex and other brain regions (for review see Reference 75). Thus, most of our knowledge of these pathologies, as well as their genesis, comes from post-mortem studies of AD brains. However, whether or not NP and NFT actually occur during normal aging — i.e., in the absence of a disease state — is currently a matter of serious debate. In this Section I will summarize our knowledge of the composition, distribution, and genesis of NP and NFT and will briefly discuss the issue of disease vs. normal aging.

Classic NPs are complex, multicellular lesions found mainly in cortical association areas, entorhinal cortex, subiculum, amygdala, pyriform cortex, and, to a lesser extent, hippocampus. They are composed of a central core of amyloid protein surrounded by dystrophic (dying) neurites (both axonal and dendritic), reactive astrocytes, and activated microglia. The NPs are thought to evolve from abnormal proteolytic processing of the β-amyloid precursor protein (βAPP). The βAPP is an integral, transmembrane protein found in all neurons and glial cells, as well as in apparently all other mammalian tissues, with especially high densities in brain and kidney. Several functions have been proposed for βAPP, including cell growth promotion, inactivation of extracellular serine proteases, and neuritic growth promotion (reviewed in Reference 75). The precise pathways and reasons for the (proposed) abnormal proteolytic processing of βAPP to the pathological amyloid β-protein (Aβ) are not yet understood but are under intense investigation. It is generally believed that an evolution of the NP may involve an initial deposition of Aβ that is not in filamentous form and is not associated with dystrophic neurites, activated astrocytes, or microglial cells. Such deposits are referred to as diffuse or preamyloid plaques and are usually far more numerous than NP in AD brains. Support for the NP evolution hypothesis derives from studies of the brains of patients with Down syndrome, which typically display classic NP and NFT after 40 to 50 years of age, but which have been found to contain only diffuse plaques at less than 3 years of age.[76] Also, the brains of some neurologically normal individuals have been found to contain diffuse plaques after about 60 years of age, and AD brains have been found to contain diffuse plaques (but not NP) in striatum and in the molecular layer of cerebellar cortex (reviewed in Reference 75). In any case, the exact origin of the initial abnormal amyloid deposit is still not known, although neurons, astrocytes, microglia, and the vascular system are all contenders.

The NFT are bundles of paired helical filaments (PHF) which, themselves, are fibrils composed of two 8 to 10-nm filaments wound around each other with a half periodicity of 80 nm. The NFT are found in perikarya and dendrites of subsets of pyramidal neurons in neocortical association areas, entorhinal cortex, subiculum, and hippocampus (especially CA1, or Sommer's sector). They appear to be closely related to neuronal death in AD brain, although not all dying neurons contain NFT. The major component of PHF is the microtubule-associated protein tau, which normally acts to stabilize the neuronal microtubules.[77,78] For as yet unknown reasons, the tau protein becomes abnormally phosphorylated and associated with ubiquitin.[79] Ubiquitin is a small protein that attaches to other cellular proteins and becomes a signal for proteolytic systems to degrade that "marked" protein. For some reason, the abnormal tau that forms the core of PHF is not or cannot be degraded. Indeed, once formed, these pathological cellular inclusions seem only to multiply and may kill the parent neuron by interfering with normal metabolic processes, such as dendritic transport.

As mentioned earlier, the presence of NP and NFT in the brains of normal aged individuals is currently being debated. It has generally been assumed that individuals who do not display neurological signs of AD or any other type of dementia are "normal." As these normal elderly individuals die from nonneural causes, their brains can serve as undemented controls for AD brains. Since these "control" brains are fairly commonly found to contain NP and NFT, some even in large enough quantities to qualify as AD brains, it has been assumed that such pathology is the result of normal aging, rather than a disease process. However, it should be pointed out that NPs do not accumulate inevitably with age, even in the very old (reviewed in Reference 80). As suggested by Miller "the 'control' brains that have NPs or NFTs, in fact, represent patients just at or below their clinical threshold."[80] In his view, the pathological lesions of AD accumulate slowly and progressively, in a long preclinical phase, to a critical threshold, which differs among individuals and beyond which clinical manifestations are evident. In any case, while AD may be caused by either genetic or environmental factors, it remains definitely age related, and it appears that the normal aging process can predispose some individuals to develop the pathologies characteristic of AD.

F. Granulovacuolar Degeneration

Pyramidal neurons in human hippocampus are uniquely susceptible to a form of neuronal pathology termed granulovacuolar degeneration (GVD), in which the cytoplasm of the cells fills with clear membrane-bound vesicles, each of which contains a single small, dense granule (reviewed in Reference 72). The vesicles may fill the neuron, distorting its shape and undoubtedly interfering with its normal cellular functions. GVD is most prominent in area CA1 of hippocampus, followed by CA2, the subiculum, and CA4. Uncommon before age 60, GVD appears in approximately 20% of brains after that age. After age 80, 75% of brains have this pathology.

There is no clear relation of GVD to the neurofibrillary tangle or the neuritic plaque. Although tangles and GVD may both exist in the same cell, this is not the rule. Further, high concentrations of GVD may occur without any plaques or tangles. The only evidence that GVD may impair cognitive function comes from its high incidence in the brains of patients with senile dementia of the Alzheimer's type. Of course, since these brains are also plagued with plaques and tangles, no firm conclusions regarding GVD can as yet be drawn.

V. CONCLUSIONS

In this chapter I have tried to provide an overview of age-related changes occurring both in peripheral sensory systems and in CNS. Two main conclusions may be drawn here.

First, in each Section I have given examples of age-related changes that definitely affect our perceptions and thus, to some extent, our cognitive abilities, but that do not involve nervous system damage as a primary cause. For example, yellowing of the lens may lead not only to changes in color perception but also to slower reading ability because of a need for more light, thus causing the individual to appear slower than average in information processing. Likewise, various changes in the auditory apparatus may lead to decreased hearing ability and thus to misperception or slower perception of speech, causing an individual to appear cognitively "slow" or to actually misinterpret what was said and thus to reply inappropriately. It is crucial that clinicians be aware of such non-nervous system-related changes in order to properly diagnose the neurological status of an individual.

Second, it is important to acknowledge that even the very old do not always show signs of dementia. While lipofuscin, cell death, astrocyte hypertrophy (and other phenomena) do seem to increase in all individuals with age, not everyone develops amyloid plaques or neurofibrillary tangles, the major hallmarks of AD as well as normal aging. From a number of *in vivo* and *in vitro* studies, we know that natural lipofuscin accumulation in cells with age does not kill the cells. Rather, a wide variety of other factors (stress, alcohol and other drug consumption, pollutants, vitamin deficiency, etc.) render neurons susceptible to cell death. However, we also know that we can lose a certain proportion of our neurons without suffering any noticeable consequences. This is most likely because other (surviving) neurons of the same population are capable of taking over the functions of the dead ones. The neurons can "sprout" new dendrites and axon collaterals, they can produce additional neurotransmitter and trophic factor substances, and thus they sustain the functional integrity of a brain system. An example of such plasticity could be loss of substantia nigra neurons with age. However, symptoms of Parkinson's disease appear only after approximately 80% of the nigra neurons have died.

A wide variety of genetic and environmental factors undoubtedly contribute to various neurological disorders as we age. Apart from genetics, we all have different environmental handicaps, such as contaminants of food, water, and air; poor diet (nutrient deficiency and excess fat); and stresses and head injuries. Provided that the genetic make-up is right and that we are careful enough, or fortunate enough, to avoid environmental hazards, we should be able to avoid dementia even into very old age.

ACKNOWLEDGMENT

The author is deeply indebted to Laurel Wheeler for her patience in typing this manuscript.

REFERENCES

1. Ordy, J. M. and Brizzee, K. R., Eds., *Sensory Systems and Communication in the Elderly*, Vol. 10, Raven Press, New York, 1979.
2. Reading, V., Visual resolution as measured by dynamic and static tests, in *Sensory Systems and Communication in the Elderly*, Vol. 10, Ordy, J. and Brizzee, K., Eds., Raven Press, New York, 1972, 20.
3. Colavita, E. B., *Sensory Changes in the Elderly*, Thomas C. Thomas, New York, 1979.
4. Kwabara, T., Age-related changes of the eye, in *Special Senses in Aging*, Han, S. and Coons, D., Eds., Institute of Gerontology, University of Michigan, Ann Arbor, 1977, 46.
5. Chylack, L. T., Jr., Aging and cataracts, in *Special Senses in Aging*, Han, S. S. and Coons, D., Eds., Institute of Gerontology, University of Michigan, Ann Arbor, 1977, 88.
6. Jani, S., The age factor in stereopsis screening, in *Sensory Systems and Communication in the Elderly*, Vol. 10, Ordy, J. and Brizzee, K., Eds., Raven Press, New York, 1966, 21.
7. Ordy, J., Brizzee, K., Beavers, T., and Medark, P., Age differences in the functional and structural organization of the auditory system in man, in *Sensory Systems and Communication in the Elderly*, Vol. 10, Ordy, J. and Brizzee, K., Eds., Raven Press, New York, 1979, 153.
8. Fozard, J., Vision and hearing in aging, in *Handbook of the Psychology of Aging*, 3rd ed., Birren, J. E. and Schaie, K. W., Eds., Academic Press, New York, 1990.
9. Feldman, M. and Vaughan, D., Changes in the auditory pathway with age, in *Special Senses in Aging*, Han, S. and Coons, D., Eds., Institute of Gerontology, University of Michigan, Ann Arbor, 1977, 143.
10. Brody, H., Organization of the cerebral cortex. III. A study of aging in the human cerebral cortex, *J. Comp. Neurol.*, 102, 511, 1955.
11. Scheibel, M. E. and Scheibel, A. B., Structural changes in the aging brain, in *Aging*, Vol. 1, Brody, H., Harman, D., and Ordy, J. M., Eds., Raven Press, New York, 1975.
12. Kenshalo, D., Changes in vestibular and somaesthetic systems as a function of age, in *Sensory Systems and Communication in the Elderly*, Vol. 10, Ordy, J. and Brizzee, K., Eds., Raven Press, New York, 1979, 269.
13. Kenshalo, D., Somaesthetic sensitivity in young and elderly humans, *J. Gerontol.*, 41(6), 732, 1986.
14. Stevens, J., Aging and spatial acuity of touch, *J. Gerontol.*, 47(1), 35, 1992.
15. Kenshalo, D., Aging effects on cutaneous and kinesthetic sensibilities, in *Special Senses in Aging*, Han, S. and Coons, D., Eds., Institute of Gerontology, University of Michigan, Ann Arbor, 1977, 189.
16. Era, P. and Heikkinen, E., Postural sway during standing and unexpected disturbances of balance in random samples of men of different ages, *J. Gerontol.*, 40(3), 287, 1985.
17. Harkins, S., Price, D., and Martelli, M., Effects of age on pain perception — thermonociception, *J. Gerontol.*, 41(1), 58, 1986.
18. Hamm, R. J. and Knisely, J. S., Environmentally induced analgesia: an age-related decline in an endogenous opioid system, *J. Gerontol.*, 40, 268, 1985.
19. Krondl, M., Smell and taste, in *Living with Sensory Loss*, National Advisory Council on Aging, 1990.
20. Van Toller, C., Dodd, G., and Billing, A., *Aging and the Sense of Smell*, Charles C Thomas, New York, 1985.
21. Doty, R., Ed., *Handbook of Olfaction and Gustation*, Marcel Dekker, New York, 1995.
22. Gilbert, A. and Wysocki, C., National Geographic's — The smell survey, in *Living with Sensory Loss*, National Advisory Council on Aging, 1990, 28.
23. Russel, M., Cummings, B., Profitt, B., Wysocki, C., Gilbert, A., and Cotman, C., Life span changes in the verbal categorization of odours, *J. Gerontol.*, 48(2), 49, 1993.
24. Moulton, D., Dynamics of cell populations in olfactory epithelium, in *Sensory Systems and Communication in the Elderly*, Vol. 10, Ordy, J. and Brizzee, K., Eds., Raven Press, New York, 1974, 247.
25. Schiffman, S., Orlandi, M., and Erikson, P., Changes in taste and smell with age: biological aspects, in *Sensory Systems and Communication in the Elderly*, Ordy, J. and Brizzee, K., Eds., Raven Press, New York, 1979, 247.
26. Brizzee, K., Klara, P., and Johnson, J., Changes in microanatomy, neurocytology and fine structure with aging, in *Aging and the Sense of Smell*, Van Toller, C., Dodd, G., and Billing, A., Eds., Charles C Thomas, New York, 1975, 43.

27. Thumfart, W., Plattig, K., and Schlicht, N., Taste and smell sensitivity in the elderly human, in *Aging and the Sense of Smell*, Van Toller, C., Dodd, G., and Billing, A., Eds., Charles C Thomas, New York, 1980, 47.

28. Ship, J. and Weiffenbach, J., Age, gender, medical treatment and medication effects on smell identification, *J. Gerontol.*, 48(1), 26, 1993.

29. Eskanazi, B., Odour perception in temporal lobe epilepsy patients with and without temporal lobectomy, in *Handbook of Olfaction and Gustation*, Doty, R., Ed., Marcel Dekker, New York, 1986, 356.

30. Murphy, C., Age related effects on the threshold, psychophysical function and pleasantness of menthol, in *Aging and the Sense of Smell*, Van Toller, C., Dodd, G., and Billing, A., Eds., Charles C Thomas, New York, 1983, 139.

31. Weiffenbach, J., Tylenda, C., and Baum, B., Oral sensory changes in aging, *J. Gerontol.*, 45(4), 121, 1990.

32. Spitzer, M., Taste acuity in institutionalized and non-institutionalized elderly men, *J. Gerontol.*, 43(3), 71, 1988.

33. Schiffman, S., Changes in taste and smell with age: psychological aspects, in *Sensory Systems and Communication in the Elderly*, Vol. 10, Ordy, J. and Brizzee, K., Eds., Raven Press, New York, 1979, 227.

34. Chauhan, J., Relationships between sour and salt taste perceptions and selected subject attributes, in *Living with Sensory Loss*, National Advisory Council on Aging, 1990, 30.

35. Langan, M. and Yearick, R., The effects of improved oral hygiene on taste perception and nutrition in the elderly, in *Living with Sensory Loss*, National Advisory Council on Aging, 1990.

36. Hall, D., Quintessence international 24(11), in *The Oral Care Report*, 5(1), 1993.

37. Spitzer, M. J., Taste perception and selected subject attributes, *J. Gerontol.*, 43(3), 71, 1988.

38. Coleman, P. D. and Flood, D. G., Neuron numbers and dendritic extent in normal aging and Alzheimer's disease, *Neurobiol. Aging*, 8, 521, 1987.

39. Brody, H., Structural changes in the aging nervous system, in *The Regulatory Role of the Nervous System in Aging*, Vol. 7, *Interdisciplinary Topics in Gerontology*, Blumenthal, H.T., Ed., S. Karger, Basel, 1970, 9.

40. Brizzee, K. R., Ordy, J. M., and Kaack, B., Early appearance and regional differences in intraneuronal and extraneuronal lipofuscin accumulation with age in the brain of a nonhuman primate (*Macaca mulatta*), *J. Gerontol.*, 29(4), 366, 1974.

41. Rogers, J. and Styren, S. D., Neuroanatomy of aging and dementia, *Rev. Biological Res. Aging*, 3, 223, 1987.

42. Hall, T. C., Miller, A. K. H., and Corsellis, J. A. N., Variations in the human Purkinje cell population according to age and sex, *Neuropathol. Appl. Neurobiol.*, 1, 267, 1975.

43. Nandy, K., Morphological changes in the cerebellar cortex of aging *Macaca nemestrina*, *Neurobiol. Aging*, 2, 61, 1981.

44. Mann, D. M. A., Yates, P. O., and Hawkes, J., The pathology of the human locus coeruleus, *Clin. Neuropathol.*, 2, 1, 1983.

45. Vijayashankar, N. and Brody, H., A quantitative study of the pigmented neurons in the nuclei locus coeruleus and subcoeruleus in man as related to aging, *J. Neuropathol. Exp. Neurol.*, 38, 490, 1979.

46. McGeer, P. L., McGeer, E. G., and Suzuki, J. S., Aging and extrapyramidal function, *Arch. Neurol.*, 34, 33, 1977.

47. Agid, Y., Javoy-Agid, F., and Ruberg, M., Biochemistry of neurotransmitters in Parkinson's disease, in *Movement Disorders 2*, Marsden, C. D. and Fahn, S. T., Eds., Butterworths, London, 1987, 166.

48. Bondareff, W. and Mountjoy, C. Q., Number of neurons in nucleus locus ceruleus in demented and non-demented patients: rapid estimation and correlated parameters, *Neurobiol. Aging*, 7, 297, 1986.

49. Chui, H. C., Mortimer, J. A., Slager, U. T., Barrow, C., Bondareff, W., and Webster, D. D., Pathological correlates of dementia in Parkinson's disease, *Arch. Neurol.*, 43, 991, 1986.

50. Fang, H. C. H., Observation of aging characteristics of cerebral blood vessels: macroscopic and microscopic features, in *Neurobiology of Aging*, Terry, R. D. and Gershon, S., Eds., Raven Press, New York, 1976.

51. Burns, E. M., Kruckeberg, T. W., Comerford, L. E., and Buschmann, M. B. T., Thinning of capillary walls and declining numbers of endothelial mitochondria in the cerebral cortex of the aging primate, *Macaca nemestrina*, *J. Gerontol.*, 34, 642, 1979.

52. Burns, E. M., Kruckeberg, T. W., Gaetano, P. K., and Shulman, L. M., Morphological changes in cerebral capillaries with age, in *Brain Aging: Neuropathology and Neuropharmacology*, Vol. 21, *Aging*, Cervos-Navarro, J. and Sarkander, H.-I., Eds., Raven Press, New York, 1983.

53. Mancardi, G. L., Perdelli, F., Rivano, C., Leonarde, A., and Bugiani, O., Thickening of the basement membrane of cortical capillaries in Alzheimer's disease, *Acta Neuropathol.*, 49, 79, 1980.

54. Dastur, D. K., Lane, M. H., Hansen, D. B., Kety, S. S., Butler, R. N., Perlin, S., and Sokoloff, L., Effects of aging on cerebral circulation and metabolism, in *Human Aging*, Birren, J. E., Butler, R. N., Greenhouse, S. W., Sokoloff, L., and Yarrow, M. R., Eds., Public Health Service Publication No. 986, 1963, 59.

55. Lassen, N. A., Cerebral blood flow and oxygen consumption in man, *Physiol. Rev.*, 39, 183, 1959.

56. Lassen, N. A., Feinberg, J., and Lane, M. H., Bilateral studies of cerebral oxygen uptake in young and aged normal subjects and in patients with organic dementia, *J. Clin. Invest.*, 39, 491, 1960.

57. Hagberg, B. and Ingvar, D. H., Cognitive reduction in presenile dementia related to regional abnormalities of the cerebral blood flow, *Br. J. Psychiatry*, 128, 209, 1976.

58. Brun, A., Gustafson, L., and Ingvar, D. H., Neuropathological findings, neuropsychiatric symptoms and regional cerebral blood flow in presenile dementia, in *Excerpta Med.*, International Congress of Neuropathology, Budapest, 1975, 10.

59. Ordy, J. M., Kaack, B., and Brizzee, K. R., Life-span neurochemical changes in the human and non-human primate brain, in *Aging*, Vol. 1, Brody, H., Harman, D., and Ordy, J. M., Eds., Raven Press, New York, 1975.

60. Minckler, T. M. and Boyd, E., Physical growth, in *Pathology of the Nervous System*, Vol. 1, Minckler, J., Ed., McGraw-Hill, New York, 1968.

61. Wisniewski, H. M. and Terry, R. D., Neuropathology of the aging brain, in *Neurobiology of Aging*, Terry, R. D. and Gershon, S., Eds., Raven Press, New York, 1976.

62. Corsellis, J. A. W., Some observations on the Purkinje cell population and on brain volume in human aging, in *Neurobiology of Aging*, Terry, R. D. and Gershon, S., Eds., Raven Press, New York, 1976.

63. Tomlinson, B. E., Blessed, G., and Roth, M., Observations of the brain of demented old people, *J. Neurol. Sci.*, 11, 205, 1970.

64. Terry, R. D. and Davies, P., Dementia of the Alzheimer type, *Annu. Rev. Neurosci.*, 3, 77, 1980.

65. Korr, H., Proliferation of different cell types in the brain of senile mice: autoradiographic studies with ^{3}H- and ^{14}C-thymidine, *Exp. Brain Res.*, Suppl. 5, 51, 1982.

66. Finch, C. E. and Morgan, D. G., RNA and protein metabolism in the aging brain, *Annu. Rev. Neurosci.*, 13, 75, 1990.

67. Harman, D., Lipofuscin and ceroid formation: the cellular recycling system, in *Lipofuscin and Ceroid Pigments*, Porta, E. A., Ed., Plenum Press, New York, 1990, 3.

68. Ivy, G. O., Kanai, S., Ohta, M., Smith, G., Sato, Y., Kobayashi, M., and Kitani, K., Lipofuscin-like substances accumulate rapidly in brain, retina and internal organs with cysteine protease inhibition, in *Lipofuscin and Ceroid Pigments*, Porta, E. A., Ed., Plenum Press, New York, 1990, 31.

69. Brizzee, K. R., Ordy, J. M., Hansche, J., and Kaack, B., Quantitative assessment of changes in neuron and glia cell packing density and lipofuscin accumulation with age in the cerebral cortex of a nonhuman primate (*Macaca mulatta*), in *Neurobiology of Aging*, Terry, R. D. and Gershon, S., Eds., Raven Press, New York, 1976, 229.

70. Bazelon, M., Fenichel, G. M., and Randall, J., Studies on neuromelanin. I. A melanin system in the human adult brain stem, *Neurology*, 17, 512, 1967.

71. Fix, J. D., A melanin-containing nucleus associated with the superior cerebellar peduncle in man, *J. Hirnforsch.*, 21, 429, 1980.

72. Ivy, G. O., MacLeod, C. M., Petit, T. L., and Markus, E. J., A physiological framework for perceptual and cognitive changes in aging, in *The Handbook of Aging and Cognition*, Craik, F. I. M. and Salthouse, T., Eds., Lawrence Erlbaum Associates, New Jersey, 1992, 273.

73. Flood, D. G. and Coleman, P. D., Cell type heterogeneity of changes in dendritic extent in the hippocampal region of the human brain in normal aging and in Alzheimer's disease, in *Neural Plasticity: A Lifespan Approach*, Petit, T. L. and Ivy, G. O., Eds., Alan R. Liss, New York, 1988, 265.

74. McWilliams, J. R. R., Age related declines in anatomical plasticity and axonal sproutings, in *Neural Plasticity: A Lifespan Approach*, Petit, T. L. and Ivy, G. O., Eds., Alan R. Liss, New York, 1988, 329.

75. Selkoe, D. J., Normal and abnormal biology of the β-amyloid precursor protein, *Annu. Rev. Neurosci.*, 17, 489, 1994.

76. Ter-Minassian, M., Kowall, N. W., and McKee, A. C., Beta amyloid protein immunoreactive senile plaques in infantile Down's syndrome, *Soc. Neurosci. Abstr.*, 18, 734, 1992.

77. Nukina, N. and Ihara, Y., Proteolytic fragments of Alzheimer's paired helical filaments, *J. Biochem. (Tokyo)*, 98, 1715, 1985.

78. Kondo, J., Honda, T., Mori, H., Hamada, Y., Miura, R., Ogawara, M., and Ihara, Y., The carboxyl third of tau is tightly bound to paired helical filaments, *Neuron*, 1, 817, 1988.

79. Mori, H., Kondo, J., and Ihara, Y., Ubiquitin is a component of paired helical filaments in Alzheimer's disease, *Science*, 235, 1641, 1987.

80. Miller, D. C., Is it time to reassess the pathological criteria for a diagnosis of Alzheimer's disease?, *Alzheimer Dis. Associated Disorders*, 7(3), 129, 1993.

Part II

NEUROTRANSMITTERS AND NEUROMODULATORS

Chapter **4**

SEROTONIN AND DOPAMINE AS NEUROTRANSMITTERS

Andrius Baskys and Gary Remington

CONTENTS

0-8493-8386-0/96/$0.00+$.50

Excitatory amino acids and gamma-aminobutyric acid (GABA) are probably the largest neurotransmitter class in the central nervous system (CNS); a smaller proportion of synaptic connections use biogenic amines dopamine, norepinephrine, histamine, and indoleamine serotonin as neurotransmitters. The last few decades have witnessed a tremendous growth in our understanding of these substances, receptor types, and their roles in psychopathology. There is a remarkably large number of publications on amine neurotransmitters, mostly inspired by hypotheses about their role in mental illness. The two major psychotropic drug classes, antidepressants and neuroleptics, are thought to influence behavior primarily by interacting with serotonin- and dopamine-mediated processes in the brain. This chapter will review information on serotonin and dopamine synthesis, uptake mechanisms, and receptor types, focusing on their role in psychopathology and drug interactions.

I. SEROTONIN

It is widely held that the diffuse serotonin (5-hydroxytryptamine, 5-HT) pathway system contributes to the regulation of a variety of psychological and biological functions. Mood, anxiety, arousal, attention, impulsivity, aggression, suicidality, and cognition are among the former, and sleep-wake cycle, appetite, pain sensation, and brain maturation are among the latter. Consequently, dysfunction of serotonin-mediated transmission has been implicated in a variety of psychiatric disorders — anxiety disorders, schizophrenia, depression, autism, alcoholism, and sociopathy, among others. The significance of serotonin and serotonin receptors in brain function and psychopathology cannot be overestimated, especially in light of recent advances in treatment of depression with a group of drugs known as serotonin selective reuptake inhibitors (SSRIs). On the other hand, interactions of antidepressant drugs with the serotonergic system in the CNS do not necessarily imply that 5-HT is involved in pathogenesis of depression but only indicate that some antidepressants may be acting on this system. In fact, clinical efficacy of SSRIs is not restricted to depression alone but may be useful in a range of disorders characterized by dysregulation of affect, anxiety, appetitive behaviors, aggression, and the appearance of intrusive thoughts.[1]

Historically, central (raphe) neurons of the brain stem with uncertain projections attracted attention of neuroanatomists since the time of Ramón y Cajal. On the other hand, a blood serum component that produced vasoconstriction (*tonus*, hence the name *serotonin*) had been well known to investigators. Identification of serotonin in brain structures had led to the proposal that it acts as a CNS neurotransmitter. Application of the histochemical fluorescence technique was crucial in further describing the location of cell bodies and axon terminals containing 5-HT, as well as other amines (for review see Reference 2).

A. Anatomical Distribution

The 5-HT immunoreactive cell bodies in the brain are divided into the rostral and caudal groups.[3] The rostral group consists of four nuclei: nucleus centralis superior, nucleus raphae dorsalis, nucleus prosupralemniscus, and hypothalamic dorsomedial nucleus. The caudal group consists of five nuclei: nucleus raphae obscurus, nucleus raphae pallidus, nucleus raphae magnus, nucleus raphae ventricularis, as well as nucleus reticularis lateralis and nucleus paragiganto cellularis lateralis. The axons leaving these nuclei form the descending pathway to the spinal cord and the ascending pathway arising within the medial forebrain bundle. Although immunoreactivity for serotonin-containing fibers has been found in virtually every part of the brain,[2] the highest density of serotonergic projections are seen in limbic structures (temporal lobe of the neocortex, amygdaloid nuclei, cingulate gyrus) and sensory centers (visual cortex, superior temporal gyrus, postcentral cortex, entorhinal cortex). The highest density of serotonin fibers in the cortex are found in layers I and IV.

B. Biosynthesis, Storage, and Release

The metabolism of serotonin has been well investigated, and detailed descriptions are available (for review see Reference 4). Serotonin is synthesized in certain nerve cells from the amino acid tryptophan, which comes from dietary protein. The enzyme tryptophan monoxigenase also known as tryptophan hydroxylase converts tryptophan to 5-hydroxytryptophan (5-HTP) which appears to be a rate-limiting step in serotonin biosynthesis.[4] Molecular cloning of tryptophan hydroxylase revealed that this enzyme is 50% homologous with tyrosine hydroxylase, the rate-limiting enzyme in catecholamine biosynthesis.[4] The next step in serotonin biosynthesis is a conversion of 5-HTP to 5-HT by the enzyme aromatic L-amino acid decarboxylase. Serotonin is found in presynaptic neurons colocalized with other neurotransmitters or neuropeptides. Substance P, thyrotropin-releasing hormone, and enkephalin have been found in the medullary serotonergic neurons. There are neurons that contain both serotonin and norepinephrine or GABA. It is believed that peptides which colocalize with serotonin modulate its postsynaptic action, and one of them, enkephalin, functions as a growth-inhibiting factor of cultured serotonergic neurons. Serotonin is released from synaptic vesicles together with the serotonin-binding protein (SPB) by way of exocytosis. The release is Ca^{2+}-dependent.

There is a great abundance of folklore among the general public and part of medical profession about the dependence of the mental symptoms on dietary components. While most of these beliefs (e.g., effects of dietary sugar or food allergies on behavior in children and adults) may be unfounded, the dependence of brain serotonin synthesis on dietary components allows tests of hypotheses about the role of diet in mood and behavior. It appears that there is experimental evidence to suggest a link between a tryptophan-depleted diet and mood and behavior in humans. Thus, removal of tryptophan from the diet significantly lowers the tryptophan and, consequently, the 5-HT level in the brain. Ingestion of a tryptophan-deficient diet lowers mood, increases aggressive behavior, and decreases the analgesic effect of morphine in normal humans (for review see Reference 5). Deficiency of folic acid in the diet can also lower brain serotonin, which results in lower mood. These findings support the role of 5-HT in regulation of mood.

In addition to the level of 5-HT in presynaptic terminals, the amount of released neurotransmitter depends on the rate of action potential generation ("firing") of the raphae neurons. The neurotransmitter release from the presynaptic terminals is elevated at the higher rates and decreased when the rates are lower. The firing rate of serotonergic neurons is controlled by somatodendritic autoreceptors which are of the 5-HT$_{1A}$ type (see below). Drugs that are full or partial agonists for these receptors (e.g., 8-hydroxy-2-(di-1-propylamino)tetralin [8-OH-DPAT], buspirone, ipsapirone, gepirone, tandospirone) reduce serotonin release from the terminals if administered acutely but desensitize them and increase serotonin release following long-term administration. Mild antidepressant and anxiolytic actions of the above compounds may be indicators of the role that serotonin plays in regulation of mood by interacting with 5-HT$_{1A}$ receptors.

C. Uptake and Degradation

The major mechanism that terminates serotonin action in the synaptic cleft is its uptake into presynaptic terminals. The uptake is an active process which depends on temperature and requires the presence of Na^+ and Cl^-. Several metabolic inhibitors, as well as Na^+, K^+-ATPase inhibitors, can nonspecifically inhibit 5-HT uptake. However, more interesting are actions of several selective inhibitors of serotonin transporter (5-HTT) protein which has been recently cloned[6,7] and was found to be similar to transporters for GABA, norepinephrine, and dopamine. Many antidepressant drugs, cocaine, and amphetamine competitively inhibit serotonin binding to 5-HTT and are potent inhibitors of serotonin uptake. Among the antidepressants, paroxetine, citalopram, and clomipramine have the highest affinity for the 5-HTT. Fluoxetine, imipramine, and amitriptyline have lower affinity, and desipramine and doxepine have lowest affinity.[7] The tertiary amine tricyclic antidepressants amitriptyline and imipramine are more potent antagonists of 5-HTT than nortriptyline and desipramine, which are more potent antagonists at the cloned noradrenaline transporter.[6] The SSRIs fluoxetine, citalopram, and paroxetine are more potent inhibitors of 5-HTT in comparison with the noradrenaline carrier.[6]

It is not at all clear how the 5-HTT inhibition is related to the potency of a given drug to act as an antidepressant as there is no correlation between the 5-HTT binding affinity of SSRIs and their antidepressant effect.[1] Moreover, the ability to inhibit 5-HTT may not be at all necessary for a particular compound to be an effective antidepressant. Thus, an antidepressant, tianeptine, enhances 5-HT reuptake,[8] suggesting that other factors may be responsible for the antidepressant action of SSRIs. There is evidence that, in addition to the inhibition of the serotonin transporter, SSRIs have numerous interactions with other serotonin receptors in the CNS. Thus, clomipramine and fluoxetine apparently interact with 5-HT_{1D} receptors, fluoxetine binds to 5-HT_{1C} receptors, and the novel SSRI antidepressant nefazodone is an antagonist for both 5-HTT and 5-HT_2 receptors,[9] suggesting that 5-HT_{1D}, 5-HT_{1C}, and 5-HT_2 receptor subtypes may be involved in antidepressant actions of SSRIs. The fact that antidepressant therapy is usually associated with a 4 to 6-week delay[10] cannot be explained by the 5-HTT inhibition followed by a transient increase in extracellular 5-HT concentration.[8,11] It is likely, therefore, that actions of antidepressants are complex and involve other neurotransmitter and second messenger systems as well. This idea is well supported by downregulation of cortical $GABA_B$, 5-HT_2 receptors, α_1-, α_2- and β-adrenoreceptors, that is observed following antidepressant treatment.[8] In experimental *in vitro* preparations serotonin enhanced N-methyl-D-aspartate (NMDA)-mediated responses,[12] suggesting that glutamate receptors may also be involved.

SSRIs are considered to be less toxic and relatively safe drugs as compared with tricyclics.[8] However, significant inhibition of cytochrome P4501A2 by fluvoxamine may be responsible for its interactions with amitriptyline, clomipramine, imipramine, theophylline, and caffeine.[13] Similarly, inhibition by paroxetine and fluvoxamine of the cytochrome P4502D6 can be responsible for their interactions with tricyclic antidepressants, neuroleptics, and antiarrhythmics.[13]

The enzyme monoamineoxidase (MAO) present on mitochondrial membranes of neurons and glia converts serotonin to 5-hydroxyindoleacetaldehyde, which can be further oxidized to 5-hydroxyindoleacetic acid (5-HIAA) or reduced to 5-hydroxytryptophol. The major metabolite of serotonin in the brain is 5-HIAA. There are MAO type A and MAO type B isoenzymes; there appears to be more MAO type B in human brain, and serotonergic cell bodies contain predominantly MAO type B.[4] Serotonin is preferentially metabolized by MAO type A. There are selective inhibitors for each type of the enzyme, e.g., clorgyline, brofaromine, cimoxatone, and moclobamide for type A and deprenyl for type B MAO. The antidepressant drugs that inhibit type A MAO also raise the amount of 5-HT in experimental animals, which could account for or contribute to their antidepressant properties.

D. Receptors

Serotonin exerts its physiological effects in the CNS by interacting with a large variety of receptors. The original classification of 5-HT receptors into 5-HT$_1$ and 5-HT$_2$ types was based on separate binding sites identified by two specific ligands, 5-[^{3}H]HT and [^{3}H]spiroperidol.[14] Application of molecular biology techniques revealed a wealth of additional 5-HT receptor subtypes. At present, it is generally believed that the family of serotonin receptors consists of at least seven distinct subtypes, termed 5-HT$_1$ to 5-HT$_7$. With the exception of the 5-HT$_3$, which is a ligand-gated ion channel, all other receptors belong to a superfamily of G protein–coupled receptors that share a seven transmembrane domain structure. These receptors show significant sequence homologies in the transmembrane regions and have characteristic features of receptors that couple with second messengers.[15] Inhibition of adenylate cyclase is a characteristic of 5-HT$_1$ receptors, except for the 5-HT$_{1C}$ subtype, which, like the 5-HT$_2$ receptor, stimulates phospholipase C and increases phosphoinositide hydrolysis. The two most recently cloned receptors, 5-HT$_6$ and 5-HT$_7$, as well as 5-HT$_4$, all stimulate adenylate cyclase. No second messenger mechanism has been identified for the 5-HT$_5$ receptor. The 5-HT$_3$ receptor is an exception because it does not have a seven transmembrane domain structure but constitutes a ligand-gated ion channel. Table 1 is an attempt to summarize information on known 5-HT–receptor types, their ligands, second messenger systems, distribution in the CNS, and possible role in behavior.

Although the physiological effects mediated by different 5-HT receptors are still a subject of ongoing investigations, links between certain subtypes and particular disorders or psychotropic drug interactions have already been identified. For example, it has been proposed that an effective antimigraine drug, sumatriptan, decreases headaches by activating an inhibitory serotonin receptor, similar to 5-HT$_{1D}$, located presynaptically on perivascular nerve fibers.[16] The net effect of this action is a blockade of neuropeptide release and impulse conduction in trigeminovascular neurons. Although serotonin has been implicated in regulation of anxiety states, to date only 5-HT$_{1A}$ agonists and some 5-HT$_2$ antagonists (e.g., ketanserin) show clear potential as anxiolytic drugs.[17] In addition, several 5-HT receptors have been associated with motor effects. Thus, "the 5-HT syndrome" can be induced by 5-HT$_{1A}$ agonists, and the head-twitch response or "wet-dog shakes" in animals are mediated by 5-HT$_2$ receptors.[15] Generation of the patterned motor output involves inhibition of spinal cord segmental inputs by 5-HT$_{2A}$ or 5-HT$_{2C}$ receptors and by extrasynaptically located 5-HT$_{1A}$, 5-HT$_{1B}$, or 5-HT$_{1C}$ receptors.[18]

The 5-HT$_3$ receptor (see Table 1) is a nonselective cation channel equally permeable to Na$^+$ and K$^+$. Because of the cation permeability modulation by Ca or Mg ions, this receptor resembles the glutamate NMDA receptor.[19] The antagonists for the 5-HT$_3$ receptor are effective antiemetics.[20] They are particularly effective when nausea is caused by cytotoxic drug therapy or radiation or follows general anesthesia. The 5-HT$_3$ receptor antagonists are ineffective against nausea associated with motion sickness or administration of apomorphine.[20] In addition, the antagonist drugs for this receptor have antinociceptive properties. Although animal studies suggest that

TABLE 1.

A Summary of Serotonin Receptor Types, Ligands, Second Messenger Systems, Distribution in the CNS and Possible Functions

Receptor	Ligands	Second Messenger	Location in CNS	Processes That May Be Influenced	Ref.
5-HT$_{1A}$	8-OH-DPAT Buspirone Ipsapirone Gepirone Tandospirone Propranolol	↓ cAMP	Midbrain raphe nuclei Dentate gyrus Lateral septum Hippocampus Cortex (deeper layers) Spinal cord (Layer I)	Anxiety Depression Sexual behavior Appetite Aggression Pain	1, 2, 22
5-HT$_{1B}$ (also known as a rat homologue of human 5-HT$_{1D}$)	5-CT Propranolol Methysergide	↓ cAMP	Globus pallidus, dorsal subiculum, substantia nigra, olivary pretectal nucleus Presynaptically localized on terminals of striatal neurons and Purkinje cells Striatum Dorsal and median raphe		2, 23–26
5-HT$_{1D\alpha}$ 5-HT$_{1D\beta}$	5-CT Metergoline CGS12066 Methysergide Sumatriptan Mianserin Yohimbine 8-OH-DPAT Spiperone	↓ cAMP Sumatriptan ↑ [Ca^{2+}]$_i$	Pyramidal layer of the olfactory tubercle, nucleus caudatus, nucleus accumbens Human cerebral cortex 5-HT$_{1D\beta}$ expressed in human and monkey frontal cortex, medulla, striatum, hippocampus, and amygdala	Vasoconstriction Appetite	27–32
5-HT$_{1C}$	Mesulergine LSD	↑ [Ca^{2+}]$_i$	Choroid plexus, much less in hippocampus	Appetite Anxiety Mood	2, 33
5-HT$_{1E}$	Methysergide Ergotamine 8-OH-DPAT 5-CT Ketanserin	↑ cAMP		?	34, 35

Receptor	Ligands	Transduction	Localization	Function	References
5-HT$_{1F}$	Sumatriptan 5-CT	↑ cAMP	Brain. Not detected in kidney, liver, spleen, heart, pancreas, testes	?	36
5-HT$_2$	Ketanserin Mianserin Haloperidol LSD Spiperone	↑ PI hydrolysis	Cortex Caudate Limbic structures Hypothalamus	Vasoconstriction Sleep Hallucinations Suicide Anxiety	1, 37, 38
5-HT$_{2B}$ 5-HT$_{2C}$	Ritanserin Methysergide Setoperone Cyproheptadine Spiperone Ketanserin	?	Spinal cord 5-HT$_{2C}$ also found in Cerebellum Intestine Heart Kidney	Regulation of patterned motor output Other ?	18, 39, 40
5-HT$_3$	Ondansetron Genisetron Tropisetron Bemesetron	Receptor-controlled ion channel	Pyriform, cingulate and enthorhinal cortex Hippocampal interneurons Amygdaloid complex, the olfactory bulb, the trochlear nerve nucleus, the dorsal tegmental region, the facial nerve nucleus, the nucleus of the spinal tract of the trigeminal nerve, and the spinal cord dorsal horn Heart	Dopamine release Emesis Pain Anxiety Aversion Psychosis	19, 20, 41–43, 95
5-HT$_4$	5-Metoxytryptamine 5-CT Cisapride Renzapride Zacopride Metoclopramide	↓ K$^+$ channels ↑ cAMP ↑ acetylcholine release	Collicular and hippocampal neurons Myenteric neurons Cardiac myocytes Adrenocortical cells	Arrhythmias Atrial fibrillation Cognitive dysfunction (?)	21
5-HT$_5$	2-Bromo-LSD Ergotamine 5-CT Methysergide	?	Cerebral cortex Hippocampus Habenula Olfactory bulb Cerebellum (granular layer)	Similar to 5-HT$_{1D}$	44

TABLE 1. (continued)

A Summary of Serotonin Receptor Types, Ligands, Second Messenger Systems, Distribution in the CNS and Possible Functions

Receptor	Ligands	Second Messenger	Location in CNS	Processes That May Be Influenced	Ref.
5-HT$_{5B}$	5-CT LSD Dihydroergotamine Methysergide Methiothepin 8-OH-DPAT	?	Medial habenulae, hippocampal CA1 neurons in adults	?	45
5-HT$_6$	Clozapine Amoxapine Clomipramine Loxapine Amitriptyline	↑ cAMP	Striatum Amygdala Cerebral cortex Olfactory tubercle	?	46
5-HT$_7$	LSD Clozapine Loxapine Amitriptyline	↑ cAMP	Hypothalamus, hippocampus, mesencephalon, cerebral cortex, olfactory bulb, olfactory tubercle	?	47, 96

Abbreviations: LSD, lysergic acid diethylamide; 5-CT, 5-carboxamidotryptamine; 8-OH-DPAT, 8-hydroxy-2-(di-1-propylamino)tetralin.

5-HT$_3$ antagonists also have anxiolytic properties, clinical data supporting these findings are still lacking. Similarly, it remains unclear whether or not drugs that interact with the 5-HT$_3$ receptor have any benefit in the treatment of psychotic symptoms or cognitive impairment.

The 5-HT$_4$ receptor (see Table 1) has not been cloned yet, and its role in CNS physiology remains unclear. It has been suggested that, because of its possible role in neurotransmitter release, this receptor could be involved in cognitive dysfunction caused by acetylcholine deficit.[21]

The remaining receptors (5-HT$_5$, 5-HT$_6$, and 5-HT$_7$, see Table 1) have been cloned only recently, and their pharmacology and roles in brain physiology are only beginning to emerge. In this regard, particularly interesting appear receptors with high affinity to the antidepressant and antipsychotic agents clozapine, loxapine, clomipramine, and nortriptyline (see Table 1). It is likely that further studies of these receptors will bring to light many mysteries related to these compounds and to the role of serotonin in major psychiatric disorders.

II. DOPAMINE

As with serotonin, dopamine (DA) has been implicated in a number of psychiatric conditions, particularly schizophrenia[48] and the affective disorders.[49] Indeed, DA is central to the current theories of schizophrenia and the action of neuroleptics. Understanding the specific role of DA is made more complicated by the identification of at least five receptor subtypes, as well as the recognition that other neurotransmitters, such as serotonin, can modulate DA activity.

A. Anatomical Distribution

Dopaminergic pathways in the brain arise from groups of cells in the midbrain and hypothalamus.[50] In the midbrain, cell groups designated A8, A9, and A10 form the DA neurons of the ventral mesencephalon, also collectively referred to as the ventral mesotelencephalic DA system because of a lack of anatomical boundaries between cell groups and overlap in projection fields.[50,51] The A8 neurons are located in the retrorubral field of the midbrain and contribute to the DA innervation of striatal and mesolimbic, but not neocortical, sites. The A9 cells constitute the lateral DA cells of the substantia nigra and ascend to provide the greatest proportion of DA innervation of the striatum. Taken together, the A8 and A9 neurons contain approximately 70% of brain dopamine and are involved in the modulation of motor behavior.[49] The A10 DA neurons arise from the ventral tegmental area and innervate both mesolimbic regions (nucleus accumbens, olfactory tubercle, septum) and mesocortical sites (cingulate, entorhinal, prefrontal, and pyriform cortices).[49,52] Evidence suggests that the neocortical DA system is much more extensive in primates than in rodents,[52] and that these neurons may be more involved in cognition, motivation, and reward.[49,52,53]

In the hypothalamus, there are four discrete groups of DA cell bodies which project to the median eminence and neurohypophysis and which modulate neuroendocrine regulation of prolactin secretion.[49,53]

It is the so-called long DA systems, that is, the nigrostriatal, mesolimbic, and mesocortical pathways, which are thought to play a prominent role in psychiatric illness.[54] The other DA system of significance in the CNS is the intermediate-length tuberoinfundibular DA pathway[54] as the effect of DA antagonism, seen with neuroleptics, can be associated with troublesome neuroendocrine side effects, e.g., amenorrhea, galactorrhea, gynecomastia.

B. Biosynthesis, Storage, and Release

The reader is referred to a number of excellent reviews of this topic.[50,52,55,56] Summarizing, tyrosine, an amino acid transported across the blood–brain barrier into the DA neuron, represents the precursor to catecholamine synthesis, including DA. Once in the neuron, L-tyrosine is converted to L-hydroxyphenylalanine (L-DOPA) by tyrosine hydroxylase (TH), which represents the rate-limiting step in DA synthesis. Depolarization of catecholaminergic terminals leads to the activation of TH, which in turn is modulated by end product inhibition by means of intraneuronal catecholamines competing at sites that bind the pterin cofactor. With depolarization and activation of TH through reversible phosphorylation, there occur kinetic changes in the enzyme which increase affinity for the pterin cofactor and decrease sensitivity to end product inhibition. Moreover, protein kinase C, as well as cAMP and Ca-calmodulin-dependent protein kinases, can all induce phosphorylation of the enzyme and, therefore, increase activity.[56]

DOPA is converted to DA by L-aromatic acid decarboxylase, and the turnover of this enzyme is so rapid that DOPA levels of the brain are negligible under normal conditions.[50] Conversion of tyrosine to L-DOPA and L-DOPA to dopamine occurs in the cytosol, with DA then being taken up into the storage vesicles. The vesicles play a dual role, storing catecholamines and mediating their release. An action potential at the nerve terminal results in the opening of Ca^{2+} channels, and increased intracellular Ca^{2+} promotes fusion of the vesicles with the neuronal membrane where exocytosis occurs.[56] The extent of DA release can be modulated by the rate and pattern of firing, in addition to presynaptic release-modulating autoreceptors.

C. Uptake and Degradation

Reuptake of DA into the nerve terminal is the primary mechanism for inactivation following release into the synaptic cleft.[55] This process is mediated by a carrier located on the outer membrane of catecholaminergic neurons,[56] with the action of DA terminated through a high-affinity uptake into presynaptic nerve terminals by means of this Na^+- and Cl^--dependent transporter protein.[50] The human DA transporter (DAT) gene has been located on chromosome 5p15.3,[57] and it appears that the carriers at DA- and norepinephrine-containing neurons are different in their specificities.[56]

The uptake process is energy dependent, reflecting a coupling process with the Na^+ gradient across the neuronal membrane. Thus, neuronal reuptake is Na^+-dependent, and drugs that inhibit Na^+ or open Na^+ channels interfere with the uptake process. Once taken up across the neuronal membrane, the amine can be stored in storage vesicles, and this uptake process requires Mg^{2+}. Catecholamines in the synaptic cleft not removed by the transport process diffuse into the extracellular space, where they can be catabolized by MAO and catechol-O-methyl-transferase (COMT) in the liver and kidney.[56]

D. Receptors

Initial classification of DA receptors distinguished two subtypes, D_1 and D_2.[58] D_1 receptors are coupled to stimulation of adenylate cyclase activity, while D_2 receptors inhibit adenylate cyclase. At this point, five DA receptors have been identified, although these can still be grouped into the original two subtypes (see Table 2).[59]

1. D_1-Like Receptors (D_1, D_5)

The D_1-like group of DA receptors comprises the D_1 and D_5 receptors, or D_{1A} and D_{1B}, coded by genes on chromosomes 5 and 4, respectively.[60-62] Examples of selective D_1 antagonists include NO 756, SCH 31966, SCH 23390, A-69024, and SK&F 83959.[63-65] SK&F 38983 and 83189 represent selective D_1 agonists.[66,67] Other commonly used agonists, such as apomorphine, influence both D_1 and D_2 receptors, exhibiting little or no selectivity between them.[67] D_1 receptor mRNA expression is high in the caudate putamen, nucleus accumbens, olfactory tubercle, and amygdala.[68,69] D_1 receptors have been implicated in negative symptomatology,[65] as well as in memory.[70] It is necessary to keep in mind, however, that D_1 and D_2 receptors are not totally independent and appear to function in an interactive fashion.[71-73]

Existing D_1 compounds suggest the capacity to discriminate between D_1 and D_5 receptors, but, to date, selective D_5 antagonists and agonists have not been identified.[74,75] There is a much lower level of expression of the D_5, vs. the D_1, receptor; in fact, the wider distributions of D_1 and D_2 receptor mRNA suggest a greater number of functions for these two receptor subtypes.[68] Highest levels of D_5 are found in the hippocampus and hypothalamus, in contrast to virtually none in the striatum.[71] There have not, as yet, been particular behaviors or psychiatric symptomatology specifically linked to this particular receptor.

2. D_2-Like Receptors (D_2, D_3, D_4)

The D_2-like receptors consist of D_2, D_3, and D_4 subtypes. The D_3 and D_4 receptors are also referred to as D_{2A} and D_{2B}, respectively, while the D_2 receptor has two variants in humans: $D_{2(long)}$ and $D_{2(short)}$, the latter missing a 29-amino acid coded by an exon.[59] The chromosomal locations in humans are as follows: D_2 (11q22-23), D_3 (3q13.3), and D_4 (11p15.5).[69] Examples of D_2 antagonists include raclopride, eticlopride, and spiperone, while (+)-PHNO, RU 24926, and quinpirole are examples of selective D_2 agonists.[66,71] D_2 receptor mRNA expression is high in the caudate putamen, nucleus accumbens, and olfactory tubercle.[68,69] The D_2 receptor has been closely associated with psychosis, as well as motor behavior and endocrinologic function, i.e., prolactin secretion.[59,76] Much of this information has been provided through the use of conventional neuroleptics, particularly the high-potency agents with their increased D_2 selectivity.

In humans, the D_3 gene has two variants, D_3 with 1200 bases and D_{3s} with 1102 bases.[59] Similar to the difficulties distinguishing D_1 and D_5 receptors, there are no selective agonists or antagonists which permit the clear distinction between D_2 and D_3 receptors. The most widely investigated D_3 antagonist to date has been (+)-UH232, which has a D_3/D_2 affinity ratio of 4:5, a ratio that is unsatisfactory in evaluating the independent role of D_3 receptors.[77] Unlike the D_2 receptor, the D_3 receptor is almost absent in the dorsal striatum but expressed selectively in the ventral striatum, especially the nucleus accumbens.[69,78] This pattern of localization, with its selectivity for the ventral, limbic DA areas suggests that a D_3 antagonist may be capable of offering antipsychotic efficacy without inducing movement disorders.[77] Further evidence supporting a possible role for D_3 in psychosis arises from data indicating that most antipsychotics display high affinities for this particular receptor.[79,80]

The D_4 receptor has eight polymorphic variants in humans, with each variant having a different number of repeat units and each repeat consisting of 16 amino acids.[59] With no selective antagonists or agonists, it is once again difficult investigating the specific functions of this receptor. D_4 receptors are relatively high in the hippocampus, thalamus, and cortex, whereas D_4 receptor mRNA is absent from the motor elements of the striatum,[69,81] a pattern of distribution which once more suggests that D_4 antagonism may lead to antipsychotic efficacy and do so without associated movement disorders. The finding that clozapine, the prototype of "atypical" neuroleptics (see Chapter 11), is approximately one order of magnitude more potent at the D_4 receptor, vs. D_2 or D_3, has garnered considerable interest, suggesting that D_4 may play a critical role in the unique clinical profile of clozapine.[59,82] Moreover, elevated D_4 receptors in post-mortem schizophrenic brains has also been reported.[83,84] The sole attribution of D_4 to the unique profile of clozapine has, however, been tempered by evidence that other neuroleptics, including conventional compounds, also have D_4-binding affinity.[85] It remains to be determined whether the D_4-binding affinity of clozapine plays an integral role in its significant efficacy in treating both the positive and negative symptomatology of schizophrenia.

3.   Autoreceptors

The notion of presynaptic autoreceptors on DA neurons in the CNS was first introduced in the early 1970s.[86] It has been established that autoreceptors exist on most portions of the cells, including soma, dendrites, and nerve terminals, with their effect, at least in part, a function of their location.[87] For example, stimulation of these autoreceptors on the somatodendritic region slows the firing rate of DA neurons, while at nerve terminals stimulation leads to an inhibition in synthesis and release.[50,88] Three functionally distinct subtypes have been identified, that is, impulse modulating, release modulating, and synthesis modulating, and it is thought that all DA autoreceptors are of the D_2 subtype.[50] It appears that DA autoreceptors can be found on striatal, mesolimbic, and mesocortical neurons.[89-92]

From a clinical standpoint, these receptors have been the subject of considerable interest, as they offer the opportunity to modulate DA activity at a different level, which may have ramifications in terms of both clinical response and side effects.[93,94]

E.   Conclusions

Table 2 summarizes our current understanding of the DA system and reflects the significant advances made during recent years, as evidenced by the identification of D_3, D_4, and D_5 receptors. Simultaneous developments in the area of neuroimaging, particularly positron emission tomography (PET), have also extended our knowledge of individual receptors, as well as their relationship to the action of various psychotropic medications. This information, in turn, has had a profound impact on theories regarding the pathophysiology of the disorders for which these medications are used.

In the case of DA, the primary focus continues to be on its role in psychosis and schizophrenia, in particular. However, its involvement in other psychiatric illnesses, such as the affective disorders, has also been well established. At this point, much attention has turned to the development of selective ligands which will offer an opportunity to elucidate the role of these newly identified receptors. In the course of this work, it is likely that other receptors will be identified, further underscoring the complexity of the CNS and the limitations of our current theories.

TABLE 2.

Dopamine Receptors

	D$_1$-Like		D$_2$-Like			Autoreceptors
	D$_1$	D$_5$	D$_2$	D$_3$	D$_4$	
Other terms	D$_{1A}$	D$_{1B}$	D$_{2(short)}$ D$_{2(long)}$	D$_{2B}$	D$_{2C}$	
Human variants			D$_{2(short)}$ D$_{2(long)}$	D$_3$ D$_{3S}$	D$_{4.2-4.10}$ (8 variants)	
Adenylate cyclase	Stimulates	Stimulates	Inhibits	Inhibits	? Inhibits	
Chromosome	5q35.1	4p15-16	11q22-23	13q13.3	11p15.5	
Anatomical distribution	Caudate putamen Nucleus accumbens Olfactory tubercle Amygdala	Hippocampus Hypothalamus	Caudate putamen Nucleus accumbens Olfactory tubercle	Hypothalamus N. accumbens Olfactory tubercle	Hippocampus Thalamus Cortex	
Role	Modulates D$_2$ (? role: EPS, positive and negative symptoms) Memory	?	Antagonism antipsychosis Side effects: movements, endocrine	Antagonism antipsychosis	Antagonism antipsychosis (? positive and negative symptoms)	Stimulation DA synthesis and release
DA pathway	Mesolimbic Mesocortical	?	Mesolimbic Nigrostriatal Tuberoinfundibular	Mesolimbic	Mesolimbic Mesocortical	Mesolimbic Nigrostriatal Mesocortical

Abbreviations: DA, dopamine; EPS, extrapyramidal symptoms.

REFERENCES

1. Dubovsky, S. L., Beyond the serotonin reuptake inhibitors: rationales for the development of new serotonergic agents, *J. Clin. Psychiatry,* 55, Suppl. 34, 1994.
2. Jacobs, B. L. and Azmitia, E. C., Structure and function of the brain serotonin system, *Physiol. Rev.,* 72, 165, 1992.
3. Azmitia, E. C., The CNS serotonergic system: progression toward a collaborative organisation, in *Psychopharmacology: The Third Generation of Progress,* Meltzer, H. Y., Ed., Raven Press, New York, 1987, 61.
4. Frazer, A. and Hensler, J. G., Serotonin, in *Basic Neurochemistry,* 5th ed., Siegel, G. J., Agranoff, A. B., Albers, R. W., and Mollnoff, P. B., Eds., Raven Press, New York, 1994, 283.
5. Young, S. N., The 1989 Borden Award Lecture: Some effects of dietary components (amino acids, carbohydrate, folic acid) on brain serotonin synthesis, mood, and behavior, *Can. J. Physiol. Pharmacol.,* 69, 893, 1991.
6. Blakely, R. D., Berson, H. E., Fremeau, R. T., Jr., Caron, M. G., Peek, M. M., Prince, H. K., and Bradley, C. C., Cloning and expression of a functional serotonin transporter from rat brain, *Nature,* 354, 66, 1991.
7. Hoffman, B. J., Mezey, E., and Brownstein, M. J., Cloning of a serotonin transporter affected by antidepressants, *Science,* 254, 579, 1991.
8. Leonard, B. E., The comparative pharmacology of new antidepressants, *J. Clin. Psychiatry,* 54, Suppl. 3, 1993.
9. Fontaine, R., Novel serotonergic mechanisms and clinical experience with nefazodone, *Clin. Neuropharmacol.,* 16, Suppl. 3, S45, 1993.
10. Rickels, K. and Schweizer, E., Clinical overview of serotonin reuptake inhibitors, *J. Clin. Psychiatry,* 51, Suppl. B, 9, 1990.
11. Blier, P. and de Montigny, C., Current advances and trends in the treatment of depression, *Trends Pharmacol. Sci.,* 15, 220, 1994.
12. Reynolds, J. N., Baskys, A., and Carlen, P. L., The effects of serotonin on N-methyl-D-aspartate and synaptically evoked depolarizations in rat neocortical neurons, *Brain Res.,* 456, 282, 1988.
13. Brosen, K., The pharmacogenetics of selective serotonin reuptake inhibitors, *Clin. Investigator,* 71, 1002, 1993.
14. Peroutka, S. J. and Snyder, S. H., Multiple serotonin receptors: differential binding of [^{3}H] 5-hydroxytryptamine, [^{3}H] lysergic acid diethylamide and [^{3}H] spiroperidol, *Mol. Pharmacol.,* 16, 687, 1979.
15. Hen, R., Of mice and flies: commonalities among 5-HT receptors, *Trends Pharmacol. Sci.,* 13, 160, 1992.
16. Moskowitz, M. A., Neurogenic versus vascular mechanisms of sumatriptan and ergot alkaloids in migraine, *Trends Pharmacol. Sci.,* 13, 307, 1992.
17. Hamon, M., Neuropharmacology of anxiety: perspectives and prospects, *Trends Pharmacol. Sci.,* 15, 36, 1994.
18. Wallis, D. I., 5-HT receptors involved in initiation or modulation of motor patterns: opportunities for drug development, *Trends Pharmacol. Sci.,* 15, 288, 1994.
19. Maricq, A. V., Peterson, A. S., Brake, A. J., Myers, R. M., and Julius, D., Primary structure and functional expression of the 5-HT$_3$ receptor, a serotonin-gated ion channel, *Science,* 254, 432, 1991.
20. Greenshaw, A. J., Behavioural pharmacology of 5-HT$_3$ receptor antagonists: a critical update on therapeutic potential, *Trends Pharmacol. Sci.,* 14, 265, 1993.
21. Bockaert, J. B., Fozard, J. R., Dumuis, A., and Clarke, D. E., The 5-HT$_4$ receptor: a place in the sun, *Trends Pharmacol. Sci.,* 13, 141, 1992.
22. Stam, N.J., Van Huizen, F., Van Alebeek, C., Brands, J., Dijkema, R., Tonnaer, J. A., and Olijve, W., Genomic organization, coding sequence and functional expression of human 5-HT$_2$ and 5-HT$_{1A}$ receptor genes, *Eur. J. Pharmacol.,* 227, 153, 1992.
23. Maroteaux, L., Saudou, F., Amlaiky, N., Boschert, U., Plassat, J. L., and Hen, R., Mouse 5-HT$_{1B}$ serotonin receptor: cloning, functional expression, and localization in motor control centers, *Proc. Natl. Acad. Sci. U.S.A.,* 89, 3020, 1992.
24. Jin, H., Oksenberg, D., Ashkenazi, A., Peroutka, S. J., Duncan, A. M., Rozmahel, R., Yang, Y., Mengod, G., Palacios, J. M., and O'Dowd, B. F., Characterization of the human 5-hydroxytryptamine1B receptor, *J. Biol. Chem.,* 267, 5735, 1992.
25. Voigt, M. M., Laurie, D. J., Seeburg, P. H. and Bach, A., Molecular cloning and characterization of a rat brain cDNA encoding a 5-hydroxytryptamine1B receptor, *EMBO Journal,* 10, 4017, 1991.
26. Hamblin, M. W., Metcalf, M. A., McGuffin, R. W., and Karpells, S. Molecular cloning and functional characterization of a human 5-HT$_{1B}$ serotonin receptor: a homologue of the rat 5-HT$_{1B}$ receptor with 5-HT$_{1D}$-like pharmacological specificity, *Biochem. Biophys. Res. Commun.,* 184, 752, 1992.

27. Van Sande, J., Allgeier, A., Massart, C., Czernilofsky, A., Vassart, G., Dumont, J. E., and Maenhaut, C., The human and dog 5-HT$_{1D}$ receptors can both activate and inhibit adenylate cyclase in transfected cells, *Eur. J. Pharmacol.*, 247, 177, 1993.

28. Bach, A. W., Unger, L., Sprengel, R., Mengod, G., Palacios, J., Seeburg, P. H., and Voigt, M. M., Structure, functional expression and spatial distribution of a cloned cDNA encoding a rat 5-HT$_{1D}$-like receptor, *J. Receptor Res.*, 13, 479, 1993.

29. Maenhaut, C., Van Sande, J., Massart, C., Dinsart, C., Libert, F., Monferini, E., Giraldo, E., Ladinsky, H., Vassart, G., and Dumont, J. E., The orphan receptor cDNA RDC4 encodes a 5-HT$_{1D}$ serotonin receptor, *Biochem. Biophys. Res. Commun.*, 180, 1460, 1991.

30. Zgombick, J. M., Borden, L. A., Cochran, T. L., Kucharewicz, S. A., Weinshank, R. L., and Branchek, T. A., Dual coupling of cloned human 5-hydroxytryptamine1D alpha and 5-hydroxytryptamine1D beta receptors stably expressed in murine fibroblasts: inhibition of adenylate cyclase and elevation of intracellular calcium concentrations via pertussis toxin-sensitive G protein(s), *Mol. Pharmacol.*, 44, 575, 1993.

31. Weinshank, R. L., Zgombick, J. M., Macchi, M. J., Branchek, T. A., and Hartig, P. R., Human serotonin 1D receptor is encoded by a subfamily of two distinct genes: 5-HT$_{1D\alpha}$ and 5-HT$_{1D\beta}$, *Proc. Natl. Acad. Sci. U.S.A.*, 89, 3630, 1992.

32. Demchyshyn, L., Sunahara, R. K., Miller, K., Teitler, M., Hoffman, B. J., Kennedy, J. L., Seeman, P., Van Tol, H. H., and Niznik, H. B., A human serotonin 1D receptor variant (5-HT1Dβ) encoded by an intronless gene on chromosome 6, *Proc. Natl. Acad. Sci. U.S.A.*, 89, 5522, 1992.

33. Julius, D., MacDermott, A. B., Axel, R., and Jessell, T. M., Molecular characterization of a functional cDNA encoding the serotonin 1c receptor, *Science*, 241, 558, 1988.

34. Zgombick, J. M., Schechter, L. E., Macchi, M., Hartig, P. R., Branchek, T. A., and Weinshank, R. L., Human gene S31 encodes the pharmacologically defined serotonin 5-hydroxytryptamine$_{1E}$ receptor, *Mol. Pharmacol.*, 42, 180, 1992.

35. McAllister, G., Charlesworth, A., Snodin, C., Beer, M. S., Noble, A. J., Middlemiss, D. N., Iversen, L. L., and Whiting, P., Molecular cloning of a serotonin receptor from human brain (5-HT1E): a fifth 5-HT1-like subtype, *Proc. Natl. Acad. Sci. U.S.A.*, 89, 1992.

36. Adham, N., Kao, H. T., Schecter, L. E., Bard, J., Olsen, M., Urquhart, D., Durkin, M., Hartig, P. R., Weinshank, R. L., and Branchek, T. A., Cloning of another human serotonin receptor (5-HT1F): a fifth 5-HT1 receptor subtype coupled to the inhibition of adenylate cyclase, *Proc. Natl. Acad. Sci. U.S.A.*, 90, 408, 1993.

37. Roth, B. L., Hamblin, M. W., and Ciaranello, R. D., Developmental regulation of 5-HT2 and 5-HT1c mRNA and receptor levels, *Brain Res. (Dev. Brain Res.)*, 58, 51, 1991.

38. Pritchett, D. B., Bach, A. W., Wozny, M., Taleb, O., Dal Toso, R., Shih, J. C., and Seeburg, P. H., Structure and functional expression of cloned rat serotonin 5-HT-2 receptor. *EMBO J.*, 7, 4135, 1988.

39. Choi, D. S., Birraux, G., Launay, J. M., and Maroteaux, L., The human serotonin 5-HT$_{2B}$ receptor: pharmacological link between 5-HT$_2$ and 5-HT$_{1D}$ receptors, *FEBS Lett.*, 352, 393, 1994.

40. Loric, S., Launay, J. M., Colas, J. F., and Maroteaux L., New mouse 5-HT2-like receptor. Expression in brain, heart and intestine, *FEBS Lett.*, 312, 203, 1992.

41. Eison, A. S. and Eison, M. S., Serotonergic mechanisms in anxiety, *Prog. Neuro-Psychopharmacol. Biol. Psychiatry*, 18, 47, 1994.

42. Tecott, L. H., Maricq, A. V., and Julius, D., Nervous system distribution of the serotonin 5-HT$_3$ receptor mRNA, *Proc. Natl. Acad. Sci. U.S.A.*, 90, 1430, 1993.

43. Butcher, M. E., Global experience with ondansetron and future potential, *Oncology*, 50, 191, 1993.

44. Plassat, J. L., Boschert, U., Amlaiky, N., and Hen, R., The mouse 5-HT5 receptor reveals a remarkable heterogeneity within the 5-HT1D receptor family, *EMBO J.*, 11, 4779, 1992.

45. Wisden, W., Parker, E. M., Mahle, C. D., Grisel, D. A., Nowak, H. P., Yocca, F. D., Felder, C. C., Seeburg, P. H., and Voigt, M. M., Cloning and characterization of the rat 5-HT$_{5B}$ receptor. Evidence that the 5-HT$_{5B}$ receptor couples to a G protein in mammalian cell membranes, *FEBS Lett.*, 333, 25, 1993.

46. Monsma, F. J., Jr., Shen, Y., Ward, R. P., Hamblin, M. W., and Sibley, D. R., Cloning and expression of a novel serotonin receptor with high affinity for tricyclic psychotropic drugs *Mol. Pharmacol.*, 43, 320, 1993.

47. Shen, Y., Monsma, F. J., Jr., Metcalf, M. A., Jose, P. A., Hamblin, M. W., and Sibley, D. R., Molecular cloning and expression of a 5-hydroxytryptamine$_7$ serotonin receptor subtype, *J. Biol. Chem.*, 268, 18200, 1993.

48. Davis, K. L., Kahn, R. S., Ko, G., and Davidson, M., Dopamine in schizophrenia: a review and reconceptualization, *Am. J. Psychiatry*, 148, 474, 1991.

49. Kapur, S. and Mann, J. J., Role of the dopaminergic system in depression, *Biol. Psychiatry*, 32, 1, 1992.

50. Roth, R. H., and Elsworth, J. D., Biochemical pharmacology of midbrain dopamine neurons, in *Psychopharmacology: The Fourth Generation of Progress,* Bloom, F. E. and Kupfer, D. J., Eds., Raven Press, New York, 1995, 227.

51. Fallon, J. H., Topographic organization of ascending dopaminergic projections, *Ann. N.Y. Acad. Sci.,* 537, 1, 1988.

52. Roth, R. H., Wolf, M. E., and Deutch, A. Y., Neurochemistry of midbrain dopamine neurons, in *Psychopharmacology: The Third Generation of Progress,* Meltzer, H. Y., Ed., Raven Press, New York, 1987, 81.

53. Creese, I., Dopamine and antipsychotic medications, in *Annu. Rev. Psychiatry,* Vol. 4, Hales, R. E. and Frances, A. J., Eds., American Psychiatric Association Press, Washington, D.C., 1985, 17.

54. Pickar, D., Neuroleptics, dopamine, and schizophrenia, *Psychiatr. Clin. North Am.,* 9, 35, 1986.

55. Bannon, M. J., and Roth, R. H., Pharmacology of mesocortical neurons, *Pharmacol. Rev.,* 35, 53, 1983.

56. Weiner, N. and Molinoff, P. B., Catecholamines, in *Basic Neurochemistry,* 5th ed., Siegel, G. J., Agranoff, A. B., Albers, R. W., and Mollnoff, P. B., Eds., Raven Press, New York, 1994, 261.

57. Vandenbergh, D. J., Persico, A. M., Hawkins, A. L., et al., Human dopamine transporter gene (DAT1) maps to chromosome 5p15.3 and displays a VNTR, *Genomics,* 14, 1104, 1992.

58. Spano, P. F., Govoni, S., and Trabucchi, M., Studies on the pharmacological properties of dopamine receptors in various areas of the central nervous system, *Adv. Biochem. Psychopharmacol.,* 19, 155, 1978.

59. Seeman, P. and Van Tol, H. H. M., Dopamine receptor pharmacology, *Curr. Opin. Neurol. Neurosurg.,* 6, 602, 1993.

60. Sunahara, R. K., Niznik, H. B., Weiner, D. M., Stormann, T. M., Brann, M. R., Kennedy, J. L., Gelernter, J. E., Rozmahel, R., Yang, Y., Israel, Y., Seeman, P., and O'Dowd, B. E., Human dopamine D_1 receptor encoded by an intronless gene on chromosome 5, *Nature,* 347, 80, 1990.

61. Zhou, Q.-Y., Grandy, D. K., Thambi, L., Kushner, J. A., Van Tol, H. H. M., Cone, R., Pribnow, D., Salon, J., Bunzow, J. R., and Civelli, O., Cloning and expression of human and rat D_1 dopamine receptors, *Nature,* 347, 76, 1990.

62. Sunahara, R. K., Guan, H.-C., O'Dowd, B. F., Seeman, P., Laurier, L. G., Ng, G., George, S., Torchia, J., Van Tol, H. H. M., and Niznik, H. B., Cloning of the gene for a human dopamine D_5 receptor with higher affinity for dopamine than D_1, *Nature,* 350, 614, 1991.

63. Waddington, J. L. and Daly, S. A., The status of "second generation" selective D_1 dopamine receptor antagonists as putative atypical antipsychotic agents, in *Novel Antipsychotic Drugs,* Meltzer, H. Y., Ed., Raven Press, New York, 1992, 109.

64. Waddington, J. L., Daly, S. A., Downes, R. P., and Deveney, A. M., Testing out the puzzle of multiple "D-1-like" vs. "D-2-like" dopamine receptor subtypes: existence and function, *Eur. Neuropsychopharmacol.,* 3, 231, 1993.

65. Kanba, S., Suzuki, E., Nomura, S., Nakaki, T., Yagi, G., Asai, M., and Richelson, E., Affinity of neuroleptics for D_1 receptor of the human brain striatum, *J. Psychiatry Neurosci.,* 19, 265, 1994.

66. Seeman, P. and Ulpian, C., Dopamine D_1 and D_2 receptor selectivities of agonists and antagonists, *Adv. Exp. Med. Biol.,* 235, 55, 1988.

67. Daly, S. A. and Waddington, J. L., D-1 receptors and the topography of unconditioned motor behavior: studies with the selective "full efficacy" benzazepine D-1 agonist SK&F 83959, *J. Psychopharmacol.,* 6, 50, 1992.

68. Mansour, A. and Watson, S. J., Dopamine receptor expression in the central nervous system, in *Psychopharmacology: The Fourth Generation of Progress,* Bloom, F. E. and Kupfer, D. J., Eds., Raven Press, New York, 1995, 207.

69. Meador-Woodruff, J. H., Update on dopamine receptors. *Ann. Clin. Psychiatry,* 6, 79, 1994.

70. Sawaguchi, T. and Goldman-Rakic, P. S., D1 dopamine receptors in prefrontal cortex: involvement in working memory, *Science,* 251: 947, 1991.

71. Walters, J. R., Bergstrom, D. A., Carlson, J. H., Chase, T. N., and Braun, A. R., D_1 dopamine receptor activation required for postsynapic expression of D_2 agonist effects, *Am. Assoc. Adv. Sci.,* 236, 719, 1987.

72. Seeman, P., Niznik, H. B., Guan, H.-C., Booth, G., and Ulpian, C., Link between D_1 and D_2 dopamine receptors is reduced in schizophrenia and Huntington diseased brain, *Proc. Natl. Acad. Sci. U.S.A.,* 86, 10156, 1989.

73. Buonamici, M., Mantegani, S., Cervini, M. A., Maj, R., Rossi, A. C., Caccia, C., Carfagna, N., Carminati, P., and Fariello, R. G., FCE 23884, substrate-dependent interaction with the dopaminergic system. I. Preclinical behavioral studies, *J. Pharmacol. Exp. Ther.,* 259, 345, 1991.

74. Niznik, H. B., and Van Tol, H. H. M., Dopamine receptor genes: new tools for molecular psychiatry, *J. Psychiatry Neurosci.,* 17, 165, 1992.

75. Monsma, F. J., Jr., Burgess, L. H., Sibley, D. R., and Civelli, O., Dopamine D1 and D5 receptors: molecular cloning and pharmacological characteristics, *Eur. Neuropsychopharmacol.,* 4, 194, 1994.

76. Schacter, M., Bedard, P., Debono, A. G., Jenner, P., Marsden, C. D., Price, P., Parkes, J. D., Keenan, J., Smith, B., Rosenthaler, J., Horowski, R., and Dorow, R., The role of D-1 and D-2 receptors, *Nature*, 286, 157, 1980.

77. Carlsson, A., Receptor research: a challenge for drug development, *Eur. Neuropsychopharmacol.*, 3, 176, 1993.

78. Schwartz, J.-C., Diaz, J., Griffon, N., Lammers, C., Levesque, D., Martres, M. P., and Sokoloff, P., The dopamine D3 receptor in nucleus accumbens: selective cellular localisation, function and regulation, *Eur. Neuropsychopharmacol.*, 4, 190, 1994.

79. Healy, D., D_1 and D_2 and D_3, *Br. J. Psychiatry*, 159, 319, 1991.

80. Sokoloff, P., Martres, M.-P., Giros, B., Bouthenet, M.-L., and Schwartz, J. C., The third dopamine receptor (D_3) as a novel target for antipsychotics, *Biochem. Pharmacol.*, 43, 659, 1992.

81. Seeman, P., Schizophrenia, antipsychotic drugs, and D_4, *Neuropsychopharmacology*, 9, 13S, 1993.

82. Seeman, P., Dopamine receptor sequences: therapeutic levels of neuroleptics occupy D_2 receptors, clozapine occupies D_4, *Neuropsychopharmacology*, 7, 261, 1992.

83. Seeman, P., Guan, H.-C., and Van Tol, H. H. M., Dopamine D_4 receptors elevated in schizophrenia, *Nature*, 365, 441, 1993.

84. Schizophrenia: D_4 receptor elevation. What does it mean?, editorial, *J. Psychiatry Neurosci.*, 19, 171, 1994.

85. Leysen, J. E., Janssen, P. M. F., Megens, A. A. H. P., and Schotte, A., Risperidone: a novel antipsychotic with balanced serotonin-dopamine antagonism, receptor occupancy profile, and pharmacological profile, *J. Clin. Psychiatry*, 55 (Suppl. 5), 5, 1994.

86. Kehr, W., Carlsson, A., Lindqvist, M., Magnusson, T., and Atack, C., Evidence for a receptor-mediated feedback control of striatal tyrosine hydroxylase activity, *J. Pharm. Pharmacol.*, 24, 744, 1972.

87. Roth, R. H., CNS dopamine autoreceptors: distribution, pharmacology, and function, *Ann. N.Y. Acad. Sci.*, 430, 27, 1984.

88. Wolf, M. E. and Roth, R. H., Autoreceptor regulation of dopamine synthesis, *Ann. N.Y. Acad. Sci.*, 604, 323, 1990.

89. Wolf, M. E., Galloway, M. P., and Roth, R. H., Regulation of DA synthesis in medial prefrontal cortex: studies in brain slices, *J. Pharmacol. Exp. Ther.*, 236, 699, 1986.

90. Talmaciu, R. K., Hoffman, I. S., and Cubeddu, L. X., Dopamine autoreceptors modulate dopamine release from the prefrontal cortex, *J. Neurochem.*, 47, 865, 1986.

91. Wolf, M. E. and Roth, R. H., Dopamine neurons projecting to the medial prefrontal cortex possess release-modulating autoreceptors, *Neuropharmacology*, 26, 1053, 1987.

92. Ereshefsky, L., Tran-Johnson, T. K., and Watanabe, M. D., Pathophysiologic basis for schizophrenia and the effect of antipsychotics, *Clin. Pharmacol.*, 9, 682, 1990.

93. Meltzer, H. Y., Relevance of dopamine autoreceptors for psychiatry: preclinical and clinical studies, *Schizophr. Bull.*, 6, 456, 1980.

94. Javitt, D. C., Weinstein, S. L., and Opler, L. A., The possible role of dopamine autoreceptors in neuroleptic atypicality, *Psychiatr. Dev.*, 1, 57, 198

95. de Bruijn, K. M., The development of tropisetron in its clinical perspective, *Ann. Oncol.*, 3(Suppl. 4), S19, 1993.

96. Ruat, M., Traiffort, E., Leurs, R., Tardivel-Lacombe, J., Diaz, J., Arrang, J.-M., and Schwartz, J.-C., Molecular cloning, characterization, and localization of a high-affinity serotonin receptor (5-HT$_7$) activating cAMP formation, *Proc. Natl. Acad. Sci. U.S.A.*, 90, 8547, 1993.

Chapter 5

ACETYLCHOLINE

Gary M. Hasey

CONTENTS

0-8493-8386-0/96/$0.00+$.50
© 1996 by CRC Press, Inc.

Acetylcholine (ACh) was the first identified neurotransmitter. Nevertheless, relatively little attention has been focused upon the possible involvement of ACh in psychiatric disorders compared with the more recently discovered monoamines, particularly serotonin in the affective disorders and dopamine in schizophrenia. The reason for this is not entirely clear as there is solid evidence implicating cholinergic mechanisms in the pathophysiology of affective disorders. Although there are less data supporting a role for ACh in schizophrenia, ACh is clearly important in the context of the side effects and toxic effects of neuroleptic drugs.

I. ANATOMY OF CENTRAL CHOLINERGIC NEURONS

Even though ACh was the first identified neurotransmitter, the anatomical distribution of central nervous system (CNS) cholinergic neurons is only now coming to light. The slow progress in this area has been largely because of the unavailability of sensitive and specific markers of cholinergic cells. Acetylcholinesterase activity and cholinergic receptor assays have frequently been employed as cholinergic markers; however, since acetylcholinesterase and cholinergic receptors may be located on postsynaptic membranes, these markers cannot differentiate actual cholinergic neurons, which synthesize and release ACh, from "cholinoceptive" neurons, which simply receive cholinergic innervation.[1] However, the recent development of immunohistochemical techniques to identify choline acetyltransferase (ChAT),[2] which is primarily, if not exclusively, found in cholinergic neurons,[3] has allowed more precise investigation into the neuroanatomical distribution of cholinergic neurons.

It is now accepted that there are three main forebrain cholinergic pathways: (1) the intrinsic striatal projection; (2) the magnocellular basal nucleus projection to the cerebral cortex and hypothalamus; and (3) the pontine tegmental projection to the extrapyramidal system, the hypothalamus, thalamus, and the cerebral cortex.[2,4] A brief discussion of these regions follows, but the reader is referred to an excellent overview by Fibiger and Vincent[4] and a comprehensive accounting of the neuroanatomy of the pontine tegmental projection by Wilson[221] for a detailed description.

A. Striatum

ChAT-labeled neurons are scattered diffusely throughout the caudate, the putamen, and the nucleus accumbens in rat brain.[2] Since these neurons do not appear to show degenerative changes when all known afferent pathways into the striatum are severed[5] and since ChAT levels outside the striatum are not decreased when the striatum is itself destroyed,[6] it is believed that these cholinergic neurons are intrinsic, i.e., the striatal cholinergic cell bodies do not project outside the striatum.[1,4] Nevertheless, these striatal cholinergic interneurons may influence CNS functioning outside the striatum by modifying the activity of striatal GABAergic, serotonergic, noradrenergic, and dopaminergic neurons[7-10] which themselves project widely outside the striatum. The interaction of cholinergic and dopaminergic neurons in the striatum is reciprocal as dopaminergic neurons are believed to exert an inhibitory effect upon cholinergic neurons in the striatum.[4] This interaction may account for the therapeutic efficacy of anticholinergic drugs in neuroleptic-induced parkinsonism.

Although the role that striatal cholinergic events might play in human psychopathology is not yet defined, there is some indirect evidence that cholinergic activity in this structure may be important in the pathophysiology of depressive illness. Positron emission tomography studies have shown that striatal glucose metabolism may be altered in depressed patients,[11] and in a study using the behavioral despair animal model of depression striatal ACh was highly correlated with hypothalamic-pituitary-adrenal (HPA) activity and the behavioral analogue of depression.[12] Alterations in the circadian rhythm of psychomotor activity, which are frequently seen in patients with major depression,[13] can be produced in rats by perturbation of cholinergic functioning in the striatum.[14]

B. Rostral Cholinergic Column of Basal Forebrain

This pathway, which provides one of the main cholinergic afferents to the cerebral cortex, appears to be involved in memory functioning[15] and is particularly vulnerable to degeneration in patients suffering from Alzheimer's disease.[16]. In all of the mammalian species examined to date, basal forebrain cholinergic neurons are arranged in a columnar fashion extending from the medial septum to the vertical and horizontal limbs of the nucleus of the diagonal band of Broca extending into the substantia innominata, also known as the basal nucleus (nucleus basalis) of Meynert.[4] Together with the magnocellular preoptic nucleus and globus pallidus, the cholinergic neurons in this region, sometimes collectively called the magnocellular basal nucleus, project to the olfactory bulb, hippocampus, amygdala and cerebral cortex.[2,4]

Many of the cholinergic neurons in the medial septal area, particularly those projecting to the hippocampus, also contain the neuropeptide galanin.[4] The amnesia produced in rats after lesions of the nucleus basalis is reversed by ACh injection into the hippocampus, but this effect is blocked by galanin, suggesting that this peptide may play a neuromodulatory role in the hippocampus by inhibiting the postsynaptic effects of ACh.[15,17] This antagonistic effect may be particularly relevant in Alzheimer's disease as galanin levels remain unchanged despite a 50 to 60% reduction in ChAT activity in the cortex and hippocampus [18]

C. Caudal Cholinergic Column of Mesencephalic
   and Pontine Reticular Formation

The second main cholinergic projection pathway to the cerebral cortex and other forebrain regions arises caudally from ChAT-staining neurons in the mesencephalic and pontine tegmentum. These cholinergic neurons, like those in the magnocellular basal nucleus in the forebrain, are arranged in a continuous column which includes the laterodorsal-tegmental (LDT) nucleus in the floor of the fourth ventricle and the pedunculopontine tegmental (PPTg) nucleus to the tip of the substantia nigra.[2,4] Axons arising in this region project to the interpeduncular nucleus, tectum, and forward to putative limbic areas, including the hypothalamus, thalamus, hippocampus amygdala, basal forebrain, and medial frontal cortex.[4]

Several neuropeptides appear to be colocalized with ACh in the neurons of the caudal cholinergic column. About one third of the neurons contain substance P and bombesin/gastrin-releasing peptide.[4] Corticotropin-releasing factor can be found in other ChAT-staining cells in this region. [4]

Cholinergic neurons in this region may be involved in higher cortical processes, such as level of consciousness,[4,5] and the integration of limbic and somatomotor activity.[5] Rats bred for hypersensitivity to cholinergic agonists show elevated pain thresholds,[19] and injections of cholinergic agonists into the PPTg nucleus suppress nociceptive responses in cats,[5] suggesting that cholinergic neurons in this region potentiate analgesic mechanisms in the brain stem.

1. Pontomesencephalic Cholinergic Neurons and Rapid Eye Movement Sleep

Pontomesencephalic cholinergic neurons project extensively into the brain stem reticular formation[20] and may represent the neuroanatomic structures mediating the profound effect of cholinergic agonists and antagonists on various states of consciousness, particularly rapid eye movement (REM) sleep. The interested reader is referred to an excellent review of this topic written by Jones.[20] Briefly, pharmacological and lesioning studies appear to confirm the critical role of pontomesencephalic cholinergic neurons in the initiation and maintenance of REM sleep. The injection of cholinergic agonists into the pontine and caudal mesencephalic tegmentum causes the appearance of REM sleep, but injection into cholinergic neurons in the region of the rostral midbrain results in awakening. This site-specific effect on sleep may explain the biphasic effect of the indirect agonist physostigmine (PHYSO), which causes first enhanced wakeful arousal then subsequent early onset of REM sleep.[20] Consistent with this, hemicholinium, which blocks the uptake of choline resulting in decreased ACh synthesis, produces decreased wakefulness as well as REM abolition.[21] The REM-inducing, and to some extent the arousing, effects of PHYSO and other cholinergic agonists have been clearly demonstrated in humans.[22,23] Other studies in both animals and humans have shown that this effect is mediated in part by muscarinic receptors as agonist-induced REM induction can be blocked by atropine or scopolamine.[24,26] There is evidence from animal studies that "tonic" REM phenomena such as electroencephalographic desynchronization and muscular atonia are, respectively, mediated by the M1 and M2 subtypes of muscarinic cholinergic receptors while "phasic" REM events, such as ponto-geniculo-occipital (PGO) electrical potentials, which correlate in animals with rapid eye movements themselves, may be mediated by nicotinic receptors.[146] It has been demonstrated, in cats, that microinjection of nicotine into the medial pontine reticular formation initiates REM sleep[148] while nicotine antagonists block PGO spikes.[147] Gillin et al.[146] found shorter REM latency, greater REM density (number of eye movements per unit time), and greater total sleep time in healthy smokers compared with nonsmokers.

Transition through the various stages of slow wave sleep and REM appears to be governed by complex interaction of monoaminergic and cholinergic neurons. In the cat, the cholinergic neurons of the LDT–PPTg complex overlap with noradrenergic cells of the locus coeruleus–parabrachial nuclei, and there is some evidence that the noradreneric neurons cease firing with the onset of REM, suggesting a permissive role for noradrenergic systems.[20] For example, it has been shown that animals treated with the noradrenaline-depleting agent reserpine do not show the initial increased wakefulness usually seen with cholinesterase treatment, but rather go directly into REM sleep.[25] Neuronal extensions from the serotonergic cells of the raphe nuclei have been discovered in the LDT, and it has been shown that serotonin can hyperpolarize cholinergic neurons,[26] thus allowing for a similar interaction between this monoaminergic system and the cholinergic elements involved in REM generation. GABA can be found in some neurons in the region of the LDT–PPTg and may modify the cholinergic effects upon REM sleep.[20]

Substance P and vasoactive intestinal peptide, which are colocalized with ACh in some neurons, are also reported to alter REM sleep[28,29] as is somatostatin,[30] which can be found near cholinergic neurons in the pontomesencephalic region.

Since shortening of REM latency is one of the most widely accepted biologic markers of major depression in patients with major depressive disorder,[31] it is possible that dysfunction of some part of the caudal cholinergic column may be involved in the pathophysiology of mood disorder.

D. Other Regions

It is universally accepted that the cerebral cortex possesses cholinergic innervation, but there is debate whether this innervation is purely extrinsic, i.e., the cholinergic elements represent the axons of cholinergic neurons whose cell bodies are located outside the cortex, or whether intrinsic cholinergic neurons that actually originate in this region exist. As mentioned earlier the cortex in rats is a major recipient of cholinergic fibers arising in the caudal cholinergic column, and extrinsic innervation has been confirmed in a post-mortem study of human brain demonstrating that cortical axons from the nucleus basalis stained positively for acetylcholinesterase.[32] However, the study also noted staining of cortical cell bodies,[32] and other investigators have identified intrinsic ChAT-staining neurons in mammalian cortex.[33] Vasoactive intestinal peptide (VIP) colocalizes with ACh in up to 80% of these intrinsic neurons, and it has been suggested that this feature may allow the differentiation of these from cholinergic fibers arising from the caudal cholinergic column, as the latter tend to contain substance P.[33] In rat cerebral cortex, VIP is reported to potentiate the stimulatory effect of ACh on phosphoinositide turnover, especially at low concentrations of ACh.[34]

Apart from the regions described above, the only areas in the CNS which stain strongly for ChAT are the motor nuclei of several cranial nerves, including the oculomotor, trochlear, abducens, motor component of the trigeminal, and facial nerve.[2]

II. CHOLINERGIC RECEPTORS

Cholinergic receptors have been historically divided into muscarinic and nicotinic types based upon the affinities for the agonists muscarine and nicotine. To date, five different subtypes of muscarinic cholinergic receptor (mAChR) have been cloned,[35] and variations in the protein structure of the nicotinic cholinergic receptor (nAChR) may indicate several functional subtypes of this receptor as well.[36] To add to the complexity, differences between peripheral and CNS receptors have been described for nAChR and, to a lesser extent, for mAChR.[35,37,38]

A great deal is known about the physical structure and pharmacology of the nAChR, but these data are largely derived from studies of receptors in peripheral tissues; the role of the nAChR in brain functioning is only recently being examined. The mAChR, on the other hand, is present in much greater numbers in the CNS and perhaps, for this reason, has been much more extensively studied in the context of, psychopathology.

A. Muscarinic Receptors

1. Structure

The mAChR belongs to the superfamily of hormone receptors which are linked to intracellular second messenger systems such as cyclic adenosine monophosphate (cAMP), cyclic guanosine monophosphate (cGMP), or the products of phosphoinositide hydrolysis through guanine nucleotide–binding regulatory proteins (G proteins). Other members of this family include dopamine receptors, alpha- and beta-adrenergic receptors, serotonin type 1, 2 and 4 receptors, and GABA-B receptors. Within this class of rhodopsin-like receptor proteins, the amino terminal, which sits outside the neuron, is connected to a series of seven alpha helical "transmembrane domains," each of which penetrates and crosses the cell membrane. The alpha helices are linked by a series of alternating intracellular and extracellular loops to form a long, snakelike molecule which winds back and forth across the cell membrane. The second extracellular loop is thought to be the site of binding with muscarinic agonists,[39] and the third intracellular loop is believed to be the site of recognition and activation of G proteins.[40]

2. G Protein Linkage

When agonists bind to the mAChR, the receptor-coupled G protein acquires increased affinity for GTP which may then displace GDP already bound to the G protein resulting in conformational changes that cause deconstruction of the heterotrimeric G protein into its alpha and beta/gamma subunits. These subunits then go on to interact with various effector enzymes, such as adenylate cyclase, phospholipase C, which initiates the hydrolysis of inositol phospholipids,[39,41,42] or guanylate cyclase, which catalyzes the formation of cGMP.

Receptor coupling with the G protein also appears to alter the affinity with which agonist ligand binds to the mAChR. While antagonist binding to the mAChR appears to be unaffected by guanine nucleotide presence, the receptor affinity for agonists varies from low to high or superhigh depending on the presence or absence of guanine nucleotides in the experimental environment.[43,44] These findings have led to a model which proposes that receptors coupled with a G protein exist in a high-affinity state for agonists while the free, i.e., uncoupled, receptor demonstrates low affinity.[39]

a. cAMP

Although there are exceptions, the effect of mAChR activation upon adenylate cyclase is generally mediated by the G_i class of G proteins and is therefore inhibitory with a resultant decrease in the accumulation of cAMP.[45] In some brain regions, such as the olfactory bulb, and in some cultured cell lines, the mAChR may be coupled with G_s as well as G_i and mAChR activation may either stimulate or inhibit adenylate cyclase.[46] In tissue cultures the direction of the effect upon adenylate cyclase activity is dependent on mAChR density and agonist concentration.[46] To further complicate matters, at least six adenylate cyclase isozymes with different activation/inhibition profiles have now been cloned.[47]

b.   Inositol Phospholipid Hydrolysis

When activated by binding with an agonist, mAChR coupled with G proteins of the G_q class initiate inositol phospholipid hydrolysis (IPH). The enzyme phospholipase C cleaves phosphatidyl inositol, 4,5-bisphosphate (PIP_2), a large molecule partly embedded in the lipids of the cell membrane, into diacylglycerol (DAG) and inositol 1,4,5-triphosphate (IP_3).[39,45,48] DAG in turn activates protein kinase C (PKC) which can alter a variety of cell processes through protein phosphorylation.[49] IP_3 stimulates the release of membrane-bound intracellular calcium,[50] which can, in turn, affect cell functioning. IP_3 may be either phosphorylated to 1,3,4,5-tetrakisphosphate, which does not liberate calcium but may have other second messenger properties,[51,52] or dephosphorylated to inositol biphosphate (IP_2) then to IP_1. IP_1 is then cycled back for reuse by conversion to inositol, an important precursor substance in this cycle,[48] by inositol-1-phosphatase (IP_1ase).

Muscarinic stimulation of IPH is dependent on agonist structure since ACh, oxotremorine-M, and carbachol activate the response to a much greater extent than oxotremorine, arecholine or pilocarpine.[53-55]

Lithium inhibits IP_1ase resulting in the accumulation of IP_1,[56-58] and with potential depletion of inositol, lithium may dampen IPH. It has been shown that lithium, in therapeutic concentrations, can block the phosphoinositide-mediated action of oxotremorine-M in hippocampal slices,[58] leading to speculation that lithium may exert its mood-stabilizing effect in patients with mood disorder at least in part through this pathway. Since other neurotransmitters, including noradrenaline, dopamine, serotonin, histamine, and peptides, such as vasopressin, substance P, angiotensin, and thyrotropin-releasing hormone, also influence IPH,[45,48] lithium may dampen excessive activity of any of several neurotransmitter systems and this may account for its ability to improve both mania and depression.

c.   cGMP

Guanylate cyclase is stimulated by the binding of ACh to the mAChR resulting in increased concentrations of cGMP.[45,59] Guanylate cyclase may be located primarily in postsynaptic rather than in presynaptic regions of the neuron, and cGMP seems to be most concentrated in the cerebellum.[45] Although the mechanism by which cGMP alters neuronal activity is unknown, injection of cGMP into *Xenopus* oocytes results in the same increase in potassium permeability produced by ACh.[39] In general, the action of cGMP appears to be opposite to that of cAMP, and cGMP may exert some of its effects by stimulating a phosphodiesterase which breaks down cAMP.[60]

3.   *Muscarinic Receptor Subtypes*

The earliest subtyping of the mAChR evolved from pharmacological studies employing the muscarinic antagonist pirenzepine, which, unlike other antagonists being used at the time, bound avidly to only a subgroup of mAChR.[61-63] Muscarinic receptors that demonstrated high affinity for this antagonist were designated M1 while those showing low affinity for pirenzepine were called M2. As of this writing there appear to be at least five different genes coding for the mAChR.[35] Two of these gene products appear to be identical to the M1- and M2-receptor subtypes defined pharmacologically using pirenzepine; however, M3, M4, and M5 receptors have also been identified. These receptors display intermediate or high affinity for pirenzepine and may have previously been designated either M1 or M2.[35] Although there may be differences in the predominance of the mAChR subtypes across various brain regions, all can be found in the CNS.

It has been suggested that in the CNS M1 and M3 receptors are mainly located on the postsynaptic neuronal membrane while M2 and M4 receptors are located presynaptically, where they serve as inhibitory autoreceptors that turn off the release of ACh when activated.[64,65] It has been suggested that in mammalian species M2 and M4 receptors are selectively coupled to G proteins of the G_i class, which inhibit cAMP, while M1, M3, and M5 receptors are coupled to those of the G_q class, which stimulate phosphoinositide hydrolysis.[35,40,66,67] However, other groups have either found no specific second messenger–receptor type linkage [55] or specificity that varies according to the brain region studied.[68,69]

4. Receptor Reserve

At some sites in the CNS, for example the striatum, there appear to be a number of mAChR which are "spare" or held "in reserve." This has been demonstrated experimentally through studies showing that a significant proportion of mAChR may be inactivated through administration of an alkylating agent without a commensurate decrease in the liberation of second messenger in response to agonist stimulation.[69] In such tissues this redundancy may mean that greater receptor loss (as might occur in Alzheimer's disease) could occur without physiological effect. In addition, it has been shown that presynaptic inhibitory mAChR autoreceptors in the hippocampus do not possess such a reserve [70] so that receptor loss in this population could quickly manifest as decreased inhibition, which actually might compensate for decreased postsynaptic mAChR numbers.

5. Muscarinic Cholinergic Neuronal Adaptation

The cholinergic nervous system possesses the capacity to adapt to changes in the biochemical environment of the CNS. This capacity, which, within certain limits, allows the maintenance of relative stability in the face of pharmacological, physiological, or pathological alteration, may be the basis for clinical phenomena such as withdrawal effects from some medications and, possibly, therapeutic effects of other treatments. These compensatory mechanisms may operate at several different levels.

a. At the Receptor Level

It is well known that receptor density, i.e., the number of mAChRs per gram of protein, can be influenced by neuronal depolarization[71] and previous levels of receptor activation.[72] Chronic exposure of cholinergic neurons to muscarinic agonists results in receptor down regulation, while chronic blockade of the mAChR with antagonists causes receptor up regulation.[44,72-80] Atropine is reported to block the increase in receptor number produced by chronic agonist treatment of neuroblastoma tissue *in vitro* [72] presumably by limiting the extent of receptor activation. Interestingly, the up regulation of mAChR seen after chronic treatment with atropine is accompanied by parallel increases in VIP-binding sites.[76] This may represent a further compensatory mechanism as VIP may potentiate the effect of ACh on phosphoinositide turnover at low concentrations of ACh.[34]

In some studies where cholinergic afferents have been cut, the expected up regulation of mAChRs has not been seen.[81,82] This may have been due to residual cholinergic input from unknown sites or may indicate that factors other than receptor activation can determine receptor number.

Down regulation in response to chronic agonist treatment appears to involve at least two processes, the first being a rapid internalization or sequestration of the receptor[83] followed by eventual receptor destruction. Receptor sequestration is suggested by studies demonstrating differences in the binding of the lipophilic ligand quinuclidinyl-benzylate (QNB) and the hydrophilic ligand N-methylscopolamine (NMS). Chronic agonist treatment of neuronal tissue *in vitro* results in greater and more rapid decreases in ^{3}H-NMS compared with ^{3}H-QNB binding.[72,80,83,84] It is proposed that ^{3}H-QNB labels more mAChR because this lipophilic ligand labels not only cell surface mAChR but also those which may have been internalized within the cell where they would be inaccessible to the hydrophilic ^{3}H-NMS, which labels only cell surface receptors.[72] Harden et al.[84] demonstrated that while such sequestered mAChRs may dissociate from the cell membrane, they appear to remain functionally linked to G proteins and may still be capable of influencing neuronal activity to the extent that ligands are able to gain access to the inside of the neuron. They suggested that this may account for the poor correlation between ^{3}H-NMS binding and measures of IPH activity. The mechanism of sequestration is unknown, but cytochalasin B, which blocks endocytosis and microtubule functioning, is reported to block receptor down regulation.[85]

The second process in receptor down regulation may involve receptor destruction. The decrease in ^{3}H-NMS after agonist treatment is rapidly reversible,[84] but the decrease in ^{3}H-QNB binding seen with longer exposure to agonist is not.[84] This suggests that some sequestered receptors can be quickly recycled to the cell membrane but others might be destroyed so that the restoration of full numbers of mAChR requires the synthesis of new receptor protein. This is supported by work showing that the protein synthesis inhibitor cycloheximide blocks this process.[86,87] Finally, studies of cardiac mAChR suggest that there may be a period of time during which newly synthesized mAChRs may not be fully functional.[88]

b.   At the Second Messenger Level

Chronic exposure to cholinergic agonist or antagonist may also lead to changes in the efficiency of signal transduction from the receptor to the second messenger system. Using mouse neuroblastoma cells exposed to carbamylcholine, Cioffi and El-Fakahany[89] found that the reversal of agonist-induced desensitization of cGMP accumulation occurs in parallel with the changes in ^{3}H-NMS binding, but the altered IPH does not. Similarly, Smith et al.[81] showed that deafferentation of the hippocampus resulted in potentiation of carbachol-stimulated IPH without any change in ^{3}H-NMS binding, and Goobar and Bartfai[90] demonstrated that chronic atropine treatment increased ^{3}H-QNB binding, but did not alter IPH. These studies suggest that IPH-associated processes are capable of uncoupling from the mAChR under certain circumstances and may present a second layer of adaptive change beyond the surface receptor. PKC may play a role in uncoupling G proteins from the mAChR as phorbol esters, which activate PKC directly, inhibit the G protein–linked carbamylcholine stimulation of cGMP accumulation without changing mAChR binding.[91]

B.   Nicotinic Receptors

The nAChR can be found in skeletal muscle, at the neuromuscular junction, in some endocrine organs (such as the adrenal chromaffin cells), in autonomic ganglia, in the electric organ of certain fish species (such as *Torpedo marmorata*), and in large numbers in many parts of the brain including mammalian neocortex. The nAChR was the first of the neurotransmitter receptors to be identified, and studies of nAChR in the electric organ and other peripheral sites have yielded a great deal of information about the structure and functioning of the nAChR outside the brain. However, since the nAChRs found in the brain differ in amino acid sequence, structure, antigenic characteristics, and pharmacology from those found in peripheral tissues,[37,214] these data may not always be safely extrapolated to nAChR located in the brain. For this reason, and because studies of cholinergic functioning in the CNS have tended to focus on the more numerous muscarinic receptors, surprisingly little is known about the role of nAChR in brain functioning.

1. Structure

Both peripheral and central nAChR belong to the ligand-regulated ion channel superfamily of receptors. When activated by ACh or other nicotinic agonists, conformational changes take place in the receptor protein to open a cation-selective transmembrane channel.[214] This channel conducts predominantly Na^+ ions, but all small cations may pass through.[92]

Muscle and electric organ nAChR are heterologous pentamers made up of two identical alpha units, and single beta, gamma, and delta units.[36] Neuronal nAChR, however, may consist of only a single alpha and a beta subunit, the agonist binding site being associated with the larger alpha subunit.[93,214] A series of hydrophilic and lipophilic amino acid sequences within the subunit polypeptides interacts with the lipid bilayer to form a transmembrane cylinder whose ends protrude on both sides.[94,95] When viewed from above using electron microscopy, the peripheral receptor has the appearance of a rosette with five electron-dense peaks equally spaced around a central pit whose axis is perpendicular to the plane of the plasma membrane.[95]

2. Subtypes

Variation in the structure of the alpha and beta subunits is used to subtype the nAChR. To date, four different genes coding for the alpha unit and two for the beta subunit have been identified.[36,214] The $alpha_1$ and $beta_1$ variants are found in skeletal muscle but not in brain, while $alpha_2$, $alpha_3$, $alpha_4$, and $beta_2$ are located in the CNS only.[38,214] The snake venom alpha bungarotoxin binds avidly to the $alpha_1$ component of peripheral nAChR, but only poorly to the $alpha_2$, $alpha_3$, and $alpha_4$ subunits of CNS nAChR.[38,96] Kappa bungarotoxin, also called neuronal bungarotoxin or Toxin F, binds tightly to CNS $alpha_3$ and $alpha_4$ but not $alpha_2$ subunits.[38,96] It has been reported that ^{125}I kappa bungarotoxin binding sites, but not ^{125}I alpha bungarotoxin binding sites, are reduced in the frontal cortex of patients with Alzheimer's disease compared with normal control brain.[38]

3. Anatomy

Binding studies to determine the anatomical distribution and functional relevance of nAChR suffer three limitations: (1) many ligands do not bind to all subtypes of nAChR, (2) there are many inactive nAChRs within the cytoplasm which will be labeled by some ligands, (3) a great number of nAChR at the cell surface, perhaps even the majority, are not electrophysiologically active.[214]

Older studies in rat brain using alpha bungarotoxin, which binds to only a small subset of brain nAChR,[38] showed greatest labeling in the hypothalamus, mammilary bodies, uncus, cortex, inferior colliculus, and hippocampus, with low concentrations in the cerebellum and caudate nucleus.[97,98] More complete nAChR labeling is achieved with ^3H-ACh after blockade of mAChR with atropine. Use of this technique in mouse brain, shows that binding is greatest in the midbrain, intermediate in the cerebral cortex and striatum, and lowest in the cerebellum, hippocampus, hypothalamus, and medulla oblongata.[99]

To my knowledge no comprehensive study of the distribution of nAChR in the human brain has been done. However, Larsson et al.[99] demonstrated the presence of substantial numbers of nAChR in human frontal cortex using ^{3}H-ACh after atropine, and Sugaya et al.[38] confirmed this using combined labeling with alpha and kappa bungarotoxin. The binding kinetics in both studies suggested that more than one subtype of nAChR is present in this region.

4. Binding Sites and Receptor Regulation

The nAChR appears to have two primary ACh-binding sites of different affinity, at least partly carried on the alpha units, which control channel gating and receptor desensitization (see References 94 and 95 for reviews). There is evidence that both sites must be occupied by ACh to effect a channel opening.[92]

From the developmental perspective there is some evidence that innervation plays an inductive role in the expression of nAChR through an unknown humoral factor. The response to ACh of cultured sympathetic neurons, for example, is increased tenfold when the cells are co-cultured with explants from the dorsal spinal cord which provide a "presynaptic" input (see Reference 214 for a review). Although this effect appears to be humoral, receptor activation is uninvolved as it occurs even in the presence of pharmacological blockade of the nAChR. Candidate molecules called acetylcholine receptor inducing activity[100] and agrin,[101] which appear to control the numbers and distribution of nAChR on myotubules, have been isolated and purified.

Denervation of skeletal muscle causes an increase in the number of muscle nAChR, but neuronal nAChR may not respond in the same way. Unexpectedly, denervation of chick ciliary ganglia results in a significant decrease in the total number of nAChR and decreased amounts of RNA coding for the nAChR, but, as there is a large population of intracellular nAChR in chick ciliary ganglion, this does not necessarily indicate decreased numbers of cell surface nAChRs.[214]

There is some evidence that the activity half-life of the nAChR, at least in the adrenal gland, is about 24 h although older inactive receptors, perhaps making up the majority of those detected by binding studies, may survive on the neuronal surface.[214] However, in the CNS, cAMP and its analogues may be capable of converting old, inactive nAChR to functionally available forms, thereby increasing the sensitivity of the neuron to nicotinic stimulation.[214]

Receptor occupancy by ACh itself and by exogenous ligands is an important determinant of subsequent nicotinic responses at the neuronal level. Prolonged exposure of the nAChR to ACh, for example, results in both decreased receptor number, or density,[102] and "desensitization," or decreased conductance response, of the ion channel to agonist binding.[94] The mechanism of desensitization is not fully understood but appears to occur through a "slow" phase, where the nAChR is stabilized in a desensitized high-affinity state for ACh, and a "fast" phase, where the nAChR assumes a shut refractory state with lower affinity for ACh.[92,94] A variety of other factors can modulate the desensitization response, including calcium ions, PKC mediated phosphorylation of parts of the receptor protein, cAMP levels, membrane hyperpolarization, and the thymic peptide thymopoetin.[95,214] This "heterosynaptic regulation" permits the integration of multiple intracellular and extracellular signals at the transmembrane receptor.

Numerous compounds, described variously as local anesthetics or noncompetitive blockers (NCB) such as chlorpromazine and phencyclidine, can decrease the time the channel remains open and/or potentiate receptor desensitization.[92,94] A single high-affinity NCB-binding site appears to exist within the ion channel itself so that ligand occupancy may sterically "plug up" the ion channel. Multiple low-affinity NCB-binding sites may be found at the point of intersection between the receptor protein and the lipid bilayer. Substance P, which is co-released with ACh by the cholinergic cells in the caudal cholinergic bundle,[33] may be an endogenous NCB which potentiates desensitization.

Somewhat unexpectedly, repeated exposure of neuronal tissue to nicotine both *in vitro* and *in vivo* appears to result in increased numbers of nAChR.[83,99] Similar increases in nAChR without change in mAChR were reported after pulsatile administration of PHYSO into the cerebral ventricles of rats.[103] Increased release of ^{3}H-ACh accompanied the increase in nAChR density seen in the hippocampus in the latter study, and it was suggested that hippocampal nAChR were presynaptic autoreceptors which potentiated the release of ACh. However, in other studies, eventual tolerance to the behavioral effects of nicotine have been reported despite increased nAChR density,[99] suggesting that compensatory decreased receptor efficacy (desensitization) takes place or that compensatory mechanisms at sites other than the nAChR are operative.

5. Activity

At this time very little is known about the way in which nicotinic cholinergic functioning influences brain activity. *In vitro* studies of slices of guinea pig cortex have shown that treatment with nicotine or the nicotinic agonist cytisine causes increased efflux of ACh.[104] Excitatory responses to nicotine have also been shown in a small number of subcortical sites; including the medial habenular nucleus,[105] interpeduncular nucleus,[106] locus coeruleus,[107] medial geniculate nucleus,[108] and the retina.[109] Using the paradigm of electrical stimulation of rat prefrontal cortex *in vitro*, Vidal and Changeux[37] showed that nicotine and other nicotinic agonists produced an increase in the excitability of cortical neurons while muscarinic agonists produced the reverse effect, i.e., caused decreased cortical excitability. They also found that, at low synaptic concentrations, ACh produces a muscarinic receptor-mediated suppression of cortical excitability but that, when present in higher concentrations, such as seen during administration of acetylcholinesterase inhibitors, ACh causes a nicotinic receptor-mediated increase in excitability. In this way, the postsynaptic consequences of the release of a quanta of ACh into the synaptic cleft can be electrophysiologically very different depending upon the relative proportions of nicotinic and muscarinic receptors present at that site and the concentration of ACh.

III. PSYCHIATRIC SYNDROMES WITH APPARENT CHOLINERGIC INVOLVEMENT

The first report clearly linking disturbed cholinergic activity with psychopathology was that of Rowntree et al.[110] in 1950. They administered the powerful cholinesterase inhibitor diisopropylfluorophosphate (DFP) by intramuscular injection to 17 patients with the diagnosis of schizophrenia, 9 with manic depressive illness, and 10 healthy control subjects. One of two euthymic manic depressives experienced mild depression, a depressed patient became much worse, and four out of six hypomanic patients were much improved. Among the control subjects "a very characteristic mental picture of depression, irritability, lassitude and apathy appeared even before the onset of unpleasant physical symptoms.... The depression was associated with a slowness, or poverty of thought, without disturbance of orientation, memory or intellectual ability".[110] Among the schizophrenic patients, 6 of 17 experienced an "activation" of psychosis or, in patients with "chronic" illness, a reemergence of acute symptoms of the type seen at illness onset.

These findings were later indirectly supported by Gershon and Shaw[111] who studied the effects on agricultural workers of organophosphorus insecticides which, like DFP, potently inhibit acetylcholinesterase. Of 16 workers who presented with toxic reactions after exposure to these chemicals, 7 presented with a depressive syndrome while 5 received the diagnosis of schizophrenia. These illnesses were severe enough to warrant treatment with antidepressants, neuroleptics, and electroconvulsive therapy (ECT), but nearly all showed complete recovery when assessed 1 year after exposure ceased.

A. Mood Disorder

Numerous studies implicating cholinergic mechanisms in the pathophysiology of mood disorder have accreted around the original reports by Rowntree et al.[110] and Gershon and Shaw.[111] These include further investigation of the psychological and physiological effects of cholinergic agonists and antagonists, examinations of cholinergic receptor numbers, measures of red blood cell choline levels, and studies of the way in which antidepressant treatments may influence cholinergic activity. The first to clearly articulate a cholinergic theory of mood disorder was Janowsky et al.[112] who proposed the cholinergic–adrenergic hypothesis of mood disorder as an extension of the earlier catecholamine hypothesis.[122] According to the cholinergic–adrenergic hypothesis, depression is the result of overactivity of cholinergic neurons relative to adrenergic neurons in specific brain regions, while mania is the result of underactivity of cholinergic neurons with respect to adrenergic neuronal functioning. Advances since that time, for example, the differentiation of numerous cholinergic and adrenergic receptor subtypes, have now rendered this hypothesis oversimplified; however, it has in the past had tremendous usefulness as a central starting point for many studies.

1. Cholinergic Agonist and Antagonist Effects on Mood

a. Acetylcholinesterase Inhibitors and ACh Precursors

The observations of Rowntree et al.[110] and Gershon and Shaw[111] that cholinergic agonists may worsen depression and improve mania in patients with affective disorders and produce depression and/or psychosis in the healthy were confirmed by a number of other investigators. Using a double-blind design, Janowsky et al.[112,113] demonstrated that intravenous physostigmine (PHYSO) significantly decreased manic symptoms even to the point of producing psychomotor retardation and depression, while saline or neostigmine, a highly potent acetylcholinesterase inhibitor which does not cross the blood–brain barrier, had no measurable effect. They also found that PHYSO induced brief relapse in euthymic remitted depressed patients.[117,118] Cohen et al.[114] reported that lecithin, a phospholipid precursor of choline, produced improvement in five out of six acutely manic patients. Tamminga,[115] attempting to relieve the symptoms of tardive dyskinesia in two patients with PHYSO or large doses of choline, found that these agents induced hopelessness, despair, suicidal ideation, and psychosis which could be reversed with atropine administration.

The psychological effects of cholinergic agonists such as PHYSO in healthy subjects is the subject of some debate. In a study comparing PHYSO with saline in normal subjects, Davis et al. [123] concluded that the syndrome of psychomotor slowing and anergia produced by PHYSO does not represent true depression, primarily because sad affect is not consistently present. Indeed, Davis et al.[123] found no significant differences in mean scores on the Bunney-Hamburg depression rating scale in PHYSO compared with saline-treated subjects. However, 2 out of their 23 subjects treated with PHYSO but none of the saline-treated subjects complained of depression. In contrast, Risch et al.[116] found that PHYSO, but not saline, produced highly significant increases in depression, tension/anxiety, anger/hostility, and anergia scores together with significant decreases in elation, friendliness and vigor scores on several self- and observer-rated psychopathology scales in healthy subjects. Like Davis et al.,[123] they found that the greatest affective changes were seen in a subgroup (7 out of 22) of the subjects tested. These findings indicate that most healthy subjects experience psychomotor slowing and anergia after PHYSO treatment but that only a subgroup will also develop significantly depressed mood. The explanation for this inconsistent affective response to PHYSO is not clear. However, very large interindividual variation in M1 mAChR density, at least in erythrocytes, has been reported in healthy subjects.[124] Since erythrocyte mAChR density is strongly correlated with such personality traits as reactive aggression and extroversion, which could influence the threshold for experiencing or reporting depression after challenge with a cholinergic agonist such as PHYSO, constitutional variance in mAChR density may underlie the differences reported in the literature. Another study in healthy subjects linking personality characteristics, such as discontent with life, irritability, and emotional lability, with robust cardiovascular and behavioral responses to PHYSO[125] further supports the notion that cholinergic activity may influence personality in ways that might influence the vulnerability or predisposition to develop depression in response to life stresses.

b.   Muscarinic Agents

In contrast to the dysphoric effects of cholinesterase inhibitors and ACh precursors, which would be expected to activate both muscarinic and nicotinic receptors, muscarinic antagonists appear to have euphoriant effects. Dilsaver[119] has collected and analyzed 30 reports involving about 100 cases describing abuse of these drugs. In addition to the psychedelic experience which these agents can produce, reasons offered for abuse of muscarinic antagonists include euphoria, improved mood, enhanced energy, and self-treatment of depression and anxiety.[119] Under controlled experimental conditions, atropine is reported to produce "marked euphoria" in healthy subjects.[110]

Studies of the effect of muscarinic antagonists upon mood in depressed patients show mixed results. Jimerson et al.[120] reported improvement in at least a subgroup of depressed patients treated for 8 weeks with trihexyphenidyl, a predominantly M1 antagonist. Kasper et al.[121] noted decreased symptoms in ten depressed patients treated for 30 days with biperiden, another M1 antagonist,[124] but Fritze et al.[125] found that biperiden did not potentiate the effects of the antidepressants mianserine or viloxazine.

c.   Nicotinic Agents

Despite evidence that nicotinic receptors are found in large numbers in the CNS,[37,99,126] nicotinic functioning has not been examined in mood-disordered patients.

Tobacco smoking, however, is extremely common in depressed patients, providing the opportunity for naturalistic study of the influence of nicotine upon mood.

The nicotine inhaled during tobacco smoking readily crosses the blood–brain barrier,[127] and both smoking and nicotine itself are reported to produce improved memory, increased arousal, decreased anxiety, and improved task performance in healthy subjects,[127,128] all of which could be beneficial in depressed patients. The increased prevalence of smoking among currently and formerly depressed individuals,[129-132] the very low rate of smoking cessation among depressed patients,[133] and the anecdotal accounts of major depression following tobacco withdrawal[132,134] may reflect a salutary effect of nicotine in patients with major depression.

Fluoxetine, desipramine, phenelzine, and phototherapy are reported to produce nicotinic subsensitivity after chronic administration,[119,135,136] and lithium may down regulate nAChR, although the latter finding has so far been noted only in skeletal muscle.[137] These data and the decreased craving for nicotine reported in abstinent smokers treated with standard doses of doxepin[138] suggest an overlap or perhaps even a similarity in the pharmacological effects of nicotine and antidepressants. In a small, retrospective study of 16 bipolar depressed patients matched for antidepressant treatment, the rate of recovery was significantly greater in patients meeting DSM-III criteria for tobacco abuse disorder than in those who did not.[139]

Nicotine also influences the activity of noradrenergic and dopaminergic neurons, both of which have been implicated in the pathophysiology of mood disorder. At the locus coeruleus, nicotine initially activates, but later suppresses, noradrenergic neuronal firing rates, as do some antidepressants.[107,140] At dopaminergic sites, activity is increased by nicotine,[141] and the incidence of idiopathic Parkinson's disease is reduced in smokers.[142]

2. Effect of Cholinergic Agonists and Antagonists on REM Sleep

Decreased REM latency (the time between sleep onset and the first REM period) and increased REM density (number of eye movements per unit of REM time) are among the most widely studied and robust biological markers of major depression.[31] As discussed earlier in this chapter, the transition from non-REM to REM sleep is strongly influenced by the relative activity of cholinergic and monoaminergic neurons in the pontomesencephalic region. The study of REM sleep in human subjects may, therefore, offer insights into the potential involvement of cholinergic mechanisms in the pathophysiology of mood disorder.

It has been shown that the onset of REM sleep may be accelerated in humans by such muscarinic agonists as arecoline or RS-86,[22,143-145] and delayed by muscarinic antagonists, such as scopolamine or biperiden.[26,146] Nicotinic mechanisms may also be involved in REM control,[147,148] and smokers are reported to have shorter REM latency and increased REM density compared with nonsmokers.[146] Depressed patients and their first-degree relatives show greater decreases in REM latency after treatment with such agonists as RS-86 or arecoline than do nondepressed psychiatric patients or healthy control subjects without family histories of affective disorder indicating supersensitivity to cholinergic stimulation.[22,143-145] This supersensitive REM response is reported to persist in depressed patients even after clinical remission in some[22] but not all studies.[23] These data suggest that depression might be associated with a persisting state of muscarinic cholinergic supersensitivity, which may be apparent even in as yet unaffected first-degree relatives. Nicotinic involvement in sleep regulation and the pathophysiology of depressive illness requires further study.

3.   ACh and the HPA Axis

Overactivity of the HPA axis is one of the most widely recognized biological abnormalities in patients with major depression.[149-151] Because HPA activation can also be induced by a variety of nonspecific stressors,[150,152-154] measures of HPA activity are of very limited clinical value. Nevertheless, findings linking glucocorticoid hypersecretion to treatment response, prognosis, inheritance patterns, and other biological markers such as REM latency[155-161] have led to the belief that this endocrine dysfunction may be related to the underlying pathophysiology of depressive illness.[162,163]

Animal studies[164] indicate that CNS cholinergic activity may be involved in HPA regulation, and this appears to be the case in humans as PHYSO induces increases in plasma cortisol and ACTH levels and failed suppression of plasma cortisol after dexamethasone[165,166] in healthy subjects. Similar changes are produced in depressed patients, but the increases in ACTH seen after PHYSO are of greater magnitude than those seen in control subjects or nondepressed psychiatric patients,[166] suggesting supersensitivity of those cholinergic systems involved in HPA regulation. Since, in another series of human studies, HPA activity was unchanged after treatment with the predominantly M1 antagonist biperiden,[125] the predominantly M1 agonist RS-86,[144] and the nonspecific mAChR antagonist scopolamine, it is possible the PHYSO effect on the HPA axis is mediated through nAChR. Indeed, nicotine is reported to alter vasopressin, prolactin, β-endorphin, and prostaglandin, as well as ACTH, levels.[167,168]

4.   Post-Mortem Studies

Many of the findings discussed in this section suggest that muscarinic receptor-mediated responses are supersensitive in depressed patients. The few attempts to examine mAChR levels directly have yielded discrepant results. Meyerson et al.[169] found increased numbers of mAChR in the brains of suicide victims compared with controls, but Kaufman et al.[170] did not.

5.   Erythrocyte Choline Levels

Choline is both the precursor and the metabolite of ACh, and erythrocyte choline (ECh) levels have been studied as a potential biochemical marker of affective disorder. Hanin et al.[171] have found that the variance of ECh is much greater in depressed compared with healthy subjects, and subgroups of both depressed patients[171] and manic patients[172,173] appear to show elevations of ECh. The manic patients with high ECh were more severely ill, less responsive to treatment, and needed greater amounts of neuroleptic than low-ECh manics.[173] These investigators point out that choline is also an important constituent of membrane phospholipids and an intermediary in single carbon metabolism, so that the changes documented do not necessarily indicate abnormalities in ACh turnover.

6.   Cholinergic Involvement in Antidepressant Treatments

a.   Tricyclic Antidepressants

The tricyclic antidepressants have prominent antimuscarinic properties,[174] which has led to some speculation that their antidepressant effect may be mediated, at least in part, through cholinergic pathways. The efficacy of the other antidepressant classes, particularly the selective serotonin reuptake inhibitors, which are essentially devoid of anticholinergic effects, clearly indicate that anticholinergic properties are not essential. This, by itself, does not necessarily preclude involvement of ACh in the pathophysiology of mood disorder. The axonal connections among cholinergic, adrenergic, and serotonergic nuclei in the pontomesencephalic region, the confluence of cholinergic and monoaminergic influences at the level of the G protein, and the process of heterosynaptic regulation discussed earlier result in great physiological interactivity such that a given pathological state might be corrected through any of several biological pathways as, for example, Parkinson's disease might be treated sucessfully with either anticholinergic or dopamine agonist drugs. In this instance the absence of dopaminergic properties in anticholergic drugs which relieve parkinsonian symptoms does not rule out the involvement of dopamine in Parkinson's disease.

b.   Electroconvulsive Therapy

Several studies have shown that electroconvulsive shock (ECS) causes an immediate drop of ACh levels in the brain in animals,[175-178] presumably because of increased neuronal release of ACh during the seizure.[179] In humans, spontaneous, chemically induced, and electrically induced seizures are all associated with elevations of cerebrospinal fluid and possibly plasma ACh levels,[180-182] suggesting that the seizure-related release of ACh is massive, resulting in the spillover of ACh into these pools. In animals, the seizure-related release of ACh is followed by an increase in ACh synthesis, as indicated by increases in brain ChAT levels, increased high affinity choline uptake, and rapid restoration of ACh levels to normal.[177,183] ACh breakdown rate, as measured by acetylcholinesterase activity, does not appear to be altered by seizure activity.[184]

The supraphysiologic cholinergic receptor stimulation resulting from repeated flooding of the CNS receptor with ACh during a course of ECT may initiate adaptive changes in cholinergic functioning which could be of significance to the mechanism of action of ECT. Studies of neuronal tissue exposed to high concentrations of cholinergic agonist indicate that increased agonist availability causes compensatory down regulation of mAChR.[44,72-80] Similarly, after repeated ECS in animals, down regulation of mAChR in the cortex and hippocampus, indicated by a decrease in the binding of the muscarinic receptor ligand ^{3}H-QNB, has been demonstrated by some[185,186] but not all groups.[187,188] Other investigators have found evidence that chronic ECS causes decreased carbachol-stimulated inositol phospholipid hydrolysis, which is indicative of cholinergic desensitization at the second messenger level.[189] Finally, studies showing that ECS blocks the cataplexy induced in animals by the ACh agonists arecoline and pilocarpine[186,190] indicate that ECS induces a functional decrease in the responsivity of the cholinergic system.

Patients who respond well to ECT reportedly demonstrate an exaggerated drop in blood pressure after the cholinergic agonist mecholyl,[191] suggesting that cholinergic supersensitivity prior to ECT is associated with a good antidepressant response. This observation is compatible with the hypothesis that at least some types of depression are due to cholinergic supersensitivity and that mAChR down regulation is an important component of the mechanism by which ECT exerts its antidepressant effects. Since atropine blocks agonist-induced mAChR down regulation in vitro[72] and ECS-induced mAChR down regulation in vivo in animals,[186] it would be expected to impair the antidepressant efficacy of ECT if this depended upon the induction of mAChR down regulation. Hasey et al.[192] tested this hypothesis in 22 patients with major depression who were randomly assigned in double-blind fashion to treatment with intravenous saline or atropine just prior to seizure induction. When depression-rating scale scores were adjusted for differences in the electrical charge delivered over the course of ECT, the response to ECT, though essentially the same the day after the last ECT in both groups, was significantly less well sustained in the atropine-treated group 1 week and 1 month post-ECT. These findings suggest that mAChR down regulation may be an important part of the antidepressant mechanism of action of ECT.

B. Schizophrenia

Several decades ago, Rowntree et al.[110] and Gershon and Shaw[111] reported that exposure to acetylcholinesterase inhibitors caused worsening of schizophrenia and the onset of a schizophrenia-like syndrome in previously normal individuals. Since that time, few studies have focused upon cholinergic functioning in schizophrenic patients, and the general ethos appears to be that cholinergic mechanisms are un-important in the pathophysiology of schizophrenia. However, the evidence impli-cating cholinergic mechanisms in memory and cognitive functioning and in mood regulation has led a few groups to consider the possibility that subgroups of schizo-phrenics, such as those with prominent cognitive impairment or those with promi-nent "negative symptoms," may show disturbances of central cholinergic functioning.

There is now substantial evidence supporting forebrain cholinergic dysfunction in patients showing senile dementia of Alzheimer's type,[16] and there is at least one report demonstrating improvement of psychotic symptoms in patients with Alzhe-imer's disease treated with PHYSO.[217] Since patients with chronic schizophrenia can have significant cognitive impairment, it has been proposed that these schizophrenics may also show abnormalities of cholinergic functioning. However, in a post-mortem study, Haroutunian et al.[218] found no differences in brain acetylcholinesterase or ChAT levels in chronically hospitalized schizophrenics compared with elderly con-trols. Earlier, Koponen et al.[219] found no differences in cerebrospinal fluid acetylcho-linesterase levels in schizophrenics with dementia compared with age-matched con-trols or patients with alcohol induced dementia.

Tandon and Greden[193] proposed that the pathophysiology of schizophrenics with prominent negative symptoms, which include symptoms with a distinct depressive flavor, such as affective blunting, emotional withdrawal, reduced energy and moti-vation, apathy, and anhedonia, may include cholinergic overactivity. As evidence for this they note that (1) the somatic symptoms described by Kraepelin[220] as being characteristic of schizophrenia, such as diminished pain sensitivity, hypersalivation, low body temperature, and diminished pupillary reflexes, are similar to the para-sympathetic activation that might be expected with increased CNS cholinergic ac-tivity; (2) antimuscarinic drugs may reduce negative symptoms and often are abused by schizophrenic patients; and (3) shortened REM latency and failure of cortisol suppression by dexamethasone, both of which may be related to cholinergic over-activity in depressed patients, are also seen in some schizophrenics. These authors propose a cholinergic–dopaminergic balance hypothesis which states that increased cholinergic activity, which ultimately leads to negative symptoms, is an attempt at compensation for the pathologically increased dopaminergic activity that produces psychosis. Persistence of some measure of cholinergic overactivity after dopaminer-gic activity normalizes may lead to the negative symptoms which remain after the acute psychotic symptoms disappear. Other studies, in animals, however, suggest that increased activity of the cholinergic neurons in the PPTg nucleus actually ap-pears to excite dopaminergic neurons, at least in the ventral tegmental area,[216] calling into some question the notion that increased cholinergic activity can universally compensate for dopaminergic overactivity.

IV. MEMORY AND ACETYLCHOLINE

Although other neurotransmitter systems are undoubtedly also involved, over the last 15 years an impressive body of literature has accumulated which suggests that CNS cholinergic mechanisms play an important role in the processes of learning and memory in lower animals and in humans.

A. Antagonist Effects

In general, cholinergic antagonists exert negative effects upon memory. Drachman[194] demonstrated that scopolamine, but not methylscopolamine (which does not cross the blood–brain barrier), interferes with memory registration and possibly retention in healthy control subjects and that this effect could be reversed by PHYSO but not by amphetamine. This adverse effect of acutely administered centrally acting cholinergic antagonists on memory registration and retention has been replicated in human subjects,[195,196] monkeys,[197] and rats.[198]

B. Agonist Effects

Cholinergic agonists appear to enhance memory functioning. The cholinergic agonist arecoline is reported to improve sentence learning in healthy subjects.[199] Oxotremorine[200] and PHYSO[201] improve memory retention and retrieval in a passive avoidance test in rats. Rats treated with very high doses of choline perform better in a radial arm maze test than control animals.[202] Memory deficits in rats or mice produced by protein synthesis inhibitors, hippocampal lesions, cholinergic denervation of the cortex and hippocampus, or inhibition of CAT activity can be corrected by cholinergic agonist treatment[203-206] or by the grafting of fetal cholinergic neurons into the damaged area.[207]

C. Neurotoxin Effects

The neurotoxin AF64 A, which selectively damages cholinergic neurons, impairs memory acquisition and retention in both a passive avoidance task and a water maze task.[208]

D. Muscarinic Receptors

While stimulation of postsynaptic mAChR facilitates memory, carbachol stimulation of what are thought to be presynaptic receptors in nuclei projecting to the cortex and hippocampus impairs memory, presumably by decreasing ACh release from the presynaptic neuron.[209] Secoverine, an antagonist thought to bind preferentially to presynaptic autoreceptors resulting in increased ACh release, is reported to facilitate the augmentation of memory retention by PHYSO.[201]

The phenomena of mAChR regulation appear also to affect memory functioning. Chronic administration of the cholinergic agonists PHYSO or oxotremorine causes decreased ^{3}H-QNB binding and, upon withdrawal, impairment of memory retention[210,211] rather than the augmentation seen after acute administration. Similarly, chronic treatment with scopolamine results in memory augmentation rather than memory impairment following withdrawal.[210] These effects suggest that memory can be altered as a consequence of the adaptive responses of the cholinergic neurons.

[3]H-QNB binding in this context is a rather poor indicator of cholinergic activity as ^{3}H-QNB–binding abnormalities may persist long after memory functioning has returned to normal following withdrawal from chronic treatment with oxotremorine.[211] This may reflect ^{3}H-QNB labelling of physiologically inactive intracellular mAChR in pretreatment assessments.

E. Nicotinic Receptors

To date few studies have examined nicotinic cholinergic effects on memory, and the results are inconsistent. Nicotine is reported to block the impairment of short-term working memory produced in healthy subjects by scopolamine,[128] while the nicotinic antagonist mecamylamine potentiates the negative effects of scopolamine on memory in rats,[212] suggesting that nicotinic activity is relevant to memory functioning. Both nicotine and arecoline are reported to have modest but statistically significant positive effects on cognition in patients with Alzheimer's disease.[215] Other investigators, however, report that muscarinic agonists can reverse the decrements in memory produced by lesions in the hippocampus or treatment with β endorphin but nicotine cannot.[205,213]

V. CONCLUSIONS AND FUTURE DIRECTIONS

To some extent ACh has become one of the poor relatives in the neurotransmitter family currently being studied in the search for the biological underpinnings of psychiatric illness. This benign neglect is unwarranted, especially in the area of the mood disorders, as the evidence for cholinergic involvement in these conditions is substantial, perhaps even compelling. In any case, the multiple layers of interaction between neurotransmitter systems are such that it is probably impossible for disordered neurotransmission to be confined to a single system. From the point of view of treatment, this means that psychiatric symptoms may well be improved by intervention at many potential sites including the site of the "primary lesion" or disturbance or at systems which modulate the effects of the neurons within the primary lesion. The efficacy of dopaminergic agonists, as well as cholinergic antagonists, in Parkinson's disease is a well-known example of this.

Even within a poor family, some members are more impoverished than others, and among the studies of cholinergic involvement in psychiatric disease, studies focusing upon nicotinic functioning are underrepresented. Current investigations of the psychotropic effects of nicotine exposure through smoking represent what, I hope, will be the beginnings of a fertile, new research direction.

Finally, modern technologies such as positron emission tomography and molecular biology bring unprecedented power to psychiatric researchers, and these might be profitably brought to bear on cholinergic neurophysiology in psychiatric illness.

REFERENCES

1. Kuhar, M., The anatomy of cholinergic neurons, in *Biology of Cholinergic Function*, Goldberg, A. M., Hanin, I., Eds., Raven Press, New York, 1976, 3.
2. Armstrong, D. M., Saper, C. B., Levey, A. I., Wainer, H., Terry, R. D., Distribution of cholinergic neurons in rat brain: demonstrated by the immunocytochemical localization of choline acetyltransferase, *J. Comp. Neurol.*, 216, 53, 1983.
3. Jenden, D. J., Chemistry and biochemical pharmacology of cholinergic neurons, in *Psychopharmacology: The Third Generation of Progress*, Meltzer, H. Y., Ed., Raven Press, New York, 1987, 233.
4. Fibiger, H. C., Vincent, S. R., Anatomy of central cholinergic neurons, in *Psychopharmacology: The third generation of progress*, Meltzer, H. Y., Ed., Raven Press, New York, 1987, 211.
5. McGeer, P. L., McGeer, E. G., Fibiger, H. C., Wickson, V., Neostriated cholineacetylase and cholinesterase following selective brain lesions, *Brain Res.*, 35, 305, 1971.
6. McGeer, E. G., Wada, J. A., Terao, A., Jung, E., Amine synthesis in various brain regions with caudate or septal lesions, *Exp. Neurol.*, 24, 277, 1969.
7. Sieklucka-Dziuba, M., Kleinrok, Z., Effects of drugs stimulating and blocking M and N cholinergic receptors on GABA level and GAD activity in rat striatum, *Acta Physiol. Pol.*, 33, 67, 1982.
8. Murrin, L. C., Kennedy, R. H., Donnelly, T. E., Effect of withdrawal from chronic propranolol treatment on high affinity choline uptake in rat brain, *J. Pharm. Pharmacol.*, 35, 677, 1983.
9. Sivam, S. P., Norns J. C., Lim D. K., Effect of acute and chronic cholinesterase inhibition with diisopropylfluorophosphate on muscarinic, dopamine, and GABA receptors of the rat striatum, *J. Neurochem.*, 40, 1414, 1983.
10. Fernando, J. C. R., Hoskins, B. H., Ho, I. K., A striatal serotonergic involvement in the behavioral effects of anticholinesterase organophosphates, *Eur. J. Pharmacol.* 98, 129, 1984.
11. Schwartz, J. M., Baxter, L. R., Jr., Mazziotta, J. C., Gerner, R. H., Phelps, M. E., The differential diagnosis of depression, *J. Am. Med. Assoc.*, 258, 1368, 1987.
12. Hasey, G., Hanin, I., The cholinergic-adrenergic hypothesis of depression reexamined using clonidine, metoprolol and physostigmine in an animal model, *Biol. Psychiatry*, 29, 127, 1991.
13. Wolff, E., Putnam, F., Post R Motor activity and affective illness. The relationship of amplitude and temporal distribution to changes in affective state, *Arch. Gen. Psychiatry*, 42, 288, 1985.
14. Sandberg, K., Hanin, I., Fisher, A., Coyle, J. T., Selective cholinergic neurotoxin: AF64A's effects in rat striatum, *Brain Res*, 293, 49, 1984.
15. Mastropaolo, J., Nadi, N. S., Ostrowski, N. L., Crawley, J. N., Galanin antagonizes acetylcholine on a memory task in basal forebrain-lesioned rats, *Proc. Natl. Acad. Sci. U.S.A.*, 85, 9841, 1988.
16. Hohmann, C., Antuono, P., Coyle, J. T., Basal forebrain cholinergic neurons and Alzheimer's disease, in *Psychopharmacology of the Aging Nervous System*, Vol. 20, Iverson, L. L., Iverson, S. D., and Snyder, S. H., Eds., Plenum Press, New York, 1988, 69.
17. Crawley, J. N., Functional interactions of galanin and acetylcholine: relevance to memory and Alzheimer's disease, *Behav. Brain Res.*, 57, 133, 1993.
18. Beal, M. F., Clevens, R. A., Chattha, G. K., MacGarvey, U. M., Mazurek, M. F., Gabriel, S. M., Galanin-like immunoreactivity is unchanged in Alzheimer's disease and Parkinson's disease dementia cerebral cortex, *J. Neurochem.*, 51, 1935, 1988.
19. Pucilowski, O., Eichelman, B., Overstreet, D. H., Rezvani, A. H., Janowsky, D. S., Enhanced affective aggression in genetically bred hypercholinergic rats, *Neuropsychobiology*, 24, 37, 1990–1991.
20. Jones, B. E., Paradoxical sleep and its chemical/structural substrates in the brain, *Neuroscience*, 40, 637, 1991.
21. Hazra, J., Effect of hemicholinicum-3 on slow wave and paradoxical sleep of cat, *Eur. J. Pharmacol.*, 11, 395, 1970.
22. Berger, M., Riemann, D., Hochli, D., Spiegel, R., The cholinergic rapid eye movement sleep induction test with RS-86, *Arch. Gen. Psychiatry*, 46, 421, 1989.
23. Nurnberger, J., Berrettini, W., Mendelson, W., Sack, D., Gershon, E. S., Measuring cholinergic sensitivity: I. Arecoline effects in bipolar patients, *Biol. Psychiatry*, 25, 610, 1989.
24. Matsuzaki, M., Okada, Y., Shuto, S., Cholinergic agents related to para-sleep state in acute brain stem preparations, *Brain Res.*, 9, 253, 1968.
25. Karczmar, A. G., Longo, V. G., Scotti de Carolis, A., A pharmacological model of paradoxical sleep; the role of cholinergic and monoamine systems, *Physiol. Behav.*, 5, 175, 1970.
26. Poland, R. E., Tondo, L., Rubin, R. T., Trelease, R. B., Lesser, I. M., Differential effects of scopolamine on nocturnal cortisol secretion, sleep architecture, and REM latency in normal volunteers, *Biol. Psychiatry*, 25, 403, 1989.
27. Leornard, C. S., Llinas, R. R., Serotonin (5-HT) inhibits mesopontive cholinergic neurons *in vitro*, *Soc. Neurosci. Abst.*, 16, 1233, 1990.

28. Riou, F., Cespalio, R., Jouvet, M., Endogenous peptides and sleep in the rat: 1 peptides decreasing paradoxical sleep, *Neuropeptides*, 2, 243, 1982.

29. Obal, F., Opp, M., Cady, A. B., Johannsen, L., Krueger, J. M., Prolactin, vasoactive intestinal peptide and peptide histidine methionine elicit selective increases in REM sleep in rabbits, *Brain Res.*, 490, 292, 1989.

30. Danguir, J., Intracerebroventricular infusion of somatostatin selectively increases paradoxical sleep in rats, *Brain Res.*, 367, 26, 1986.

31. Kupfer, D. J., The sleep EEG in diagnosis and treatment of depression, in *Depression: Basic Mechanisms, Diagnosis and Treatment*, Rush, A. J., Altshuler, K. Z., Eds., Guilford Press, New York, 1986, 102.

32. Mesulam, M. M., Geula, C., Mash, D., Brimijoin, S., Immunocytochemical demonstration of axonal and perikaryal acetylcholinesterase in human cerebral cortex, *Brain Res.*, 539, 233, 1991.

33. Eckenstein, F., Baughman, R. W., Two types of cholinergic innervation in cortex, one co-localized with vasoactive intestinal polypeptide, *Nature*, 309, 153, 1984.

34. Raiteri, M., Marchi, M., Paudice, P., Vasoactive intestinal polypeptide (VIP) potentiates the muscarinic stimulation of phosphoinositide turnover in rat cerebral cortex, *Eur. J. Pharmacol.*, 133, 127, 1987.

35. Bonner, T. I., The molecular basis of muscarinic receptor diversity, *TINS*, 12, 148, 1989.

36. Steinbach, J. H., Ilfune, C., How many kinds of nicotinic acetylcholine receptor are there?, *TINS*, 12, 3, 1989.

37. Vidal C., Changeux, J. P., Pharmacological profile of nicotinic acetylcholine receptors in the rat prefrontal cortex: an electrophysiological study in a slice preparation, *Neuroscience*, 29, 261, 1989.

38. Sugaya, K., Giacobini, E., Chiappinelli, V. A., Nicotinic acetylcholine receptor subtypes in human frontal cortex: changes in Alzheimer's disease, *J. Neurosci. Res.*, 27, 349, 1990.

39. Schmerlik, M. I., Structure and regulation of muscarinic receptors, *Annu. Rev. Physiol.*, 51, 217, 1989.

40. Bluml, K., Mutshler, E., Wess, J., Identification of an intracellular tyrosine residue critical for muscarinic receptor-mediated stimulation of phosphatidylinositol hydrolysis, *J. Biol. Chem.*, 269, 402, 1994.

41. Fisher, S. K., Agranoff, B. W., Enhancement of the muscarinic synaptosomal phospholipid labeling effect by the ionophore A231187, *J. Neurochem.*, 37, 968, 1981.

42. Worley, P. F., Heller, W. A., Snyder, S., Baraban, J. M., Lithium blocks a phosphoinositide-mediated cholinergic response in hippocampal slices, *Science*, 239, 1428, 1988.

43. Ehlert, F. J., Roeske, W. R., Yamamura, H. I., Muscarinic receptor:regulation by guanine neucleotides, ions on N-ethylmalleimide, *Fed. Proc.*, 40, 153, 1981.

44. Spencer, D. G., Jr., Horvath, E., Traber, J., GTP effects in rat brain slices support the non-interconvertability of M_1 and M_2 muscarinic acetylcholine receptors, *Life Sci.*, 42, 993, 1987.

45. McGeer, P. L., Eccles, J. C., McGeer, E. G., Principles of synaptic biochemistry. in *Molecular Neurobiology of the Mammalian Brain*, 2nd ed., Plenum Press, New York, 1987, 151.

46. Dittman, A. H., Weber, J. P., Hinds, T. R., Choi, E. J., Migeon, J. C, Nathanson, N. M., Storm, D. R., A novel mechanism for coupling of m4 muscarinic acetylcholine receptors to calmodulin-sensitive adenylyl cyclases: crossover from G protein-coupled inhibition to stimulation, *Biochemistry*, 33, 943, 1994.

47. Choi, E. J., Zia, Z., Villacres, E. C., Storm, D. R., *Curr. Opinion Cell Biol.*, 5, 269, 1993.

48. Berridge, M. J., Inositol trisphosphate and diacylglycerol as second messengers, *Biochem. J.*, 220, 345, 1984.

49. Nishizuka, Y., Studies and perspectives of protein kinase C, *Science*, 233, 305, 1986.

50. Streb, H., Irvine, R. F., Berridge, M., Schultz, I., Release of Ca^{2+} from a nonmitochondrial intracellular store in pancreatic acinar cells by inositol-1,4,5-trisphosphate, *Nature*, 306, 67, 1983.

51. Drummond, A. H., Lithium and inositol lipid-linked signaling mechanisms, *TIPS*, 8, 129, 1987.

52. Worley, P. F., Barabon, J. M., Snyder, S. H., Beyond receptors: multiple second-messenger systems in brain, *Ann. Neurol.*, 21, 217, 1987.

53. Gongora, J. L., Sierra, A., Mariscal, S., Physostigmine stimulates phosphoinositide breakdown in the rat neostriatum, *Eur. J. Pharmacol.*, 155, 49, 1988.

54. Sokolovsky, M., Cohen-Armon, M., Cross talk between receptors: muscarinic receptors, sodium channels, and guanine nucleotide-binding protein(s) in rat membrane preparations and synaptoneurosomes, in *Advances in Second Messenger and Phospho-Protein Research*, Vol. 21, Adelstein, R., Klee, C., Rodbell, M., Eds., Raven Press, New York, 1988, 11.

55. Baumgold, J., White, T., Pharmacological differences between muscarinic receptors coupled to adenylate cyclase inhibition, *Biochem. Pharmacol.*, 38, 1605, 1989.

56. Kendall, D. A., Nahorski S. R., Depolarization-evoked release of acetylcholine can mediate phosphoinositide hydrolysis in slices of rat cerebral cortex, *Neuropharmacology*, 26, 513, 1987.

57. Batty, I., Nahorski, S. R., Differential effects of lithium on muscarinic receptor stimulation of inositol phosphates in rat cerebral cortex slices, *J. Neurochem.*, 45, 1514, 1985.

58. Worley, P. F., Baraban, J. M., McCarren, M., Cholinergic phosphatidylinositol modulation of inhibitory, G protein-linked, neurotransmitter actions: electrophysiological studies in rat hippocampus, *Proc. Natl. Acad. Sci. U.S.A.*, 84, 3467, 1987.

59. Hanley, M. R., Iversen, L. L., Muscarinic cholinergic receptors in rat corpus striatum and regulation of guanosine cyclic 3', 5'-monophosphate, *Mol. Pharmacol.*, 14, 246, 1977.

60. Hartzell, H. C., Fischmeister, R., Opposite effects of cyclic GMP and cyclic AMP on Ca^{2+} current in single heart cells, *Nature*, 323, 273, 1986.

61. Hammer, R., Berrie, C. P., Birdsall, N. J. M., Pirenzepine distinguishes between different subclasses of muscarinic receptors, *Nature*, 283, 90, 1980.

62. Watson, M., Vickroy, T. W., Roeske, W. R., Subclassification of muscarinic receptors based upon the selective antagonist pirezepine, *TIPS Suppl.*, 1, 9, 1984.

63. Fukuda, K., Kubo, T., Akiba, I., Molecular distinction between muscarinic acetylcholine receptor subtypes, *Nature*, 327, 623, 1987.

64. Richards, M. H., Rat hippocampal muscarinic autoreceptors are similar to the M2(cardiac) subtype: comparison with hippocampal M1, atrial M2 and ileal M3 receptors, *Br. J. Pharmacol.*, 99, 753, 1990.

65. McKinney, M., Miller, J. H., Aagard, P. J., Pharmacological characterization of the rat hippocampal muscarinic autoreceptor, *J. Pharmacol. Exp. Ther.*, 264, 74, 1993.

66. Aagaard, P., McKinney, M., Pharmacological characterization of the novel cholinomimetic L-689,660 at cloned and native brain muscarinic receptors, *J. Pharmacol. Exp. Ther.*, 267, 1478, 1993.

67. Peralta, E. G., Ashkenazi, A., Winslow, J. W., Ramachandran, J., Capon, D. J., Differential regulation of PI hydrolysis and adenylyl cylase by muscarinic receptor subtypes, *Nature*, 334, 434, 1988.

68. Fisher, S. K., Bartus, R. T., Regional differences in the coupling of muscarinic receptors to inositol phospholipid hydrolysis in guinea pig brain, *J. Neurochem.*, 45, 1085, 1985.

69. Fisher, S. K., Snider, M., Differential receptor occupancy requirements for muscarinic cholinergic stimulation of inositol lipid hydrolysis in brain and in neuroblastomas, *Mol. Pharmacol.*, 32, 81, 1987.

70. Vickroy, T. W., Malphurs, W. L., Defiebre, N. C., Absence of receptor reserve at hippocampal muscarinic autoreceptors which inhibit stimulus-dependent acetylcholine release, *J. Pharmacol. Exp. Ther.*, 267, 1198, 1993.

71. Lugami, Y. A., Bradford, H. F., Birdsall, N. J. M., Hulme, E. C., Depolarisation-induced changes in muscarinic cholinergic receptors in synaptosomes, *Nature*, 277, 481, 1979.

72. El-Fakahany, E. E., Lee, J. H., Agonist-induced muscarinic acetylcholine receptor down-regulation in intact rat brain cells, *Eur. J. Pharmacol.*, 132, 21, 1986.

73. Laduron, P., Axoplasmic transport of muscarinic receptors, *Nature*, 286, 287, 1980.

74. Ben-Barak, J., Gazit, H., Silman, I., Dudai, Y., *In vivo* modulation of the number of muscarinic cholinergic receptors in the rat brain by cholinergic ligands. *Eur. J. Pharmacol.*, 74, 73, 1981.

75. McKinney, M. T., Coyle, J. T. Regulation of neocortical muscarinic receptors: effects of drug treatment and lesions, *J. Neurosci.*, 2, 97, 1982.

76. Hedlund, B., Abens, J., Bartfai, T., Vasoactive intestinal polypeptide and muscarinic receptors: supersensitivity induced by long-term atropine treatment, *Science*, 220, 519, 1983.

77. Luthin, G. R., Wolfe, B. B., Comparison of [³H] pirenzepine and [³H] quinuclidinylbenzilate binding to muscarinic cholinergic receptors in rat brain, *J. Pharmacol. Exp. Ther.*, 228, 648, 1984.

78. Feigenbaum, P., El-Fakahany, E. E., Regulation of muscarinic cholinergic receptor density in neuroblastoma cells by brief exposure to agonist: possible involvement in desensitization of receptor function, *J. Pharmacol. Exp. Ther.*, 233(1), 134, 1985.

79. Marchi, M., Raiteri, M., Interaction acetylcholine-glutamate in rat hippocampus: involvement of two subtypes of M-2 muscarinic receptors, *J. Pharmacol. Exp. Ther.*, 248, 1255, 1988.

80. Shaw, C., van Huizen, F., Cynader, M. S., A role for potassium channels in the regulation of cortical muscarinic acetylcholine receptors in an *in vitro* slice preparation, *Mol. Brain Res.*, 5, 71, 1989.

81. Smith, C. J., Court, J. A., Keith, A. B., Increases in muscarinic stimulated hydrolysis of inositol phospholipids in rat hippocampus following cholinergic deafferentation are not paralleled by alterations in cholinergic receptor density, *Brain Res.*, 485, 317, 1989.

82. Raulli, R. E., Arendash, G., Crews, F. T., Effects of nBM lesions on muscarinic stimulation of phosphoinositide hydrolysis, *Neurobiol. Aging*, 10, 191, 1989.

83. Maloteaux, J.-M., Gossuin, A., Pauwels, P. J., Laduron, P. M., Short-term disappearance of muscarinic cell surface receptors in carbachol-induced desensitization, *FEBS*, 156, 103, 1983.

84. Harden, T. K., Petch, L. A., Traynelis, S. F., Waldo, G. L., Agonist-induced alteration in the membrane form of muscarinic cholinergic receptors, *J. Biol. Chem.*, 260, 13060, 1985.

85. Siman, R. G., Klein, W. L., Cholinergic activity regulates muscarinic receptors in central nervous system cultures, *Proc. Natl. Acad. Sci. U.S.A.*, 76, 4141, 1979.

86. Klein, W. L., Nathanson, N., Nirenberg, M., Muscarinic acetylcholine receptor regulation by accelerated rate of receptor loss, *Biochem. Biophys. Res. Commun.*, 90, 506, 1979.

87. Shifrin, G. S., Klein, W. L., Regulation of muscarinic acetylcholine receptor concentration in cloned neuroblastoma cells, *J. Neurochem.*, 34, 993, 1980.

88. Hunter, D. D., Nathanson, N. M., Biochemical and physical analyses of newly synthesized muscarinic acetylcholine receptors in culture embryonic chicken cardiac cells, *J. Neurosci.*, 6, 3739, 1986.

89. Cioffi, C. L., El-Fakahany, E. E., Differential sensitivity of phosphoinositide and cyclic GMP responses to short-term regulation by a muscarinic agonist in mouse neuroblastoma cells, *Biochem. Pharmacol.*, 38, 1827, 1989.

90. Goobar, L., Bartfai, T., Long-term atropine treatment lowers the efficacy of carbachol to stimulate phosphatidy breakdown in the cerebral cortex and hippocampus of rats, *Biochem. J.*, 250, 727, 1988.

91. Lai, W. I. S., El-Fakahany, E. E., Phorbol ester-induced inhibition of cyclic GMP formation mediated by muscarinic receptors in murine neuro-blastoma cells, *J. Pharmacol. Exp. Ther.*, 241, 366, 1987.

92. Stroud, R. M., Finer-Moore, J., Acetylcholine receptor structure, function and evolution, *Annu. Rev. Cell Biol.*, 1, 317, 1985.

93. Oblas, B., Singer, R. H., Boyd, N. D., Location of a polypeptide sequence within the alpha subunit of the acetylcholine receptor containing the cholinergic binding site, *Mol. Pharmacol.*, 29, 649, 1986.

94. Changeux, J. P., Revah, F., The acetylcholine receptor molecule: allosteric sites and the ion channel, *TINS*, 10, 245, 1987.

95. Changeux, J. P., Giroudot, J., Dennis, M., The nicotinic acetylcholine receptor: molecular architecture of a ligand-regulated ion channel, *TIPS*, 8, 459, 1987.

96. Watson, M., Roeske, W. R., Yamamura, H. I., Cholinergic receptor heterogeneity, in *Psychopharmacology: The Third Generation of Progress*, Meltzer, H. Y., Ed., Raven Press, New York, 1987, 241.

97. Morley, B. J., Robinson, G. R., Brown, G. B., Kemp, G. E., Bradley, R. J., Effects of dietary choline on nicotinic acetylcholine receptors in brain, *Nature*, 266, 848, 1977.

98. Volpe, B. T., Francis, A., Gazzaniga, M. S., Schechter, N., Regional concentration of putative nicotinic cholinergic receptor sites in human brain, *Exp. Neurol.*, 66, 737, 1979.

99. Larsson, C., Lundberg, P. A., Halen, A., Adem, A., Nordberg, A., *In vitro* binding of 3H-acetylcholine to nicotinic receptors in rodent and human brain, *J. Neural Transm.*, 69, 3, 1987.

100. Usdin, T. B., Fischbach, G. D., Purification and characterization of a polypeptide from chick brain that promotes the accumulation of acetycholine receptors in chick myotubes, *J. Cell Biol.*, 103, 493, 1986.

101. Nitkin, R. M., Smith, M. A., Magill, C., Fallon, J. R., Yao, Y. M., Wallace, B. G., McMahan, U. J., Identification of agrin, a synaptic organizing protein from torpedo electric organ, *J. Cell Biol.*, 105, 2471, 1987.

102. Schwartz, R. D., Kellar, K. J., Nicotinic cholinergic receptor binding sites in the brain: regulation *in vivo*, *Science*, 220, 214, 1983.

103. DeSarno, P., Giacobini, E., Modulation of acetylcholine release by nicotinic receptors in the rat brain, *J. Neurosci. Res.*, 22, 194, 1989.

104. Beani, L., Bianchi, C., Nilsson, L., Nordberg, A., Romanelli, L., Sivilotti, L., The effect of nicotine and cytisine on 3H-acetylcholine release from cortical slices of guinea-pig brain, *Naunyn-Schmiedebergs Arch. Pharmacol.*, 331, 293, 1985.

105. McCormick, D. A., Connors, B. W., Lighthall, J. W., Prince, D. A., Comparative electrophysiology of pyramidal and sparsely spiny stellate neurons in the neocortex, *J. Neurophysiol.*, 54, 782, 1985.

106. Brown, J. H., Brown, S. L., Agonists differentiate muscarinic receptors that inhibit cyclic AMP formation from those that stimulate phosphoinositide metabolism, *J. Biol. Chem.*, 259, 3777, 1984.

107. Egan, T. H., North, R. A., Actions of acetylcholine and nicotine on rat locus coeruleus neurons *in vitro*, *Neuroscience*, 25, 565, 1986.

108. Tebecis, A. K., Properties of cholinoceptive neurons in the medial geniculate nucleus, *Br. J. Pharmacol.*, 38, 117, 1970.

109. Lipton, S. A., Aizenman, E., Loring, R. H., Neural nicotinic acetylcholine responses in solitary mammalian retinal ganglion cells, *Pfluegers Arch. Gesamte Physiol.*, 410, 37, 1987.

110. Rowntree, D. W., Nevin, S., Wilson, A., The effects of diisopropylfluorphosphonate in schizophrenia and manic depressive psychosis, *J. Neurol. Neurosurg. Psychiatry*, 13, 47, 1950.

111. Gershon, S., Shaw, F. H., Psychiatric sequelae of chronic exposure to organophosphorus insecticides, *Lancet*, I, 1371, 1961.

112. Janowsky, D. S., El-Yousef, M. K., Davis, J. M., Sckerke, H. J., A cholinergic-adrenergic hypothesis of mania and depression, *Lancet*, 2, 632, 1972.

113. Janowsky, D. S., El-Yousef, M. K., Davis, J. M., Parasympathetic suppression of manic symptoms by Physo, *Arch. Gen. Psychiatry*, 28, 542, 1973.

114. Cohen, B. M., Lipinski, J. F., Altesman, R. I., Lecithin in the treatment of mania: double-blind, placebo-controlled trials, *Am. J. Psychiatry*, 139, 1162, 1982.

115. Tamminga, C., Depression associated with oral choline, *Lancet*, 2, 905, 1976.

116. Risch, S. C., Janowsky, D. S., Kalin, N. H., Cholinergic bendophin hypersensitivity associated with depression, in *Biological Markers in Psychiatry and Neurology*, Usid, E., Hanin, I., Eds., Pergamon Press, Oxford, 1982, 269.

117. Janowsky, D. S., El-Yousef, M. K., Davis, J. M., Acetylcholine and depression, *Psychosom. Med.*, 36, 248, 1974.

118. Janowsky, D. S., Risch, S. C., Parker, D., Huey, L., Judd, L., Increased vulnerability to cholinergic stimulation in affective disorder patients, *Psychopharmacol. Bull.*, 16, 29, 1980.

119. Dilsaver, S. C., Antimuscarinic agents as substances of abuse: a review, *J. Clin. Psychopharmacol.*, 8, 14, 1988.

120. Jimmerson, D. C., Nurnberger, J. I., Simmons, S., Gershon, E. S., Anticholinergic treatment for depression, in *Proceedings of the Annual Meeting of the American Psychiatric Association Syllabus*, The Association, Toronto, 1982, 218.

121. Kasper, S., Moises, H. W., Beckman, H., The anticholinergic biperiden in depressive disorders, *Pharmacopsychiatrica*, 14, 95, 1981.

122. Bunney, W. E., Davis, J. M., Norepinephrin in depressive states. A review. *Arch. Gen. Psychiatry*, 13, 484, 1965.

123. Davis, K. L., Hollister, L. E., Overall, J., Johnson, A., Train, K., Physostigmine: effects of cognition and affect in normal subjects, *Psychopharmacology*, 51, 23, 1976.

124. Moises, H. W., Bering, B., Muller, W. E., Personality factors predisposing to depression correlate significantly negatively with M1-muscarinic and B adrenergic receptor densities on blood cells, *Eur. Arch. Psychiatr. Neurol. Sci.*, 237, 209, 1988.

125. Fritze, J., Lanczik, M., Sofic, E., Struck, M., Riedere, R., Cholinergic neurotransmission seems not to be involved in depression but possibly in personality, *J. Psychiatr. Neurosci.*, 20, 39, 1995.

126. Martino-Barrows, A. M., Kellar, K. J., [^{3}H] Acetylcholine and [^{3}H](–)nicotine label the same recognition site in rat brain, *Mol. Pharmacol.*, 31, 169, 1986.

127. Pomerleau, O. F., Nicotine as a psychoactive drug: anxiety and pain reduction, *Psychopharmacol. Bull.*, 22, 865, 1986.

128. Wesnes, K., Revell, A., The separate and combined effects of scopolamine and nicotine on human information processing, *Psychopharmacology*, 84, 5, 1984.

129. Hughes, J. R., Hatsukami, D. K., Mitchell, J. E., Haskell, D., Prevalence of smoking among psychiatric outpatients, *Am. J. Psychiatry*, 143, 993, 1986.

130. Kandel, D. B., Davies, M., Adult sequelae of adolescent depressive symptoms, *Arch. Gen. Psychiatry*, 43, 255, 1986.

131. Glassman, A. H., Stetner, F., Walsh, T., Raizman, P. S., Fleiss, J. L., Cooper, T. B., Covey, L. S., Heavy smokers, smoking cessation, and clonidine. *J. Am. Med. Assoc.*, 259, 2963, 1988.

132. Glassman, A. H., Helzer, J. E., Covey, L. S., Cottler, L. B., Stetner, F. S., Tipp, J. E., Johnson, J., Smoking, smoking cessation and major depression, *J. Am. Med. Assoc.*, 264, 1546, 1990.

133. Anda, R. F., Williamson, D. F., Escobedo, L. G., Mast, E. E., Giovino, G. A., Remington, P. L., Depression and the dynamics of smoking, *J. Am. Med. Assoc.*, 264, 1541, 1990.

134. Flanagan, J., Maany, I., Smoking and depression. *Am. J. Psychiatry*, 139, 541, 1982.

135. Dilsaver, S. C., Davidson, R. K., Fluoxetine subsensitizes a nicotinic mechanism involved in the regulation of core temperature, *Life Sci.*, 41, 1165,1987.

136. Dilsaver, S. C., Harihan, M., Desipramine subsensitizes nicotinic mechanism involved in regulating core temperature, *Psychiatry Res.*, 25, 105, 1988.

137. Pestronk, A., Drachman, D. B., Lithium reduces the number of acetylcholine receptors in skeletal muscle, *Science*, 210, 342, 1980.

138. Edwards, N. B., Murphy, J. K., Downs, A. D., Ackerman, B. J., Rosenthal, T. L., Doxepin as an adjunct to smoking cessation: a double blind pilot study, *Am. J. Psychiatry*, 146, 373, 1989.

139. Hasey, G. M., Cooke, R., Warsh, J. J., Bonello, A., Jorna, T., Nicotine use and recovery from depressive illness, in *Proceedings of the Annual Meeting of the Canadian Psychiatric Association*, Toronto, Ontario, September 1990.

140. Olpe, H. R., Jones, R. S. G., Steinmann, M. W., The locus coeruleus: actions of psychotropic drugs, *Experientia*, 39, 242, 1983.

141. Clarke, P. B. S., Hommer, D., Pert, A., Skirboll, L., Electophysiological actions of nicotine on subtantia nigra single units, *Br. J. Pharmacol.*, 85, 827, 1985.

142. Baron, J. A., Cigarette smoking and Parkinson's disease, *Neurology*, 36, 1490, 1986.

143. Sitaram, N., Dube, S., Jones, R., Pohl, R., Gershon, S., Acetylcholine and alpha adrenergic sensitivity in the separation of depression and anxiety, *Psychopathology*, 17 (Suppl. 3), 24, 1984.

144. Kreig, J. C., Berger, M., REM sleep and cortid response to the cholinergic challenge with RS 86 in normals and depressives, *Acta Psychiatri. Scand.*, 76, 600, 1987.

145. Schreiber, W., Lauer, C. K., Krumrey, K., Hosboer, F., Kreig, J. C., Cholinergic REM sleep induction test in subjects at high risk for psychiatric disorders, *Biol. Psychiatry*, 32, 79, 1992.

146. Gillin, J. C., Salin-Pascual, R., Valazquez-Moctezuma, J., Shiromani, P., Zoltoski, R., Cholinergic receptor subtypes and REM sleep in animals and normal controls, in *Progress in Brain Research*, Cuello, A. C., Ed., Elsevier, New York, 1993, 379.

147. Hu, B., Bouhassira, D., Steriade, M., Deschenes, M., The blockage of PGO waves in the cat LGN by nicotinic antagonists, *Brain Res.*, 473, 394, 1988.

148. Velazquez-Moctezuma, J., Shalauta, M. D., Gillin, J. C., Shiromani, P. J., Microingestions of nicotine in the medial pontine reticular formation elicits REM sleep, *Neurosci. Lett.*, 115, 265, 1990.

149. Sachar, E. J., Hellman, L., Roffwang, H. P., Halpern, F. S., Fukushimo, D. K., Gallagher, T. F., Disrupted 24-hr patterns of cortisol secretion in psychotic depression, *Arch. Gen. Psychiatry*, 28, 19, 1973.

150. Carroll, B. J., Feinberg, M., Greden, J. F., Tarika, J., Albala, A., Haskelt, R., James, N. M., Kronfol, Z., Lohr, N., Steiner, M., de Vigne, J. P., Young, E., A specific laboratory test for the diagnosis of melancholia — standardization, validation and clinical utility, *Arch. Gen. Psychiatry*, 38, 15, 1981.

151. Charlton, B. G., Leake, A., Wright, C., Griffiths, H. W., Ferrier, I. N., A combined study of cortisol, ACTH and dexamethasone concentrations in major depression: multiple time-point sampling, *Br. J. Psychiatry*, 150, 791, 1987.

152. Edelstein, C. K., Roy-Byrne, P., Fawzy, F. I., Effect of weight loss on the dexamethasone suppression test, *Am. J. Psychiatry*, 140, 338, 1987.

153. Baumgartner, A., Graf, K. J., Kurten, I., The dexamethasone suppression test, in schizophrenia, and during experimental stress, *Biol. Psychiatry*, 20, 675, 1985.

154. Ceulemans, D. L. S., Westenbery, H. G. M., van Praag, H. M., The effect of stress on the dexamethasone suppression test, *Psychiatry Res.*, 14, 189, 1985.

155. Brown, W. A., Johnston, R., Mayfield, D., The 24 hour dexamethasone suppression test in a clinical setting: relationship to diagnosis, symptoms and response to treatment, *Am. J. Psychiatry*, 136, 543, 1979.

156. Asnis, G. M., Halbreich, V., Sachar, E. J., Nathan, R. S., Ostrow, L. C., Novacenko, H., Davis, M., Endicott, J., Puig-Antich, J., Plasma cortisol secretion and REM period latency in adult endogenous depression, *Am. J. Psychiatry*, 140, 750, 1983.

157. Schlessor, M. A., Winokur, G., Sherman, B. M., Hypothalamic-pituitary-adrenal axis activity in depressive illness. Its relationship to classification, *Arch. Gen. Psychiatry*, 37, 737, 1980.

158. Frazer, A. R., Choice of antidepressant based on the dexamethasone suppression test, *Am. J. Psychiatry*, 140, 786, 1983.

159. Ames, D., Burrows, G., Davies, B., Maguire, K., Norman, T., A study of the dexamethasone suppression test in hospitalized depressed patients, *Br. J. Psychiatry*, 14, 311, 1984.

160. Grunhaus, L., Zelnik, T., Albala, A. A., Rabin, D., Haskett, R. F., Zus, A. P., Greden, J. F., Serial dexamethasone suppression tests in depressed patients treated only with electroconvulsive therapy, *J. Affective Disord.*, 13, 233, 1987.

161. Bowie, P. C. W., Beani, A. Y., Bowie, L. J., The prognosis of primary depressive illness: its relationship to the dexamethasone suppression test, *Br. J. Psychiatry*, 150, 787, 1987.

162. Gerken, A., Maier, W., Holsboer, F., Weekly monitoring of dexamethasone suppression response in depression: its relationship to change of body weight and psychopathology, *Psychoneuroendocrinology*, 10, 261, 1985.

163. Lu, R.-B., Ho, S.-L., Ho, B. T., Leu, S.-Y., Shian, L.-R., Chen, W.-L., Correlation between plasma cortisol and CSF catecholamines in endogenous depressed dexamethasone nonsuppressors, *J. Affective Disord.*, 10, 177, 1986.

164. Hasey, G. M., Hanin, I., Hypothalamic-pituitary-adrenal axis activity is increased and correlated with brain acetylcholine levels in physostigmine but not neostigmine treated rats, *Psychoneuroendocrinology*, 15, 357, 1990.

165. Carroll, B. J., Greden, J. F., Rubin, R. T., Haskett, R., Feinberg, M., Schteingart, D., Neurotransmitter mechanism of neuroendocrine disturbance in depression, *Acta Endocrinol. (Suppl.)*, 220, 40, 1978.

166. Risch, S. C., Janowsky, D. S., Gillin, J. C., Rauch, J. L., Loevinger, B. L., Huey, L. Y., Muscarinic supersensitivity of anterior pituitary ACTH release in major depressive illness, adrenal cortical dissociation, *Psychopharmacol. Bull.*, 19, 343, 1983.

167. Marty, M. A., Erwin, V. G., Cornell, K., Zgombick, J. M., Effects of nicotine on beta-endorphin, alpha-MSH and ACTH secretion by isolated perfused mouse brains and pituitary glands, *in vitro*, *Pharmacol. Biochem. Behav.*, 22, 317, 1985.

168. Fuxe, K., Andersson, K., Eneroth, P., Harfstrand, A., Agnati, L. F., Neuroendocrine actions of nicotine and of exposure to cigarette smoke: medical implications, *Psychoneuroendocrinology*, 14, 19, 1989.

169. Meyerson, L. R., Wennogle, L. P., Abel, M. S., Coupet, J., Lippa, A. S., Rauh, C. E., Beer, B., Human brain receptor alterations in suicide victims, *Pharmacol. Biochem. Behav.*, 17, 159, 1982.

170. Kaufman, C. A., Gillin, J. C., Hill, B., O'Laughlin, T., Phillips, I., Kleinman, J. E., Wyatt, R. J., Muscarinic binding in suicides, *Psychiatry Res.*, 12, 47, 1984.

171. Hanin, I., Kopp, U., Spiker, D., Neil, J. F., Shaw, D. H., Kupfer, D. J., RBC and plasma choline levels in control and depressed individuals: a critical evaluation, *Psychiatry Res.*, 3, 345, 1980.

172. Hanin, I., Cohen, B. M., Kopp, U., Lipinski, J. F., Erythrocyte and plasma choline in bipolar psychiatric patients: a follow up study, *Psychopharmacol. Bull.*, 18, 186, 1982.

173. Stoll, A. L., Cohen, B. M., Snyder, M. B., Hanin, I., Erythrocyte choline concentration in bipolar disorder: a predictor of clinical course and medication response, *Biol. Psychiatry*, 29, 1171, 1991.

174. Snyder, S. H., Yamamura, H. I., Antidepressants and the muscarinic acetylcholine receptor, *Arch. Gen. Psychiatry*, 34, 236, 1977.

175. Richter, D., Crossland, J., Variation in acetylcholine content of the brain with physiological state, *Am. J. Physiol.*, 159, 247, 1949.

176. Takahashi, R., Nasu, T., Tamura, T., Relationship of ammonia and acetylcholine levels to brain excitability, *J. Neurochem.*, 7, 103, 1961.

177. Longoni, R., Mulas, A., Novak, B. O., Effect of single and repeated electroshock applications on brain acetylcholine levels and choline acetyltransferase activity in the rat, *Neuropharmacology*, 15, 283, 1976.

178. Spignoli, G., Pepeu, G., Oxiracetam prevents electroshock-induced decrease in brain acetylcholine and amnesia, *Eur. J. Pharmacol.*, 126, 253, 1986.

179. Adams, H. E., Hoblit, P. R., Sutker, P. B., Electroconvulsive shock, brain acetylcholinesterase activity and memory, *Physiol. Behav.*, 4, 113, 1969.

180. Tower, D. B., McEachern, D., Acetylcholine neuronal activity, *Can. J. Res.*, 27, 105, 1949.

181. Fink, M., Cholinergic aspects of convulsive therapy, *J. Nerv. Ment. Dis.*, 142, 475, 1966.

182. Sevastyanova, G. A., Tretyakova, K. A., The state of the acetylcholine-cholinesterase system in tonic meso-diencephalic convulsive fits, *Zh. Nevropatol. Psikhiatr., im. S. S. Korsakova*, 69, 1811, 1969.

183. Atterwill, C. K., Lack of effect of repeated electroconvulsive shock on [3H] spiroperidol and [3H] 5-hydroxytryptamine binding and cholinergic parameters in rat brain, *J. Neurochem.*, 35, 729, 1980.

184. Pryor, G. T., Effect of repeated ECS on brain weight and brain enzymes, in *Psychobiology of Convulsive Therapy*, Fink, M., Kety, S., McGaugh, J., Eds., V. H. Winston and Sons, Washington, D.C., 1974, 171.

185. Dasheiff, R. M., Savage, D. D., McNamara, J. O., Seizures down-regulate muscarinic cholinergic receptors in hippocampal formation, *Brain Res.*, 235, 327, 1982.

186. Lerer, B., Studies on the role of brain cholinergic systems in the therapeutic mechanisms and adverse effects of ECT and lithium, *Biol. Psychiatry*, 20, 20, 1985.

187. Deakin, J. F. W., Owen, F., Cross, A. J., et al., Studies on possible mechanisms of action of electroconvulsive therapy: effects of repeated electrically induced seizures on rat brain receptors for monamines and other neurotransmitters, *Psychopharmacology*, 73, 345, 1981.

188. Kellar, K. J., Cascio, C. S., Begstrom, D. A., et al., Electroconvulsive shock and reserpine: effects on B-adrenergic receptors in rat brain, *J. Neurochem.*, 37, 830, 1981.

189. Newman, M. E., Miskin, I., Lerer, B., Effects of single and repeated electroconvulsive shock administration on inositol phosphate accumulation in rat brain slices, *J. Neurochem.*, 49, 19, 1987.

190. Green, A. R., Bloomfield, M. R., Aterwill, C. K., et al., Electroconvulsive shock reduces the cataleptogenic effect of both haloperidol and arecoline in rats, *Neuropharmacology*, 18, 447, 1979.

191. Funkenstein, D. H., Greenblatt, M., Solomon, H.C., Autonomic nervous system changes following electric shock treatment, *J. Nerv. Ment. Dis.*, 108, 409, 1948.

192. Hasey, G. M., Warsh, J. J., Cooke, R. G., Smith, I., Martin, B., Goldbloom, D., Atropine blocks the efficacy of ECT, in *Proceedings of the Annual Meeting of the American Psychiatric Association*, The Association, San Francisco, 1993.

193. Tandon, R., Greden, J. F., Cholinergic hyperactivity and negative schizophrenic symptoms, *Arch. Gen. Psychiatry*, 46, 745, 1989.

194. Drachman, D. A., Memory and cognitive function in man: does the cholinergic system have a specific role? *Neurology*, 27, 783, 1977.

195. Drachman, D. A., Doffsinger, D., Sahakian, B.J., Kurdziel, S., Fleming, P., Aging, memory and the cholinergic system: a study of dichotic listening, *Neurobiol. Aging*, 1, 39, 1980.

196. Gelenberg, A. J., Van Putten, T., Lovori, P. W., et al., Anticholinergic effects on memory: benztropine vs. amantadine, *J. Clin. Psychopharmacol.*, 9, 180, 1989.

197. Penetar, D. M., The effects of atropine, benactyzine, and Physo on a repeated acquisition baseline in monkeys, *Psychopharmacology*, 87, 69, 1985.

198. Richter-Levin, G., Segal, M., Spatial performance is severely impaired in rats with combined reduction of serotonergic and cholinergic transmission, *Brain Res.*, 477, 404, 1989.

199. Sitaram, N., Weingartner, H., Human serial learning: enhancement with arecoline and choline and impairment with scopolamine, *Science*, 201, 274, 1978.

200. Altman, H. J., Stone, W. S., Ogren, S. O., Evidence for a possible functional interaction between serotonergic and cholinergic mechanisms in memory retrieval, *Behav. Neural Biol.*, 48, 49, 1987.

201. Gower, A. J., Enhancement by secoverine and Physo of retention of passive avoidance response in mice, *Psychopharmacology*, 91, 326, 1987.
202. Meck, W.H., Smith, R.A., Williams, C.L., Pre- and postnatal choline supplementation produces long-term facilitation of spatial memory, *Dev. Psychobiol.*, 21, 339, 1988.
203. Micheau, J., Destrade, C., Jaffard, R., Physostigmine reverses memory deficits produced by pre-training electrical stimulation of the dorsal hippocampus in mice, *Behav. Brain Res.* 15, 75, 1985.
204. Nabeshima, T., Noda, Y., Itoh, K., Kameyama, T., Role of cholinergic and GABAergic neuronal systems in cycloheximide-induced amnesia in mice, *Pharmacol. Biochem. Behav.* 31, 405, 1988.
205. Tilson, H. A., Harry, G. J., McLamb, R. L., Peterson, N. J., Rodgers, B. C., Pediaditakis, P., Ali, F., Role of dentate gyrus cells in retention of a radial arm maze task and sensitivity of rats to cholinergic drugs, *Behav. Neurosci.*, 102, 835, 1988.
206. Ueki, A., Miyoshi, K., Effects of cholinergic drugs on learning impairment in ventral globus pallidus-lesioned rats, *J. Neural. Sci.*, 90, 1, 1989.
207. Gage, F. H., Bjorklund, A., Cholinergic septal grafts into the hippocampal formation improve spatial learning and memory in aged rats by an atropine-sensitive mechanism, *J. Neurosci.*, 6, 2837, 1986.
208. Walsh, T. J., Tilson, H. A., Dehaven, D. L., Mailman, R. B., Fisher, A., Hanin, I., AF64A, a cholinergic neurotoxin, selectively depletes acetylcholine in hippocampus and cortex, and produces long-term passive avoidance and radial-arm maze deficits in the rat, *Brain Res.*, 321, 91, 1984.
209. Robinson, S. E., Hambrecht, K. L., Lyeth, B. G., Basal forebrain carbachol injection reduces cortical acetylcholine turnover and disrupts memory, *Brain Res.*, 445, 160, 1987.
210. Loullis, C. C., Dean, R. L., Lippa, A. S., et al., Chronic administration of cholinergic agents: effects on behavior and calmodulin, *Pharmacol. Biochem. Behav.*, 18, 601, 1983.
211. Wehner, J. M., Upchurch, M., The effects of chronic oxotremorine treatment on spatial learning and tolerance development in mice, *Pharmacol. Biochem. Behav.*, 32, 543, 1987.
212. Levin, E. D., McGurk, S. R., South, D., Butcher, L. L, Effects of combined muscarinic and nicotinic blockade on choice accuracy in the radial-arm maze, *Behav. Neural. Biol.*, 51, 270, 1989.
213. Introini, I. B., Baratti, C. M., The impairment of retention induced by β-endorphin in mice may be mediated by a reduction of central cholinergic activity, *Behav. Neural. Biol.*, 41, 152, 1984.
214. Berg, D. T., Boyd, R. T., Halvorsen, S. W., Jacob, M. H., Margiotta, J. F., Regulating the number and function of neuronal acetylcholine receptors, *Trends Neurosci.*, 12, 16, 1989.
215. Sunderland, T., Tariot, P. N., Newhouse, P. A., Differential responsivity of mood, behavior and cognition to cholinergic agents in elderly neuropsychiatric populations, *Brain Res. Rev.*, 13, 371, 1988.
216. Yeomans, J. S., Mathur, A., Tampakeras, M., Rewarding brain stimulation: role of tegmental cholinergic neurons that activate dopamine neurons, *Behav. Neurosci.*, 107, 1077, 1993.
217. Cummings, J. L., Gorman, D. G., Shapira, J., Physostigmine ameliorates the delusions of Alzheimer's disease, *Biol. Psychiatry*, 33, 536, 1993.
218. Haroutunian, V., Davidson, M., Kanof, P. D., Perl, D. P., et al., Cortical cholinergic markers in schizophrenia, *Schizophrenia Res.*, 12, 137, 1994.
219. Koponen, H., Reikkinen, P. J., Cerebrospinal fluid acetylcholinesterase in patients with dementia associated with schizophrenia or chronic alcoholism, *Acta Psychiatr. Scand.*, 83, 441, 1991.
220. Kraepelin, E., *Dementia praecor and paraphrenia*, Translated by Barclay, R. M., Edinburgh, Livingstone, 1919. Reprinted by Krieger Publishing Co., Huntington, New York, 1971.
221. Wilson, P., A photographic perspective on the origins, form, course and relations of the acetylcholinesterase-containing fibres of the dorsal tegmental pathway in the rat brain, *Brain Res. Rev.*, 10, 85, 1985.

Chapter **6**

THE PHARMACOLOGY OF THE GAMMA-AMINOBUTYRIC ACID SYSTEM

M. Frances Davies

CONTENTS

0-8493-8386-0/96/$0.00+$.50
© 1996 by CRC Press, Inc.

I. INTRODUCTION

Gamma-aminobutryric acid (GABA) is the most important inhibitory neurotransmitter in the central nervous system (CNS); it is found ubiquitously in all areas of the brain. It has been estimated that about 40% of all terminals in the cortex are GABAergic.[1] Excitatory transmission is largely responsible for sending messages throughout the CNS to direct the performance of everyday tasks, but within the CNS its effects must be balanced by inhibitory forces to prevent ubiquitous uncontrolled excitation. This hyperexcitable state can result in seizures being elicited. Reduction of the strength of the GABAergic system by certain agents reliably provokes convulsant activity, and, alternatively, increase of its transmission usually has anticonvulsant activity. However, the GABAergic system should not be thought of as merely a braking system for the excitatory system. It is more realistic to think of it as being involved with shaping, integrating, and refining the information transmitted by the excitatory system. Because GABA is found in every brain region, it is likely that GABAergic transmission is involved to some extent in every CNS function. As will be described below, increasing in the strength of GABAergic transmission has effects on such diverse CNS functions as vigilance and conciousness, anxiety, thermoregulation, learning, food consumption, hormone regulation, motor control, and pain control.

GABA is found in small, local interneurons in almost every brain region, where it is the neurotransmitter used to send local inhibitory signals. There are also GABAergic projection neurons in which the cell bodies are found in one brain region but the axons project to a distant region. An example of this is the GABAergic neurons of neostriatum whose axons project to the substantia nigra.[2]

II. METABOLISM OF GABA

A. Synthesis

GABA synthesis is a highly regulated process in the CNS. GABA is synthesized in GABAergic terminals by the removal of a carboxyl group from glutamate by the enzyme glutamic acid decarboxylase (GAD). The principal supply of glutamate comes from the tricarboxylic acid (TCA) cycle and, therefore, is ultimately derived from glucose. GAD is a cytosolic enzyme that requires the cofactor pyridoxal phosphate (vitamin B_6) to be active. The amount of GAD in the CNS far exceeds the amount required for the normal synthetic demands of new GABA; therefore, there must be a controlling mechanism that limits the rate of GABA synthesis. The proportion of the enzyme supply that is associated with pyridoxal phosphate appears to be one point of regulation, as at least 50% of the GAD present in the brain is not bound to the cofactor and is therefore inactive.[3] The interconversion of cofactor-associated to nonassociated GAD is strongly enhanced by adenosine triphosphate (ATP), while inorganic phosphates antagonize the effects of ATP.[3]

B. Uptake and Catalysis

Once the GABAergic terminal is depolarized by an incoming action potential, the synthesized GABA that is kept in synaptic vesicles is then released into the extracellular space (Figure 1). There it interacts with all GABA receptors located on adjacent neurons or on the GABAergic terminal itself. The effect of GABA is terminated by reuptake into nerve terminals or into glial cells by high-affinity uptake systems.[4] Once taken up into nerve terminals or glial cells, GABA is metabolized to succinic semialdehyde by GABA-transaminase (GABA-T), a mitochondrial aminotransferase that also binds pyridoxal phosphate. Succinic semialdehyde is then oxidized to succinic acid by succinic semialdehyde dehydrogenase. Succinic acid is ultimately returned to the TCA cycle for further metabolic uses.[3]

III. DRUGS THAT AFFECT THE GABA SYNTHETIC PATHWAY

A. GAD Inhibitors

Some compounds that prevent the synthesis of GABA do so by either complexing with the cofactor of GAD, pyridoxal phosphate, as is the case for semicarbazide,[5] while others, such as 3-mercaptopropionic acid, compete with glutamate for the active site.[6] Administration of these agents to animals reduces the *in vivo* GABA concentration in the CNS and can lead to seizure activity.

B. GABA-T Inhibitors

Because GABA is the major inhibitory transmitter, attempts have been made to increase synaptic concentrations of GABA and, therefore, its overall inhibitory strength by preventing the catabolism of GABA. GABA-T has been targeted for this purpose since it is the major catalytic enzyme in the breakdown pathway of GABA. Administration of GABA-T inhibitors, such as the irreversible inhibitor γ-vinyl GABA (vigabatrin), does cause a large increase in the GABA concentration. It has been found clinically to be an effective anticonvulsant, especially in the treatment of partial epilepsies.[7] One compound, valproic acid, used clinically as an anticonvulsant, does inhibit GABA-T, but it is unclear whether or not this is the mechanism by which it prevents seizure activity.[8]

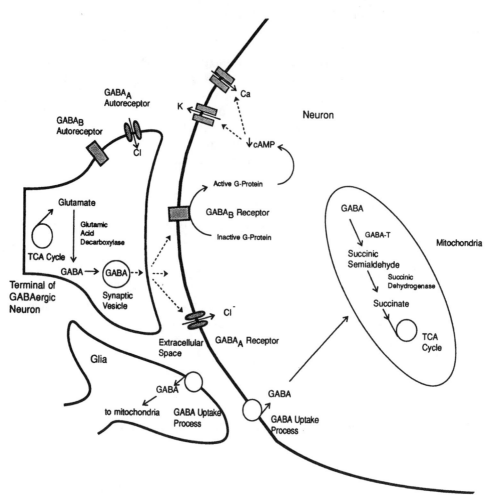

Figure 1
Schematic presentation of the synthetic pathways and receptor activation of the GABAergic system.

C. GABA-Uptake Inhibitors

Any effective strategy proposed to increase GABAergic transmission must do so by increasing the concentration of GABA around its receptor sites. Administration of GABA is not a viable solution because it has limited CNS penetrability and is rapidly removed from the synaptic region by uptake processes. Blockade of the GABA-uptake processes should, however, increase the amount of GABA in the synapse. Early uptake blockers, such as nipecotic acid and guvacine, were moderately effective *in vitro* but do not easily enter the CNS.[9] Newer compounds, such as SKF 89976A and SKF 100330A that are actually derivatives of nipecotic acid and guvacine, are potent inhibitors of GABA uptake and are behaviorally active after systemic administration. These compounds possess anticonvulsant activity but can also cause catalepsy.[10]

IV. GABA RECEPTORS

A. Major Divisions

There are two major classes of GABA receptors, termed $GABA_A$ and $GABA_B$, and each is a member of a different class of receptor superfamily. The newly described $GABA_C$ class of receptors seems be closely related to the $GABA_A$ receptor structurally but has a different pharmacological profile.[11] Because of limited knowledge of this receptor, it will not be discussed further.

The $GABA_A$ receptor is a member of the ligand-activated ion channel family. Other members of this family are the acetylcholine nicotinic receptor,[12] the glycine receptor, and some of the glutaminergic receptors, such as the N-methyl-D-aspartate receptor.[12] These receptors directly and rapidly open the ion channel intimately associated with the receptor. On postsynaptic neurons, those that receive a GABAergic signal from other neurons, $GABA_A$ receptors mediate fast inhibitory postsynaptic potentials (ipsps) that reach their peak within a few milliseconds and typically last a few tens of milliseconds.

$GABA_B$ receptors, like adrenergic and peptide receptors, are members of the large G protein–coupled receptor superfamily. In this family the signal transduction pathway is slow but can be of long duration because the agonist binding to the receptor causes a cascade of events that eventually affect membrane potassium and calcium channels. Because of the multiple steps involved in the signal transduction pathway, $GABA_B$ receptors mediate slow ipsps that reach their peak after about 100 ms but can also last hundreds of milliseconds. Both types of GABA receptors are also found on the terminals of GABAergic neurons and act as autoreceptors to control the amount of GABA being released by the terminals.[13]

B. GABA_B Receptors

Although the $GABA_B$ receptor has not been cloned, it has the functional characteristics of being a member of the G protein–coupled receptor superfamily because activation of this receptor is blocked by pertussis toxin and has been shown to be coupled to the Gi and/or Go subclasses of G proteins.[14] Activation of this receptor inhibits adenylyl cyclase and possibly directly affects calcium and potassium channels. In addition to being sensitive to GABA itself, these receptors are activated by selective agonists, such as baclofen and 3-aminopropyl phosphinic acid, but not by selective $GABA_A$ agonists, such as muscimol or 4,5,6,7-tetrahydroiso-xazole [5,4-C]pyridin-3-ol (THIP). Selective $GABA_B$ antagonists, such as phaclofen, 2-OH saclofen, and CGP 35348, have now been developed. $GABA_B$ receptors have yet to be cloned, so it is not definitively known whether multiple forms of this receptor exist. However, some evidence for $GABA_B$ receptor heterogeneity has been obtained with pharmacological and neurochemical studies.[13]

1. Distribution of GABA_B Receptors

$GABA_B$ receptors are widely distributed throughout the CNS and are also present on peripheral organs. Their pattern of distribution differs from that of $GABA_A$ receptors. In many brain regions both receptor subtypes exist together; however, there are areas where $GABA_B$ sites are found but very few $GABA_A$ receptors are present and vice versa.[14a] $GABA_B$ receptors have been demonstrated on both neurons and glial cells.[13] On neurons, $GABA_B$ receptors are found on the dendrites and soma, where they are likely to have postsynaptic actions, and on the terminals of both GABAergic and non-GABAergic neurons, where they may physiologically modulate neurotransmitter release.[13]

2. Physiological Role of GABA$_B$ Receptors

Activation of GABA$_B$ receptors has been shown to modify hormonal levels: reducing corticotropin-releasing hormone, melanocyte-stimulating hormone, prolactin-releasing factor, and luteinizing hormone, but increasing androgen production.[13] There are many behavioral effects of GABA$_B$ activation including epileptogenesis, hypotension, increased gastric motility, bronchiolar relaxation, reduced memory retention and consolidation, and muscle relaxation.[13] However, the antinociceptive activity of baclofen may be the most interesting effect, clinically. In rodents, there is substantial evidence that baclofen is antinociceptive.[15] In humans, the evidence is less definitive although trigeminal neuralgia has been successfully treated with baclofen.[16]

Selective GABA$_B$ antagonists may prove to be useful therapeutic agents for the treatment of epilepsy. There is also speculation, based on animal studies, that they may be useful for their cognitive-enhancing, anxiolytic, and neuroprotective properties.[13]

C. GABA$_A$ Receptors

GABA$_A$ receptors are localized on a complex, multisubunit structure that forms the chloride ion channel. This complex also has binding sites some important drug classes. Activation of this complex by GABA causes Cl$^-$ ions to flow through the channel. The structure of the GABA$_A$ complex is now being elucidated by the techniques of molecular biology. It is composed of five subunits arranged around a central pore that functions as the Cl$^-$ channel.[12] The concept that multiple, functionally different GABA$_A$ receptors exist within the mammalian brain is now firmly established. Five different subunit classes have been described, termed α, β, γ, δ, and ρ classified on the basis of sequence homology.[17] Within some of the subunit classes more than one form of the subunit has been detected so that six α, three β, three γ, one δ, and two ρ have now been described in mammals.[18-20] Each subunit has a unique distribution within the CNS.[21,22] This uneven distribution, coupled with newly emerging evidence that the subunit composition determines the functional and pharmacological characteristics of the receptor, indicates that the GABA$_A$ receptors may be tailored to meet the particular functional demands of a particular synapse.[17] A functional GABA$_A$ receptor/Cl$^-$ ionophore can be produced in cells that express only the α and β subunits, although these receptors lack the ability to be modulated by benzodiazepines. The species of α subunit primarily determines the benzodiazepine pharmacology,[23] yet high-affinity benzodiazepine binding requires the presence of a γ subunit in the complex,[24] with the subunit γ_2 usually being the most potent in bestowing benzodiazepine efficacy.[25] The contributions of the β, δ, or ρ subunits to receptor pharmacology are not yet well understood.

The determination of the stoichiometry of subunits within the GABA$_A$ complex has proved to be a challenging task. Initially, it was thought that, because α and β subunits could form a GABA/Cl$^-$ ionophore and because the γ subunits were only required for benzodiazepine sensitivity, 2α 2β 1γ would be a preferred configuration in native receptors. Recent studies have indicated that 2α 1β 2γ may be a more preferred combination.[26] As yet, the stoichiometry of δ- or ρ-containing receptors has not been determined.

Immunoprecipitation studies have shown that the subunit composition of native GABA$_A$ receptors may, in fact, be more complicated than originally appreciated. It has now been demonstrated that, while many receptors have one variant of a subunit, some receptors have more than one type of α subunit.[27,28]

In light of the large number of subunit varieties present in the CNS that can be combined in multiple ways to make a pentameric complex, the theoretical number of possible GABA receptor subtypes is staggering. It is unlikely with the existing experimental tools that the functional and pharmacological characteristics of every GABA$_A$ receptor subtype will be elucidated. However, some progress has been made in understanding the individual contributions of some of the subunits to the characteristics of the GABA$_A$ receptors, especially in terms of their sensitivity to psychoactive drugs.

1. Selective GABA$_A$ Receptor Antagonists

The ability of the GABA$_A$ receptor complex to allow the flow of Cl ions through its channel in response to GABA can be blocked in two ways. The first class of compounds are competitive inhibitors, such as bicuculline, that compete with GABA for its binding site.[29] The second class of compounds inhibit in a noncompetitive manner by physically blocking the Cl⁻ channel, and the classic example of such a noncompetitive inhibitor is picrotoxin.[30] The most obvious result of substantially reducing GABAergic transmission is to elicit convulsions, and both bicuculline and picrotoxin are convulsants.

2. Selective GABA$_A$ Receptor Agonists

GABA binds and activates both A and B types of GABA receptors and is rapidly taken up into nerve terminals and glia; therefore, its usefulness as an experimental tool is limited. Compounds that selectively bind to and activate the GABA$_A$ receptor, such as muscimol, THIP, isoguvacine, and isonipecotic acid, have been found useful in understanding the GABA system.[31]

These compounds were also explored as candidate therapeutic agents for the treatment of seizure disorders, based on the hypothesis that shifts of the excitation–inhibition balance that favored excitation would tend to cause convulsions, while increasing overall inhibitory strength would be anticonvulsive. Unfortunately, while some were found to suppress seizure activity in animals,[32] none has proved to be clinically useful. In humans, the major drawback has been that THIP and muscimol can produce psychotomimetic responses, sedation, and ataxia.[33]

V. OTHER SITES ON THE GABA$_A$ RECEPTOR

In addition to having binding sites for GABA on the GABA$_A$/Cl ion channel complex, there exist binding sites for many other classes of compounds that, when activated, can change the functional characteristics of the complex. These include sites for benzodiazepines, barbiturates, picrotoxin, neurosteroids, and possibly anesthetic agents. The characteristics and clinical relevance of each site will be described below. All of the sites on the GABA receptor described, except the picrotoxin and the cation sites, are targets for therapeutic agents.

A. Benzodiazepine-Binding Site

This binding site was originally characterized as the receptor site for the 1,4-benzodiazepine class of drugs. This chemical class includes the clinically important antianxiety drugs diazepam, chlordiazepoxide, and triazolam. More recently, many other compounds of diverse chemical classes have been found to bind to this same site. This site is a unique modulatory site in that it can either enhance or reduce the GABA$_A$ receptor function, depending on the type of ligand.[34] Activation of this receptor by a benzodiazepine agonist shifts the GABA dose–response curve to the left but does not change the maximal response. At the channel level, benzodiazepines increase the probability of chloride channel opening,[35] therefore, the effect of benzodiazepines would only be detected at submaximal concentrations of GABA, and they would not have any effect if GABA were either absent or present at supersaturating concentrations. This ability of benzodiazepines to merely enhance the actions of a natural neurotransmitter may explain why their behavioral effects are relatively mild. Compounds that decrease the GABA sensitivity of the receptor complex are called inverse agonists.[36] The benzodiazepine site can also be occupied but not activated by benzodiazepine receptor antagonists, such as flumazenil (Ro15-1788), which can antagonize the actions of both agonists and inverse agonists.[37] Since benzodiazepine agonists block the effects of both agonists and inverse agonists, the three types of ligands are thought to have overlapping domains on the GABA$_A$ receptor complex.

The multiple behavioral actions of benzodiazepine receptor agonists make them clinically useful anxiolytics, anticonvulsants, muscle relaxants, and hypnotic agents. It is also known that they can affect food and water consumption[38] and reduce body temperature.[39]

1. Benzodiazepine Receptor Heterogeneity vs. Partial Agonism Hypothesis of Behavioral Selectivity

It was originally assumed that all benzodiazepine receptor agonists had similar pharmacological profiles, acting through a single benzodiazepine receptor to produce every observed behavioral action, although not necessarily in the same dose range.[40] Behavioral selectivity was thought to be because of partial agonist activity for the benzodiazepine receptor.[41] Evidence for partial vs. full agonist activity has been obtained in many receptor systems. Not all agonists produce the same maximal effect. Those that cause a maximal change in receptor function are called full agonists, and those whose effects reach a submaximal level of receptor activation are called partial agonists. In the benzodiazepine receptor system, evidence for some partial agonists has been found.[42] It has been postulated that partial agonists would be behaviorally selective, reducing anxiety and having anticonvulsant activity but being minimally sedative or causing little ataxia.[41] *In vitro* evidence for partial agonist activity for some benzodiazepine receptor ligands has been found,[42] although there is still some debate whether or not benzodiazepine receptor heterogeneity also plays a role in behavioral selectivity.

The concept that all benzodiazepine effects are mediated by a single benzodiazepine receptor has been challenged by recent behavioral, receptor binding, and molecular biology studies. For example, in behavioral studies, the traditional 1,4-benzodiazepines have anticonvulsant and anxiolytic effects at lower doses than are required to produce sedation,[40] whereas the novel benzodiazepine compound zolpidem produces sedative effects at doses much lower than those required to cause anticonvulsant or anxiolytic effects.[43] Both binding studies and the techniques of molecular biology provide direct evidence for multiple benzodiazepine receptor subtypes. Early receptor binding studies had detected two central benzodiazepine receptors,[44,45] and on this basis these subtypes were called Types I and II. However, more than two pharmacologically relevant subtypes of benzodiazepine receptors have been found in the spinal cord,[46] and evidence for multiple $GABA_A$/benzodiazepine receptor subtypes have also been provided by molecular biology techniques that have characterized many $GABA_A$ receptor subunit variants.[47] Expression of combinations of these subunits in various transfected cell systems produces benzodiazepine receptors that have different ligand affinities[23] and varying ability to be activated by benzodiazepine receptor ligands.[25] For behavioral selectivity to occur, the benzodiazepine receptor subtypes within the CNS must be localized to specific regions, and it is known that GABA receptor subunit variants are not distributed evenly throughout the CNS.[21] Behavioral evidence for benzodiazepine receptor heterogeniety has been found.[48] The existence of this heterogeneity opens the door for creating therapeutic agents that would be designed to have specific effects without unwanted side effects.

2. Behavioral Effects of Benzodiazepine Receptor Inverse Agonists

The benzodiazepine receptor inverse agonists have behavioral actions which are in many instances opposite to those observed for the agonists. The most potent of the inverse agonists is DMCM, a β-carboline compound that is a very good convulsant.[49] Other inverse agonists such as β-CCM or FG-7142 are only proconvulsants because they cannot elicit seizures themselves but can reduce the seizure threshold for other compounds. These compounds also increase the anxiety state of animals and humans.[50]

B. Barbiturate-Binding Site

Barbiturates and other compounds, such as etomidate and etazolate, bind to a separate site on the $GABA_A$ receptor complex and modulate the function of GABA. At low concentrations barbiturates increase channel-opening time, but at higher concentrations they can directly activate the chloride channel without having any GABA bound to its site.[35] These two properties may contribute to the more potent sedative/hypnotic potential of barbiturates compared with the effects of benzodiazepines. The barbiturate pentobarbital reliably potentiates GABA responses in almost every recombinant receptor combination tested, whether or not a γ subunit is present.[51,52]

Barbiturates have an important place in clinical neurological practice as anticonvulsant and anesthetic agents and as sedative/hypnotics.[53] Phenobarbital, the primary anticonvulsant barbiturate, is effective for partial, complex partial, and secondarily generalized seizures. Mysoline has also been shown to be useful in the treatment of essential tremor and several other movement disorders.

C. Picrotoxin-Binding Site

The flow of Cl ions through the $GABA_A$/Cl channel is inhibited in a noncompetitive manner by several convulsants such as picrotoxin, TBPS (t-butylbicyclophosphorothionate), and pentylenetetrazol.[54] At present, there are no therapeutic agents which are known to have their actions by interacting with the picrotoxin site, although it may be a target for toxic agents.

D. Neurosteroid-Binding Site

Some 3α-hydroxylated, 5-reduced pregnane steroids are known to have anxiolytic, anticonvulsant, and sedative, effects that may be mediated by the $GABA_A$ receptor.[55] The steroid general anesthetic alphaxalone also appears to have its action by enhancing the flow of ions through the GABA receptor complex following binding to the neurosteroid site.[56] These neurosteroids probably increase the open duration of the channel,[57] and, like barbiturates, high concentrations of neurosteroids (>1 μM) can directly activate $GABA_A$ receptors.[58,59] The pharmacological profiles of neurosteroids and hypnotic barbiturates are remarkably similar,[59,60] but they appear to act at distinct sites.[61]

Some steroids that are known to interact with GABA are found endogenously in the brain, and it has been hypothesized that the $GABA_A$ receptor may provide a link between the endocrine and nervous systems. For instance, on a behavioral level ovarian endocrine status has been shown to modulate the anxiolytic potency of diazepam.[62]

E.   Anesthetic Agents

Recent evidence has indicated that general anesthetic agents such as propofol and the inhalation anesthetics have their principal action by interacting with the $GABA_A$ receptor and, perhaps, the glycine receptor.[63] Effects of anesthetics on other molecular targets, such as voltage-gated ion channels and other ligand-gated channels, and changes in membrane fluidity have been observed; however, the concentrations required are in general greater than could be achieved during the anesthetic state.[64] Further work is needed to determine if all $GABA_A$ receptor subtypes are equally sensitive to anesthetics and to elucidate how the anesthetic molecule interacts with these receptor complexes.

F.   Propofol Site

Propofol, a novel short-acting anesthetic, is known to enhance GABAergic transmission and has actions similar to pentobarbital and alphaxalone. However, these three drugs appear to have separate sites of action. A direct interaction at the level of benzodiazepine recognition sites has also been excluded.[56]

G. Cation Site

The $GABA_A$ receptor complex is sensitive to the divalent cation zinc (Zn^{2+})[65,66] and the trivalent cation lanthanum (La^{3+}).[67] The subunit composition of the $GABA_A$ receptor can have a profound effect on the sensitivity to modulation by these transition metal ions. Inclusion of the γ subunit dramatically alters the sensitivity of the resulting $GABA_A$ receptor to Zn^{2+} and La^{3+}. When α and β subunits were singly expressed or coexpressed, the conductance through the resulting receptor/channel complex was completely inhibited by micromolar concentrations of Zn^{2+}.[65] However, when a γ subunit was also incorporated, the complex was virtually insensitive to the blocking action of Zn^{2+}.[65,66] Evidence has now been found for two such populations of $GABA_A$/Cl⁻ channel receptors in the adult rat cortex and cerebellum.[68] Interestingly, the effects of La^{3+} are qualitatively opposite to those of Zn^{2+}. Micromolar concentrations of La^{3+} dose-dependently stimulated the Cl⁻ currents of $GABA_A$ receptors composed of α, β, γ subunits; yet, when only the α and β subunits were expressed, this ion only slightly potentiated the GABA response.[67]

VI. PHYSIOLOGICAL AND PHARMACOLOGICAL FACTORS THAT MODULATE GABA$_A$ RECEPTOR SUBUNIT COMPOSITION

A. Development

The subunit composition of GABA$_A$ receptors is now known to change under many physiological conditions. During development, the levels of GABA$_A$ receptor subunit mRNA expression change, and each subunit exhibits a unique regional and temporal developmental profile.[69-71] These changes may underlie the increased sensitivity of fetal brain to GABA and decreased sensitivity to benzodiazepine.[72] It has been proposed that during fetal brain development, when there are few well-developed synapses present, GABA acts as a neurotropic factor that promotes neuronal growth and differentiation, synaptogenesis, and synthesis of its own receptors.[71] In the fetal brain, GABA would be not be released into a specialized postsynaptic junction and would have to diffuse to distant receptors where the concentration of GABA would be much lower. A GABA$_A$ receptor with a greater sensitivity to GABA would be capable of being activated even at such low concentrations. Early in fetal development, subunits that have a high affinity for GABA are found, and only later do the subunits that confer low affinity for GABA increase.[71]

B. Aging

Changes in subunit composition of GABA$_A$ receptors have also been found in aged animals.[73] Changes in GABA$_A$ receptor subunit composition may underlie some of the changes in drug sensitivity to benzodiazepines and barbiturates observed in the elderly.

C. Long-Term Drug Exposure

It should be of particular interest to the clinician that chronic exposure to drugs that interact with the GABA$_A$ receptor can also change the expression pattern of GABA$_A$ receptor subunits and, consequently, change the pharmacological characteristics of the receptor. Changes in subunit composition have been demonstrated with the chronic exposure of the animal or cultured neurons to benzodiazepine receptor agonists, GABA, ethanol, neurosteroids, and pentobarbital.[74-78] In many cases the subunit changes make the GABA$_A$ receptor less sensitive to the chronically present pharmacological agent. These subunit changes may be more subtle. An example of this is the neurosteroid pregnanolone that, when chronically present, has been shown to uncouple the allosteric interactions of steroids and benzodiazepines.[79] Changes in the composition of GABA$_A$ receptors may be the underlying cause for some behaviorally observable phenomena, such as drug tolerance, physical dependence, and problems that occur with drug withdrawal. These modifications in subunit composition may be beneficial to the patient if they reduce incidence of sedation or ataxia, but they may also reduce the effectiveness of the drug during long-term treatment, as may be the case for the anticonvulsant effects of benzodiazepines. The fact that chronically administered drugs can change parameters of CNS function either on a short- or long-term basis should be kept in mind by the clinician.

D. Intracellular Processes

It is now appreciated that the function of $GABA_A$ receptors is modulated by the intracellular milieu. Cl⁻ ion movement through the channel is known to be affected by the phosphorylation state of the receptor,[80,81] the intracellular free calcium,[82] and cyclic GMP[83] concentrations. Most cloned $GABA_A$ receptor subunits have consensus sequences for phosphorylation by a variety of protein kinases,[84] and the ability of different kinases to phosphorylate $GABA_A$ receptors has been documented.[80] The functional consequences of $GABA_A$ receptor phosphorylation are not clear because the results from different laboratories are contradictory.[85] The diversity in phosphorylation effects may in fact be a function of the brain region under study, because every region has a unique $GABA_A$ subunit population and protein kinase population, both of which can also change with age.[86]

The susceptibility of the $GABA_A$ receptor to undergo phosphorylation and also dephosphorylation by phosphatases may allow modulation of this receptor's function by agonists acting upon G protein–coupled receptors. One of the principal effects of activation of these receptors is to change second messenger systems that modulate protein kinase and phosphatase activity. In cerebellar Purkinje cells, potentiation of $GABA_A$–mediated inhibition has been shown to be modulated by activation of the β-adrenergic receptor, a G protein–coupled receptor.[87] This interaction between the two major classes of receptors, the neurotransmitter-gated ion channel receptors and the G protein–coupled receptors, may prove to be important for understanding the mechanism of action of many psychoactive drugs.

VII. SUMMARY

The GABAergic system is the main inhibitory neurotransmitter system in the CNS. Suppression of its activity can result in seizure activity, whereas enhancement has numerous behavioral effects, such as causing an increase in the seizure threshold, analgesia, anxiolysis, and sedation. This system is modulated at many steps along the GABA synthesis pathway, at its release site, at the uptake site, and at its receptor sites. In addition, the functional characteristics of these receptors can change during development and aging or be dynamically changed by chronic drug administration and by the actions of other neurotransmitters. Even though the complexity of the receptor systems may be daunting, it is just this built-in structural diversity of the GABA receptors that offers the exciting possibility of creating new, behaviorally selective therapeutic agents.

REFERENCES

1. Bloom, F. E., Iversen, L. L., Localizing ³H-GABA in nerve terminals of rat cerebral cortex by electron microscopic autoradiography, *Nature*, 229, 628, 1971.
2. Vincent, S. R., Nagy, J. I., Fibiger, H. C., Increased striatal glutamate decarboxylase after lesions of the nigrostriatal pathway, *Brain Res.*, 143, 168, 1978.
3. Martin, D. L., Rimvall, K., Regulation of γ-aminobutyric acid synthesis in the brain, *J. Neurochem.*, 60, 395, 1993.

4. Iversen, L. L., Kelly, J. S., Uptake and metabolism of γ-aminobutyric acid by neurones and glial cells, *Biochem. Pharmacol.*, 24, 933, 1975.

5. Holtz, P., Palm, D., Pharmacological aspects of vitamin B$_6$, *Pharmacol. Rev.*, 16, 113, 1964.

6. Wu, J. Y, Roberts, E., Properties of brain L-glutamate decarboxylase inhibition studies, *J. Neurochem.*, 23, 759, 1974.

7. Sabers, A., Gram, L., Pharmacology of vigabatrin, *Pharmacol. Toxicol.*, 70, 237, 1992.

8. Macdonald, R. L., Kelly, K. M., Mechanisms of action of currently prescribed and newly developed antiepileptic drugs, *Epilepsia*, 35 (Suppl. 4), S41, 1994.

9. Krogsgaard-Larsen, P., Falch, E., Larsson, O. M., Schousboe, A., GABA uptake inhibitors, relevance to antiepileptic drug research, *Epilepsy Res.*, 1, 77, 1987.

10. Karbon, E. W., Enna, S. J., Ferkany, J. W., Biochemical and behavioral studies following subchronic administration of GABA uptake inhibitors in mice, *Neuropharmacology*, 30, 1187, 1991.

11. Shimada, S., Cutting, G., Uhl, G. R., γ-Aminobutyric acid A or C receptor? γ-Aminobutyric acid rho 1 receptor RNA induces bicuculline-, barbiturate-, and benzodiazepine-insensitive gamma-aminobutyric acid responses in *Xenopus* oocytes, *Mol. Pharmacol.*, 41, 683, 1992.

12. Schofield, P. R., Darlinson, M. G., Fujita, N., Burt, D. R., Stephenson, F. A., Rodriquez, H., Rhee, L. M., Ramachandran, J., Reale, V., Glencorse, T. A., Seeburg, P. H., Barnard, E. A., Sequence and functional expression of the GABA$_A$ receptor shows a ligand-gated receptor super-family, *Nature*, 328, 221, 1987.

13. Bowery, N. G., GABA$_B$ receptor pharmacology, *Annu. Rev. Pharmacol. Toxicol.*, 33, 109,1993.

14. Morishita, R., Kato, K., Asano, T., GABA$_B$ receptors couple to G proteins G$_o$, G$_o$* and G$_{i1}$ but not to G$_{i2}$, *FEBS Lett.*, 271, 231, 1990.

14a. Bowery, N. G., Hudson, A. L., Price, G. W., GABA$_A$ and GABA$_B$ receptor site distribution in the rat central nervous system, *Neuroscience*, 20, 365, 1987.

15. Aley, K. O., Kulkarni, S. K., Baclofen analgesia in mice, a GABA$_B$-mediated response, *Meth. Findings Exp. Clin. Pharmacol.*, 13, 681, 1991.

16. Green, M. W., Selman, J. E., Review article, the medical management of trigeminal neuralgia, *Headache*, 31, 588, 1991.

17. Wisden, W., Seeburg, P. H., GABA$_A$ receptor channels: subunits to functional entities, *Curr. Opinion Neurobiol.*, 2, 263, 1992.

18. Olsen, R. W., Tobin, A., Molecular biology of GABA$_A$ receptors, *FASEB J.*, 4, 1469, 1990.

19. Burt, D. R., Kamatchi, G. L., GABA$_A$ receptor subtypes, from pharmacology to molecular biology, *FASEB J.*, 5, 2916, 1991.

20. Cutting, G. R., Curristin, S., Zoghbi, H., O'Hara, B., Seldin, M. F., Uhl, G. R., Identification of a putative γ-aminobutyric acid (GABA) receptor subunit rho2 cDNA and colocalization of the gene encoding rho2 (GABRR2) and rho1 (GABRR1) to human chromosome 6q14-q21 and mouse chromosome 4, *Genomics*, 12, 801, 1992.

21. Wisden, W., Laurie, D. J., Monyer, H., Seeburg, P. H., The distribution of 13 GABA$_A$ receptor subunit mRNAs in the rat brain. I. Telencephalon, diencephalon, mesencephalon, *J. Neurosci.*, 12, 1040, 1992.

22. Laurie, D. J., Seeburg, P. H., Wisden, W., The distribution of 13 GABA$_A$ receptor subunit mRNAs in the rat brain. II. Olfactory bulb and cerebellum, *J. Neurosci.*, 12, 1063, 1992.

23. Pritchett, D. B., Luddens, H., Seeburg, P. H., Type I and type II GABA$_A$-benzodiazepine receptors produced in transfected cells, *Science*, 245, 1389, 1989.

24. Pritchett, D. B., Sontheimer, H., Shivers, B., Ymer, S., Kettenmann, H., Schofield, P. R., Seeburg, P. H., Importance of a novel GABA$_A$ receptor subunit for benzodiazepine pharmacology, *Nature*, 338, 582, 1989.

25. Puia, G., Vicini, S., Seeburg, P. H., Costa, E., Influence of recombinant γ-aminobutyric acid$_A$ receptor subunit composition on the action of allosteric modulators of γ-aminobutyric acid-gated Cl⁻ currents, *Mol. Pharmacol.*, 39, 691, 1991.

26. Backus, K. H., Arigoni, M., Drescher, U., Scheurer, L., Malherbe, P., Mohler, H., Benson, J. A., Stoichiometry of a recombinant GABA$_A$ receptor deduced from mutation-induced rectification, *Neuroreport*, 5, 285, 1993.

27. Luddens, H., Killisch, I., Seeburg, P. H., More than one alpha variant may exist in a GABA$_A$/benzodiazepine receptor complex, *J. Rec. Res.*, 11, 535, 1991.

28. Mertens, S., Benke, D., Mohler, H., GABA$_A$ receptor populations with novel subunit combinations and drug binding profiles identified in brain by α$_5$- and δ-subunit-specific immunopurification, *J. Biol. Chem.*, 268, 5965, 1993.

29. Mohler, H., Okada, T., GABA receptor binding with ³H (+) bicuculline methiodide in rat CNS, *Nature*, 267, 65, 1977.

30. Ticku, J. K., Maksay, G., Convulsant/depressant site of action at the allosteric benzodiazepine–GABA receptor complex, *Life Sci.*, 33, 2363, 1983.

31. Krogsgaard-Larsen, P., Falch, E., Jacobsen, P., GABA agonists, structural requirements for interaction with the GABA-benzodiazepine receptor complex, in *Actions and Interactions of GABA and Benzodiazepines*, Bowery, N. G., Ed., Raven Press, New York, 1984, 109.
32. Loscher, W., Frey, H. H., Reiche, R., Schultz, D., High anticonvulsant potency of γ-aminobutyric acid (GABA)mimetic drugs in gerbils with genetically determined epilepsy. *J. Pharmacol. Exp. Ther.*, 226, 839, 1983.
33. Lloyd, K. G., Morselli, P. L., Psychopharmacology of GABAergic drugs, in *Psychopharmacology: The Third Generation of Progress*, Meltzer, H. Y., Ed., Raven Press, New York, 1987, 183.
34. Hommer, D. W., Skolnick, P., Paul, S. M., The benzodiazepine/GABA receptor complex and anxiety, in *Psychopharmacology: The Third Gereration of Progress*, Meltzer, H. Y., Ed., Raven Press, New York, 1987, 977.
35. Study, R. E., Barker, J. L., Diazepam and (–) pentobarbital, fluctuation analysis reveals different mechanisms for potentiation of γ-aminobutyric acid responses in cultured chick neurons, *Proc. Natl. Acad. Sci. U.S.A.*,78, 7180, 1981.
36. Polc, P., Bonetti, E. P., Schaffner, R., Haefely, W., A three-state model of the benzodiazepine receptor explains the interactions between the benzodiazepine antagonist Ro 15-1788, benzodiazepine tranquilizers, beta-carbolines, and phenobarbitone, *Naunyn Schmiedebergs Arch. Pharmacol.*, 321, 260, 1982.
37. Polc, P., Laurent, J. P., Scherschlicht, R., Haefely, W., Electrophysiological studies on the specific benzodiazepine antagonist Ro 15-1788, *Naunyn Schmiedebergs Arch. Pharmacol.*, 316, 317, 1981.
38. Cooper, S. J., Benzodiazepines as appetite-enhancing compounds, *Appetite*, 1, 7, 1980.
39. Jackson, H. C., Nutt, D. J., Body temperature discriminates between full and partial benzodiazepine receptor agonists, *Eur. J. Pharmacol.*, 185, 243, 1990.
40. Gardner, C. R., Interpretation of the behavioral effects of benzodiazepine receptor ligands, *Drugs Future*, 14, 51, 1989.
41. Haefely, W. E., Novel anxiolytics that act as partial agonists at benzodiazepine receptors, *Trends Pharmacol. Sci.*, 11, 452, 1990.
42. Knoflach, F., Drescher, U., Scheurer, L., Malherbe, P., Mohler, H., Full and partial agonism displayed by benzodiazepine receptor ligands at recombinant gamma-aminobutyric acidA receptor subtypes, *J. Pharmacol. Exp. Ther.*, 266, 385, 1993.
43. Depoortere, H., Zivkovic, B., Lloyd, K. G., Sanger, D. J., Perrault, G., Langer, S. Z., Bartholini, G., Zolpidem, a novel nonbenzodiazepine hypnotic. I. Neuropharmacological and behavioral effects, *J. Pharmacol. Exp. Ther.*, 237, 649, 1986.
44. Klepner, C. A., Lippa, A. S., Benson, D. I., Sano, M. C., Beer, B., Resolution of two biochemically and pharmacologically distinct benzodiazepine receptors, *Pharmacol. Biochem. Behav.*, 11, 457, 1979.
45. Unnerstall, J. R., Kuhar, M. J., Niehoff, D. L., Palacios, J. M., Benzodiazepine receptors are coupled to a subpopulation of γ-aminobutyric acid (GABA) receptors, evidence from a quantitative autoradiographic study, *J. Pharmacol. Exp. Ther.*, 218, 797, 1981.
46. Maguire, P. A., Davies, M. F., Villar, H. O., Loew, G. H., Evidence for more than two central benzodiazepine receptors in rat spinal cord, *Eur. J. Pharmacol.*, 214, 85, 1992.
47. Levitan, E. S., Schofield, P. R., Burt, D. R., Rhee, L. M., Wisden, W., Kohler, M., Fujita, N., Rodriguez, H. F., Stephenson, A., Darlinson, M. G., Barnard, E. A., Seeburg, P. H., Structural and functional basis for GABA$_A$ receptor heterogeneity, *Nature*, 335, 76, 1988.
48. Davies, M. F., Onaivi, E. S., Chen, S. W., Maguire, P. A., Tsai, N. F., Loew, G. H., Evidence for central benzodiazepine receptor heterogeneity from behavior tests, *Pharmacol. Biochem. Behav.*, 49, 47, 1994.
49. Petersen, E. N., DMCM, a potent convulsive benzodiazepine receptor ligand, *Eur. J. Pharmacol.*, 94, 117, 1983.
50. Skolnick, P., Crawley, J. N., Glowa, J. R., Paul, S. M., β-Carboline-induced anxiety states, *Psychopathology*, 17 (Suppl. 3), 52, 1984.
51. Porter, N. M., Angelotti, T. P., Twyman, R. E., Macdonald, R. L., Kinetic properties of $\alpha_1\beta_1$ γ-aminobutyric acid$_A$ receptor channels expressed in Chinese hamster ovary cells, regulation by pentobarbital and picrotoxin, *Mol. Pharmacol.*, 42, 872, 1992.
52. Malherbe, P., Sigel, E., Baur, R., Persohn, E., Richards, J. G., Mohler, H., Functional characteristics and sites of gene expression of the $\alpha_1\beta_1\gamma_2$ isoform of the rat GABA$_A$ receptor, *J. Neurosci.*, 10, 2330, 1990.
53. Smith, M. C., Riskin, B. J., The clinical use of barbiturates in neurological disorders, *Drugs*, 42, 365, 1991.
54. Olsen, R. W., The GABA postsynaptic membrane receptor–ionophore complex. Site of action of convulsant and anticonvulsant drugs, *Mol. Cell Biochem.*, 39, 261, 1981.
55. Gee, K. W., Lan, N. C., Bolger, M. B., Wieland, S., Belelli, D., Chen, J. S., Pharmacology of a GABA$_A$ receptor coupled steroid recognition site, in *GABAergic Synaptic Transmission*, Biggio, G., Concas, A., Costa, E., Eds., Raven Press, New York, 1992, 111.

56. Concas, A., Santoro, G., Serra, M., Sanna, E., Biggio, G., Neurochemical action of the general anaesthetic propofol on the chloride ion channel coupled with GABA$_A$ receptors, *Brain Res.*, 542, 225, 1991.

57. Twyman, R. E., Macdonald, R. L., Neurosteroid regulation of GABA$_A$ receptor single-channel kinetic properties of mouse spinal cord neurons in culture, *J. Physiol.*, 456, 215, 1992.

58. Puia, G., Santi, M. R., Vicini, S., Pritchett, D. B., Purdy, R. H., Paul, S. M., Seeburg, P. H., Neurosteroids act on recombinant human GABA$_A$ receptors, *Neuron*, 4, 759, 1990.

59. Barker, J. L., Harrison, N. L., Lange, G. D., Owen, D. G., Potentiation of γ-aminobutyric acid-activated chloride conductance by a steroid anaesthetic in cultured rat spinal neurons, *J. Physiol.*, 386, 485, 1987.

60. Harrison, N. L., Majewska, M. D., Harrington, J. W., Barker, J. L., Structure-activity relationships for steroid interaction with the γ-aminobutyric acid$_A$ receptor complex, *J. Pharmacol. Exp. Ther.*, 241, 346, 1987.

61. Turner, D. M., Ransom, R. W., Yang, J. S., Olsen, R. W., Steroid anesthetics and naturally occurring analogs modulate the aminobutyric acid receptor complex at a site distinct from barbiturates, *J. Pharmacol. Exp. Ther.*, 248, 960, 1989.

62. Bitran, D., Purdy, R. H., Kellogg, C. K., Anxiolytic effect of progesterone is associated with increases in cortical allopregnanolone and GABA$_A$ receptor function, *Pharmacol. Biochem. Behav.*, 45, 423, 1993.

63. Franks, N. P., Lieb, W. R., Molecular and cellular mechanisms of general anaesthesia, *Nature*, 367, 607, 1994.

64. McGivern, J., Scholfield, C. N., General anaesthetics and field currents in unclamped, unmyelinated axons of rat olfactory cortex, *Br. J. Pharmacol.*, 101, 2172, 1990.

65. Draguhn, A., Verdoorn, T. A., Ewert, M., Seeburg, P. H., Sakmann, B., Functional and molecular distinction between recombinant rat GABA$_A$ receptor subtypes by Zn^{2+}, *Neuron*, 5, 781, 1990.

66. Smart, T., A novel modulatory binding site for zinc on the GABA$_A$ receptor complex in cultured rat neurons, *J. Physiol.*, 447, 587, 1992.

67. Im, M. S., Hamilton, B. J., Carter, D. B., Im, W. B., Selective potentiation of GABA-mediated Cl⁻ current by lanthanum ion in subtypes of cloned GABA$_A$ receptors, *Neurosci. Lett.*, 144, 165, 1992.

68. Davies, M. F., Maguire, P. A., Loew, G. H., Zinc selectively inhibits flux through benzodiazepine-insensitive γ-aminobutyric acid chloride channels in cortical and cerebellar microsacs, *Mol. Pharmacol.*, 44, 876, 1993.

69. Brooks-Kayal, A. R., Pritchett, D. B., Developmental changes in human γ-aminobutyric acid$_A$ receptor subunit composition. *Ann. Neurol.*, 34, 687, 1993.

70. Beattie, C. E., Siegel, R. E., Developmental cues modulate GABA$_A$ receptor subunit mRNA expression in cultured cerebellar granule neurons, *J. Neurosci.*, 13, 1784, 1993.

71. Laurie, D. J., Wisden, W., Seeburg, P. H., The distribution of thirteen GABA$_A$ receptor subunit mRNAs in the rat brain. III. Embryonic and postnatal development, *J. Neurosci.*, 12, 4151, 1992.

72. Chisholm, J., Kellogg, C., Lippa, A., Development of benzodiazepine binding subtypes in three regions of rat brain, *Brain Res.*, 267, 388, 1983.

73. Mhatre, M. C., Fernandes, G., Ticku, M. K., Aging reduces the mRNA of alpha 1 GABA$_A$ receptor subunit in rat cerebral cortex, *Eur. J. Pharmacol.*, 208, 171, 1991.

74. Kang, I., Miller, L. G., Decreased concentrations following chronic lorazepam administration, *Br. J. Pharmacol.*, 103, 1285, 1991.

75. Montpied, P., Ginns, E. I., Martin, B. M., Roca, D., Farb, D. H., Paul, S. M., γ-Aminobutyric acid (GABA) induces a receptor-mediated reduction in GABA$_A$ receptor α subunit messenger RNAs in embryonic chick neurons in culture, *J. Biol. Chem.*, 266, 6011, 1991.

76. Montpied, P., Morrow, A. L., Karanian, J. W., Ginns, E. I., Martin, B. M., Paul, S. M., Prolonged ethanol inhalation decreases γ-aminobutyric acid$_A$ receptor α subunit mRNAs in the rat cerebral cortex, *Mol. Pharmacol.*, 39, 157, 1991.

77. Shingai, R., Sutherland, M. L., Barnard, E. A., Effects of subunit types of the cloned GABA$_A$ receptor on the response to a neurosteroid, *Eur. J. Pharmacol.*, 206, 77, 1991.

78. Tseng, Y. T., Miyaoka, T., Ho, I. K., Region-specific changes of GABA$_A$ receptors by tolerance to and dependence upon pentobarbital, *Eur. J. Pharmacol.*, 236, 23, 1993.

79. Friedman, L., Gibbs, T. T., Farb, D. H., γ-Aminobutyric acid$_A$ receptor regulation, chronic treatment with pregnanolone uncouples allosteric interactions between steroid and benzodiazepine recognition sites, *Mol. Pharmacol.*, 44, 191, 1993.

80. Leidenheimer, J. N., Machu, T. K., Endo, S., Olsen, R. W., Harris, R. A., Browning, M. D., Cyclic AMP-dependent protein kinase decrease γ-aminobutyric acid$_A$ receptor-mediated ^{36}Cl⁻ uptake by brain microsacs, *J. Neurochem.*, 57, 722, 1991.

81. Stelzer, A., Kay, A. R., Wong, R. K. S., GABA$_A$-receptor function in hippocampal cells is maintained by phosphorylation factors, *Science*, 241, 339, 1988.

82. Marchenko, S. M., Mechanism of modulation of GABA-activated current by internal calcium in rat central neurons, *Brain Res.*, 546, 355, 1991.

83. Bradshaw, D. J., Simmons, M. A., Intracellular cyclic GMP modulates $GABA_A$ receptor function, *Soc. Neurosci. Abstr.*, 19, 1138, 1993.

84. Swope, S. L., Moss, S. J., Blackstone, C. D., Huganir, R. L., Phosphorylation of ligand-gated ion channels, possible mode of synaptic plasticity, *FASEB J.*, 6, 2514, 1992.

85. Lanius, R. A., Pasqualotto, B. A., Shaw, C. A., Age-dependent expression, phosphorylation and function of neurotransmitter receptors, pharmacological implications, *Trends Pharmacol. Sci.*, 14, 403, 1993.

86. Hanson, P. I., Schulman, H., Neuronal Ca^{2+}/calmodulin-dependent protein kinases, *Annu. Rev. Biochem.*, 61, 559, 1992.

87. Sessler, F. M., Mouradian, R. D., Cheng, J. T., Yeh, H. H., Liu, W. M., Waterhouse, B. D., Noradrenergic potentiation of cerebellar Purkinje cell responses to GABA, evidence for mediation through the β-adrenoceptor-coupled cyclic AMP system, *Brain Res.*, 499, 27, 1989.

Chapter **7**

GLUTAMATE RECEPTORS

John F. MacDonald, J. Martin Wojtowicz, and Andrius Baskys

CONTENTS

I. IONOTROPIC GLUTAMATE RECEPTORS

The neurons of mammalian central nervous system (CNS) are responsible for a vast array of behaviors ranging from simple reflexes to cognitive function in humans. Most neuroscientists believe that the connections between neurons, the synapses, play a significant role in the genesis and expression of brain function. In turn, dysfunction of neurons and synapses is likely responsible for the pathogenesis underlying many if not all neurological and psychiatric disorders. Synaptic transmission between neurons is largely via chemical synapses, although some neurons are physically coupled to each other by means of junctions that permit the direct transfer of electrical signals between the cells (electrical synapses). Chemical synapses require the transformation of an electrical signal in the presynaptic neuron into a chemical signal(s) that is subsequently either (1) transduced back to an electrical signal in order to modulate the excitability of the postsynaptic neuron or (2) transformed into a metabolic signal which can modify the activity of enzymes in the postsynaptic cell and potentially lead to changes in the synthesis and/or functions of cellular proteins (genetic modifications). In the inactive or resting state, the presynaptic neuron releases from its terminal a low but steady level of transmitter. The function of this release is unknown, but it likely has a trophic action to maintain the function of the synapse. When the presynaptic terminal is electrically activated by the invasion of an action potential, the probability of transmitter release increases by as much as 200,000 times. The neurotransmitter diffuses across the synaptic cleft and binds to various transmembrane glycoproteins which serve as postsynaptic receptors for the transmitter (see Chapter 1). Some of these receptors are ion channels responsible for generating electrical signals in the postsynaptic neuron while others are coupled to signal transduction molecules which cause complex enzymatic changes in the postsynaptic neuron. The flow of charged ions through channels generates a new electrical signal, which is either excitatory or inhibitory — an excitatory postsynaptic potential (epsp) or an inhibitory postsynaptic potential (ipsp), respectively. Synapses provide the potential for complex processing of information. Initially, it was thought that excitatory transmission between neurons in simple reflex circuits was relatively constant while inhibitory pathways sculpted this activity to form more elaborate and plastic behavior.[1] This hypothesis underestimated the sophistication of excitatory transmission. For example, if afferent presynaptic axons to the hippocampus are stimulated briefly at a high frequency, the epsps recorded in postsynaptic CA1 pyramidal neurons are dramatically enhanced for many hours or even for days. This phenomenon is called long-term potentiation (LTP), and it may form the basis of some kinds of learning.[2]

A large number of endogenous molecules have been identified as central neurotransmitters. However, excitatory transmission in the mammalian CNS is primarily mediated by L-glutamic acid. The earliest indication that this simple amino acid was the key excitatory transmitter came from the work of Hyashi.[3] Hyashi demonstrated that applications of L-glutamate to the surface of the cerebral cortex of dogs caused powerful convulsantlike activity reminiscent of grand mal seizures in humans. These observations were somewhat surprising as this amino acid plays a central role in the intermediate metabolism of most cells and is found in high concentrations in both neurons and glial cells. It did not seem entirely logical that this essential intracellular component could at the same time cause such tremendous excitation when applied exogenously to the CNS. Even more unexpected was the much later discovery that extracellular L-glutamate could be highly toxic to central neurons. Approximately 10 years after these initial experiments, Curtis and colleagues[4,5] examined the excitability of single central mammalian neurons using newly developed microelectrode recording techniques. In addition, drugs could be applied in close proximity to the recorded neuron using the technique of microiontophoresis. Using this technique, they screened an impressive number of potential transmitter candidates and observed that L-glutamic acid and related dicarboxylic amino acids caused a dramatic increase in the firing rates of every neuron they could record from in the CNS. This excitation was associated with a depolarization of the neurons which mimicked the epsp. Ironically, Curtis erroneously concluded that the remarkable ubiquity of excitation by L-glutamate made it very unlikely that it was a neurotransmitter. However, we now recognize that functional glutamate receptors are found on most central neurons, as well as many glia cells, and that L-glutamate plays a major and ubiquitous role in excitatory synaptic function within the mammalian CNS. Other excitatory amino acids, including L-aspartate, some sulfonated analogues of glutamate such as L-homocysteic acid, and the tripeptide NAAG, may also act as excitatory transmitters.[6-8]

The presence of specific postsynaptic glutamate receptors on central neurons has been demonstrated in a number of ways. For example, by using tissue cultured neurons, a high density of receptors has been traced to the subsynaptic membrane, and the presence of labeled receptors (using antibodies, high affinity antagonists, etc.) has been demonstrated on postsynaptic neurons in the CNS. Ionotropic glutamate receptors are also found within a specialized postsynaptic structure located directly beneath the presynaptic terminal called the postsynaptic density. The postsynaptic density is also associated with a number of important enzymes including protein kinases and phosphatases. L-Glutamic acid is the primary excitatory amino acid transmitter, and it is a true mixed agonist. That is, it has high affinity and efficacy at all subtypes of glutamate receptors. Thus, diversity of function is largely determined by presence of multiple subtypes of postsynaptic receptors. L-Aspartate, in contrast, has a reasonably high affinity and selectivity for the N-methyl-D-aspartate (NMDA) receptor subtype, suggesting that it may play a more restricted role in excitatory transmission.

Excitatory amino acid agonists optimally require the presence of a basic amino as well as two acidic groups in molecule. This mimics the dicarboxylic acid structure of L-glutamic acid. The amino group should lie beside the α acid group and not at the second one (the omega group). The backbone chain length should be from three to six atoms in length in order to provide the appropriate separation of negatively charged groups. A wide variation in the omega group is still possible. The affinity of excitatory amino acids characteristically displays a strong dependence upon chirality, and a high degree of flexibility of the glutamate molecule is likely when interacting with its postsynaptic receptors.[8] These receptors can be divided into at least four major categories on the basis of conventional pharmacological, structural, and functional criteria.[9] Glutamate receptors that are ion channels or ionotropic glutamate receptors fall into at least three major groups based upon selectivity of agonists: (1) the NMDA receptor, (2) the (α-amino-3-hydroxy-5-methyl-4-isoxazolepropionic acid, or AMPA, receptor, and (3) a high-affinity kainate receptor (relative to the affinity of kainate for AMPA receptors). In contrast, metabotropic glutamate receptors or mGluRs form the fourth category (see below) and instead act as signal transduction effectors. Recently, a substantial number of glutamate receptor proteins have been cloned using expression techniques. Each homomeric receptor subunit may form a native receptor, or alternatively it may have to be associated with another known, or as yet unknown, receptor subunit. This molecular biological approach has confirmed the existence of each of the glutamate receptor subtypes.[10]

The majority of glutamate-mediated epsps are biphasic in time course consisting of an initial rapid peak followed by a secondary and slowly decaying component. This dual aspect of the epsp arises because released glutamate initiates the rapid activation and desensitization of AMPA receptors followed by a much slower activation and longer-lasting action of NMDA receptors. Thus, the currents which underlie the epsp consist of a fast ($\tau = 3$ to 5 ms) and slow ($\tau = 50$ to 200 ms) component generated through activation of AMPA and NMDA channels, respectively (see Chapter 1).[11-13] Although not well established, kainate receptors may also contribute to the rapid phase of the epsp provided they are actually present at the synapse.[14] Whether or not mGluRs contribute directly to the epsps is not yet clear.

A. The NMDA Receptor

The NMDA receptor is unique among ligand-gated cation channels in that it is permeable to Ca^{2+}, while it is blocked by Mg^{2+}.[15] Mg^{2+} ions cause a voltage-dependent block of the channel by interacting with a binding site thought to be located within the channel pore.[16] In the presence of physiological concentrations of Mg^{2+} (1 to 3 mM), this blockade underlies the voltage dependence of the NMDA receptor–mediated component of epsps. When single epsps are evoked, the NMDA component is largely blocked by extracellular Mg^{2+}. It is not until a series of summating epsps produce a larger depolarization that the Mg^{2+} blockade is gradually relieved. The resulting influx of the second messenger Ca^{2+} through NMDA channels then stimulates a variety of calcium-dependent enzymes, including protein kinase C, calcium, calmodulin-dependent kinases, calcium-dependent proteases, phosphatases, etc. Such a mechanism is thought to underlie the induction of LTP. The influx of Ca^{2+} also contributes to the fade of NMDA responses in the continued presence of transmitter (desensitization or inactivation). However, desensitization shows an extremely slow onset relative to AMPA and kainate receptors, and several different processes are known to contribute to this phenomenon. When glutamate is rapidly applied to these channels, they demonstrate, relative to AMPA channels, very slow kinetics of channel activation and they tend to open in long-lasting bursts.[13] These basic properties account for the slow onset and offset of the NMDA-mediated component of the epsp.

The NMDA channel is also blocked by a variety of arylcyclohexylamines, including phencyclidine (PCP), dizocilpine (MK801), TCP, ketamine, and cyclazocine. These compounds cause a voltage-dependent block and likely interact at a site corresponding to or at least overlapping with the Mg^{2+} site.[17] These agents also induce "psychotic" states and do so at serum concentrations which suggest a selective blockade of this receptor. Therefore, it has been proposed that agonists for NMDA receptors might provide a potential new class of antipsychotics.[18] Another unique feature of the NMDA receptor is that it possesses a binding site for the inhibitory transmitter glycine, although this site is entirely unrelated to inhibitory transmission. For example, glycine has a much higher affinity for the NMDA receptor, and this site is pharmacologically distinct from the inhibitory one (e.g., strychnine has no effect on glycine binding to the NMDA receptor). Occupancy of this glycine site by an agonist appears to be an absolute requirement. Simply put, if there is no glycine present, NMDA channels will not open. The CNS appears to have saturating or near saturating concentrations of glycine (about 1 to 3 μM) in the synaptic cleft. At this time there is little evidence that changes in levels of glycine modulate NMDA receptor function, although transporters for glycine have been identified in the CNS, suggesting that removal of glycine from the cleft might provide such a mechanism. A number of competitive antagonists of the glycine site have been identified, including 7-chloro-kyrunenate, and this compound strongly depresses NMDA responses. Glycine has two distinct actions on NMDA receptors: (1) it greatly potentiates peak NMDA currents, and (2) it greatly speeds the rate of recovery from desensitization.[16]

Unlike other types of glutamate receptors, high-affinity and highly selective competitive antagonists were identified for NMDA receptors over 10 years ago (e.g., they likely bind at or near the binding site for NMDA). These competitive antagonists displace the binding of glutamate to subsynaptic densities which contain the native receptors. Among the most common antagonists are AP5 (2-amino-5-phosphono-valeric acid) and CPP [3-((+))-2-carboxypiperazine-4-yl)-propyl-*l*-phosphonic acid]. The most potent competitive antagonists are longer chained acidic amino acids. A D configuration and interacidic group chain length of four to six atoms is the preferred conformation, and the order of potency for substitutions at the omega or nonaminated terminal is $PO_3H_2 > CO_2H > SO_3H$.[8]

Endogenous polyamines such as spermine can both enhance and block NMDA receptors.[19] The blockade is voltage dependent and resembles that observed with Mg^{2+}, suggesting that it is because of channel block (binding to site within the channel pore). The enhancement occurs at lower concentrations than the block, and it has not been observed in all cells studied, likely because of variations in the composition of expressed subunits. This enhancement is manifest in two ways: (1) a glycine-insensitive component — spermine enhances NMDA currents in saturating concentrations of glycine and (2) a glycine-sensitive component — spermine enhances the affinity of the NMDA receptor for glycine and, therefore, reduces glycine-sensitive desensitization.[19] These effects are mimicked by Mg^{2+} and Ca^{2+},[20] and likely by protons. Reducing and oxidization agents also modulate NMDA channel function likely by acting on cysteine residues in one of the NMDA receptor subunits.[21] No physiological role has yet been demonstrated for these actions, but such evidence suggests that a wide variety of positively charged ions and drugs can regulate NMDA receptor function at this site or sites.

The NMDA receptor is likely a pentameric structure composed of two classes of subunits which combine to form the native receptor. [16,22,23] The first subcategory, NMDAR1 (rat clones with nine splice variants) or NMDARε (mouse clones), responds with relatively small currents when expressed as a heteromer, but combinations including a NMDAR2 subunit (rat clones: NMDAR2A,B,C,D) or NMDARε (mouse clones: NMDARε1,ε2,ε3,ε4) produce large and robust responses. NMDAR2 subunits show low homology with NMDAR1 subunits, and homomeric NMDAR2 subunits cannot form functional channels. Recombinant NMDA receptors demonstrate all of the pharmacological properties of the native receptors although different combinations of subunits differ quantitatively in their properties. For example, different combinations of subunits show varying degrees of permeability to Ca^{2+} and sensitivity to blockade by Mg^{2+}. Recent evidence suggests that ionotropic glutamate receptors possess three transmembrane regions with the carboxy terminus located intracellularly.[24,25] The amino terminus is extracellular and contributes to the agonist recognition site while the region previously named the TM3 region is thought to form a extracellular loop involved in the binding of agonist and, in the case of NMDA receptors, the binding of its co-agonist, glycine.[26] The region originally termed TM2 is thought to form a short blind intracellular loop, and this region of each class of subunit is believed to contribute to lining each channel pore. Several features of the channel pores (e.g., Ca^{2+} permeability and Mg^{2+} block in the case of NMDA receptors) are influenced by mutations made to specific sites within the TM2 region.

B. The AMPA Receptor

AMPA binds to this class of receptors with a high affinity (nanomolar range). However, kainate also binds with a relatively low affinity (high micromolar range) but is nevertheless very effective at activating this receptor. A number of relatively specific competitive antagonists have also been identified, including quinoxalinediones such as CNQX and NBQX. These antagonists have little effect on NMDA receptors, although in higher concentrations they can block the glycine site. They do, however, block high-affinity kainate receptors. In contrast, AMPA receptors, but not kainate receptors, are selectively blocked by GYKI 52466 (a 2,3-benzodiazepine).[27,28] AMPA receptors incorporate a nonspecific cation channel, and some receptors also demonstrate low calcium permeability. For example, in cultures of hippocampal neurons kainate currents can be classified into two groups — those with little calcium permeability and those with significant permeability.[29,30] The inclusion of one subunit, the GluR2 subunit, eliminates Ca^{2+} permeability, and most neurons apparently possess native receptors containing this subunit. Four major AMPA receptor subunits have been cloned (GluR1,2,3,4 or GluRA,B,C,D) and a number of alternatively spliced variants identified.[23]

Applications of glutamate or AMPA to central neurons demonstrate that the response is characterized by an extremely rapid rate of desensitization. This rapid desensitization of AMPA receptors can be dramatically slowed by treatment with a number of compounds, including the lectins concancavalin A and wheat germ agglutinin, by the nootropic drug aniracetam (see Chapter 13 of this volume), and by benzothiazides such as diazoxide and cyclothiazide.[27,28] In at least some cases the decay of the rapid phase of the epsp can also be significantly slowed, suggesting that decay of epsps is determined at least in part by AMPA receptor desensitization.[31-33] This contrasts with the much slower desensitization of NMDA receptors. It has also been suggested that agents that reduced AMPA receptor desensitization may enhance learning capacity and memory. AMPA receptor function can also be enhanced by low concentrations of Zn^{2+}, by sulfhydryl agents, and depressed by barbiturates such as pentobarbital. They are also antagonized by a number of toxins from spiders and wasps (e.g., philanthotoxin).[23]

C. The Kainate Receptor

It has been known for some time that high-affinity binding sites for kainate exist in the CNS.[34] Binding data give values for affinities in the low nanomolar range whereas AMPA demonstrates little affinity for such sites. This binding can be competitively antagonized by quinoxalinediones such as CNQX (although with slightly lower potency than for AMPA/kainate receptors). A distinctive distribution of these kainate-binding sites has also been demonstrated within the CNS, and a number of kainate-binding proteins have been cloned, further suggesting a functional role for these sites. Recent evidence has demonstrated that at least some of these sites, as well as functional kainate channels, are found on glia cells of the cerebellum,[35] but their function is unknown. The existence of pure kainate receptors in the CNS has been surmised from observations that low (nanomolar) concentrations of kainate can cause a powerful excitation of hippocampal neurons. A pure population of kainate receptors has also been identified on mammalian C fibers and in dorsal root ganglia neurons (DRGs).[36] More recently, it has been shown that cultured hippocampal neurons possess high-affinity responses to kainate which rapidly desensitize. Desensitization of these kainate receptors is insensitive to the benzothiazides which block the desensitization of AMPA receptors, and they are not blocked by GYKI 52466 and related compounds.[28,37,38] These kainate responses are, however, selectively blocked by NS102.[39] These high-affinity kainate channels may contribute to the rapid phase of at least some epsps in cultured hippocampal neurons.[14,38] Cloned receptors with similar functional and pharmacological properties have also been identified (GluR4,5,6 and KA1,A2) although the contribution of these subunits to the native receptors has yet to be delineated.

II. METABOTROPIC GLUTAMATE RECEPTORS

In 1979 Eccles and McGeer[40] introduced the concept of "metabotropic" neurotransmission in the brain. In their view, an essential requirement for a neurotransmitter to be considered metabotropic was its indirect action on membrane conductances mediated by a chemical reaction or a series of reactions in the postsynaptic cells, as opposed to direct opening of ion channels by "ionotropic" transmission. This implies that, at least in theory, any neurotransmitter can have ionotropic and metabotropic modes of action depending on its respective interaction with ionotropic or metabotropic receptors.

The concept of metabotropic glutamate receptors has centered around the observation that excitatory amino acids stimulated the phosphoinositol breakdown cascade,[41] leading to the production of several intracellular messengers including 1,3,5-inositol trisphosphate and 1,2-diacylglycerol.[42] For example, in cultured striatal neurons quisqualate, L-glutamate, NMDA, and kainate stimulated phosphoinositide (PI) hydrolysis with the following order of potency: quisqualate > glutamate > NMDA = kainate.[41] Stimulation of PI hydrolysis by ibotenate also occurred in slices of corpus striatum, frontal cortex, hypothalamus,[43] and in spinal cord slices,[44] although the extent of it was highest in the hippocampus. NMDA and kainate were found effective in some studies[41,45] but not in others.[43,44] Quinolinate was ineffective.[44] A significant step in understanding metabotropic glutamate receptors was the discovery that the earlier identified glutamate agonist with an "unusual" pharmacological profile, *trans*-ACPD, was a selective metabotropic glutamate receptor agonist that potently stimulated PI hydrolysis.[46-48] However, studies of metabotropic receptors have been continuously hampered by the lack of specific antagonists. Recently, it has been reported that newly synthesized phenylglycine derivatives, (S)-4-carboxyphenylglycine[49] and (RS)-alpha-methyl-4-carboxyphenylglycine (MCPG), at high concentrations could act as metabotropic antagonists.[50] The questions of whether or not these compounds act as competitive antagonists and, if so, then at which receptor subtype (see below) await further clarification.

Molecular studies confirmed the existence of a metabotropic glutamate receptor family consisting of at least seven distinct receptor subtypes, termed mGluR1 to mGluR7 (Table 1) (for review see Reference 51). It appeared that metabotropic glutamate receptors are larger than other G protein–coupled receptors and have a common structural architecture consisting of seven transmembrane domains with a large sequence domain at the N terminal and a C terminal domain longer than that of any known receptor with seven transmembrane domains. The mGluRs have sequence similarity with each other of approximately 40%, but no sequence homology is observed with any other members of the G protein–coupled receptor family.

From these studies it became apparent that a metabotropic glutamate receptor that was originally hypothesized as mediating PI hydrolysis in response to glutamate agonist stimulation possesses several subtypes with different agonist selectivity and specific distribution in the brain. Expression of these subtypes in artificial systems, such as frog oocytes (see Chapter 1), revealed that they are linked to two main second messenger systems: stimulation of PI hydrolysis and inhibition of forskolin-stimulated cAMP production. It remains to be seen whether or not these interactions represent "true" second messenger links inherent to these receptors in native cells.

TABLE 1.
Metabotropic Glutamate Receptor Subtypes, Distribution in the CNS and Second Messenger Coupling

Subtype	Length	mRNA Distribution	Coupling	Pharmacology
mGluR1α	1199	Cerebellum, olfactory bulb, thalamus, diencephalon, mesencephalon, brain stem, cerebral cortex, striatum	PI hydrolysis	quis>ibo>glu>BMAA>*trans*-ACPD MCPG-antagonist
mGluR1β	906	Hippocampus, cerebral cortex, striatum	PI hydrolysis	
mGluR1c	897	Similar to mGluR1α	PI hydrolysis	
mGluR2	872	Olfactory bulb, cerebral cortex	Inhibition of cAMP	glu>*trans*-ACPD>ibo>quis
mGluR3	879	Glial cells of corpus callosum, anterior commisurae, neurons of cerebral cortex	Inhibition of cAMP	glu>*trans*-ACPD>ibo>quis
mGluR4	912	Granule cells of cerebellum, olfactory bulb, thalamus, pontine nucleus	Inhibition of cAMP	L-AP4>L-SOP>glu
mGluR5	1171	Striatum, hippocampus, olfactory bulb, olfactory tubercle, accumbens, lateral septum	PI hydrolysis	quis>glu ≥ ibo>*trans*-ACPD
mGluR6	871	Inner nuclear layer of the retina	Inhibition of cAMP	L-AP4>L-SOP>glu>D-AP4
mGluR7	915	Thalamus, neocortex, hypothalamus, hippocampus, olfactory bulb	Inhibition of cAMP	L-AP4>L-SOP>glu>quis>1S,3R-ACPD

Abbreviations: quis, quisqualate; ibo, ibotenate; glu, L-glutamate; L-AP4, L-2-amino-4-phosphonobutric acid; L-SOP, L-serine-O-phosphate; *trans*-ACPD, *trans*-1-aminocyclopentane-1,3-dicarboxylic acid; BMAA, β-N-methylamino-L-alanine; MCPG, (RS)-alpha-methyl-4-carboxyphenylglycine.

There have been numerous efforts to understand the functional role of metabotropic receptors. Over 15 different response types to putative metabotropic agonists were documented in neurons by using electrophysiological techniques (see Reference 51 for review). The absence of selective antagonists makes the identification of metabotropic receptor function a very difficult task. It appears that metabotropic glutamate receptors mediate essentially two types of responses. Suppression of a strong inhibitory current known as a Ca-dependent K^+ current increases membrane excitability and facilitates continuous neuronal firing. This action can be seen as neuromodulatory and can create favorable conditions for synaptic potentiation, e.g., LTP (see Chapter 1 of this volume), to take place. Presynaptic inhibition of excitatory synaptic neurotransmission is another prominent feature of metabotropic receptor agonists. In the hippocampus it shows a strong negative correlation with the age of the animal, being most pronounced in early postnatal (approximately 12 to 25 days) preparations and virtually absent in slices prepared from adult (3-month-old) rats.[52] These findings correspond well with stronger PI stimulation by metabotropic agonists observed in younger animals[43,53] and indirectly suggest that PI hydrolysis in the presynaptic terminals may be involved in the mechanism of the metabotropic receptor–induced regulation of synaptic transmission. The physiological significance of this strong inhibitory effect on excitatory synaptic transmission could probably be best understood in light of the fact that postsynaptic inhibition in the hippocampus may be weak or insufficient at the early stages of development.[54] The metabotropic receptor–mediated inhibition at that stage may serve as a temporary negative feedback mechanism, reducing synaptic excitation and preventing nerve tissue from seizures and damage.

It has been well established that glutamate receptors are involved in neurotoxicity and hypoxic ischemic brain damage.[55,56] It is likely that, as regulators of excitatory synaptic transmission, metabotropic glutamate receptors play an important role in the underlying pathological processes. However, it is not at all clear at present what this role is. There are a number of studies indicating that stimulation of metabotropic glutamate agonists can be neuroprotective.[57-59] Other reports suggest the opposite: that metabotropic receptor activation leads to seizures and neuronal damage.[60,61] It is therefore appropriate to discuss processes involved in excitotoxic nerve cell death in greater detail.

III. EXCITOTOXICITY

Paradoxically, nerve cells "at rest" are highly electrically charged by uneven distributions of ions across the cellular membrane. The ionic gradients are continuously being maintained by ionic pumps and carriers. Thus, given an adequate supply of energy, the concentrations of Na^+, K^+, Cl^-, Ca^{2+}, HCO^-, and H^+ ions are maintained inside the cells at levels appropriate to support the metabolic processes of the cell. It follows that at rest the nerve cells are in a state of dynamic homeostasis which is essential for life. The homoestasis is self-adjusting; that is, enzymes which produce energy and pump the ions require an appropriate ionic milieu to function. A brief disruption of this steady-state situation can be tolerated by nerve cells because of an elaborate system of "buffers" which can temporarily absorb disruption, such as a change in the concentration of an ion, but in the long term there is a limit beyond which the homeostasis cannot be pushed without fatal results.

The toxic effects of the excitatory amino acids can be understood in terms of their disruptive effects on the homeostasis and, in particular, by their actions on ionic fluxes. Under normal circumstances the excitatory amino acids serve as efficient transmitters by taking advantage of the charged state of the neurons and by opening certain ionic channels. By doing so they dissipate the energy accumulated in the ionic gradients. Each time a sodium or a calcium channel is opened, the cell loses some of its homeostatic energy and needs to compensate for the loss as soon as possible.

The ionic fluxes are, of course, necessary for electrical signaling in the nervous system, but excessive changes in the ionic gradients have an obvious potential for cell destruction. During intense electrical activity nerve cells can be seen as performing a delicate balancing act between life and death.

Although the concept of excitotoxicity is clear, its mechanistic interpretation still requires further studies. A large body of evidence suggests that excessive increase in the intracellular level of calcium is a trigger for subsequent cell death.[62] Calcium is a particularly powerful indicator of the disruption in homeostasis because of its large concentration gradient across the cellular membrane. Normal levels of calcium in nerve cells are thought to be around 0.1 μM, i.e., 20,000 times less than the external concentration of 2 mM in the extracellular fluid. Consequently, any disruption of delicate metabolic systems responsible for buffering and pumping of these ions can cause relatively large changes in the internal concentration. However, recent findings indicate that large changes in the intracellular levels of calcium are not lethal unless calcium enters through specific ionic channels.

For example, experiments on excitotoxic effects of calcium in cultured neurons show that increases of intracellular calcium caused by calcium influx through voltage-dependent channels are less lethal than comparable increases caused by glutamate-gated calcium channels.[63] In particular, glutamate exerts its toxic effects through activation of the NMDA type of channels known to have relatively high permeability to calcium ions. It has been proposed that the lethal effect of calcium entering through the NMDA channels may be because of colocalization of these channels with particularly sensitive intracellular targets.

In addition to the locus of action, the neurotoxic effects of calcium may depend on some other poorly understood factors. This is evident from the recent findings on the physiological effects of metabotropic glutamate receptors. Certain subtypes of this class of receptors are linked to the inositol phosphate second messenger cascades and would be expected to increase the intracellular levels of calcium[64] and possibly lead to cell death. Although some studies support the idea that metabotropic agonists are neurotoxic,[60,61] there have been reports that they can act as neuroprotective agents.[57,65-67] Such neuroprotective action is particularly puzzling in view of enhanced NMDA responses following metabotropic receptor activation.[68] The exact intracellular targets for calcium toxicity have not yet been identified, but a variety of enzymes, such as calcium-sensitive kinases, phosphatases, phospholipases, and proteases, are known to be activated during the NMDA channel activity and could serve as targets. Possible excessive activation of these enzymes may lead to excitotoxic cell death or turn on the programmed cell death, or apoptosis,[78] machinery.

These *in vitro* results should be extended with caution to the *in vivo* situation since the homeostasis in isolated cultured neurons may be considerably altered because of artificial incubation conditions and the lack of normal interactions with other neurons, glia, and blood circulation. Nevertheless, the excitotoxic neuronal death does occur *in vivo* and has been implicated in the ischemic/anoxic cell death.[62] Large increases in the extracellular concentrations of excitatory amino acids, glutamate and aspartate, have been observed following cerebral ischemia (see Reference 69 for review), in epilepsy,[70] and following a head injury.[71] One important difference between the *in vitro* and *in vivo* excitotoxic effects is a much more prolonged time course of the effects in the latter case. This delay could be due to the time required for synthesis of new channel proteins or expression of latent channels triggered by the initial glutamate insult. Such phenomena have been reported in a number of studies. For example, in cultured neurons repeated applications of glutamate produced enhancement of calcium currents. This increase was suppressed by cycloheximide, a blocker of protein synthesis.[72] In this case, however, the addition of KCl as a depolarizing agent mimicked the effects of glutamate, unlike in the excitotoxicity study[63] discussed above. Nevertheless, a blocker of NMDA receptors, APV, blocked the enhancement of calcium currents caused by glutamate and KCl, suggesting specific involvement of NMDA receptors. Hence, even though the link between glutamate and anoxic injury is tentative, the majority of experimental strategies for neuroprotection during anoxia/ischemia involve a blockade of calcium channels or NMDA channels which are likely to be opened by the excessive concentration of the excitatory amino acids in the extracellular space.[73]

In addition to possible localized excitotoxic cell death during focal ischemia, such as stroke, the lethal effects of amino acid transmitters may contribute to a variety of neurodegenerative and neurological diseases. These include Huntington's disease (HD), Alzheimer's disease, Parkinson's disease, epilepsy, and some others.[74,75]

In case of the HD, for example, the evidence rests largely on the ability of the NMDA receptor agonists to mimic neuropathology of HD patients. This is characterized by neuronal loss in the striatum.[76] These findings, combined with a genetic defect resulting in overexpression of an abnormal protein, lead to a tentative hypothesis of HD pathology. According to this hypothesis, the abnormal protein (called huntingtin[77]) is overexpressed in some striatal neurons or glia. The role of this protein is not known at present, but it is thought to possess multiple glutamine residues which could interfere with normal metabolism of the neurotransmitter glutamate. Possible excess of glutamate (or deficiency of an inhibitory transmitter GABA) could, in turn, lead to degeneration of some neurons in the striatum. It has been determined that certain GABAergic neurons are particularly vulnerable to degeneration during early stages of HD. Why this process occurs only in the striatum and affects only certain types of neurons remains a mystery.

In summary, research of excitatory amino acid transmitters and receptors is rapidly advancing and holds promise of novel pharmacological approaches to treat a whole series of neuropsychiatric conditions for which no treatment is currently available.

REFERENCES

1. Eccles, J. C., *The Physiology of Synapses*, Springer-Verlag, Berlin, 1964, 1-316.
2. Bliss, T. V. P. and Collingridge, G. L., A synaptic model of memory: long-term potentiation in the hippocampus, *Nature*, 361, 31, 1993.
3. Hyashi, T., Effects of sodium glutamate on the nervous system, *Keio J. Med.*, 3, 183, 1954.
4. Curtis, D. R., Phillis, J. W., and Watkins, J. C., Chemical excitation of spinal neurons, *Nature*, 183, 611, 1959.
5. Curtis, D. R. and Watkins, J. C., Acidic amino acids with strong excitatory actions on mammalian neurons, *J. Physiol.*, 166, 1, 1960.
6. Foster, A. C. and Fagg, G. E., Acidic amino acid binding sites in mammalian neuronal membranes: their characteristics and relationship to synaptic receptors, *Brain Res. Rev.*, 7, 103, 1984.
7. Cotman, C. W., Monaghan, D. T., and Ganong, A. H., Excitatory amino acid neurotransmission: NMDA receptors and Hebb-type synaptic plasticity, *Annu. Rev. Neurosci.*, 11, 61, 1988.
8. Watkins, J. C., Krogsgaard-Larsen, P., and Honore, T., Structure activity relationships in the development of excitatory amino acid receptor antagonists and competitive antagonists, *Trends Pharmacol. Sci.*, 11, 25, 1990.

9. Gasic, G. P. and Hollmann, M., Molecular neurobiology of glutamate receptors, *Annu. Rev. Physiol.*, 54, 507, 1992.

10. Seeburg, P.H., The TIPS/TINS lecture: The molecular biology of mammalian glutamate receptor channels, *Trends Pharmacol. Sci.*, 14, 297, 1993.

11. Clark, K. A. and Collingridge, G. L., Synaptic potentiation of dual-component excitatory postsynaptic currents in the rat hippocampus, *J. Physiol.*, 482, 39, 1995.

12. Clements, J. D., Lester, R. A. J., Tong, G., Jahr, C. E., and Westbrook, G. L., The time course of glutamate in the synaptic cleft, *Science*, 258, 1498, 1992.

13. Lester, R. A., Clements, J. D., Westbrook, G. L., and Jahr, C. E., Channel kinetics determine the time course of NMDA receptor-mediated synaptic currents, *Nature*, 346, 565, 1990.

14. Lerma, J., Paternain, A. V., Naranjo, J. R., and Mellström, B., Functional kainate-selective glutamate receptors in cultured hippocampal neurons, *Proc. Natl. Acad. Sci. U.S.A.*, 90, 11688, 1993.

15. Ascher, P., Divalent cations and the NMDA channel, *Biomed. Res.*, 9, Suppl. 2, 31, 1988.

16. McBain, C. J. and Mayer, M. L., N-Methyl-D-aspartic acid receptor structure and function, *Physiol. Rev.*, 74, 723, 1994.

17. MacDonald, J. F. and Nowak, L. M., Mechanisms of blockade of excitatory amino acid receptor channels, *Trends Neurosci.*, 11, 167, 1990.

18. Toru, M., Kurumaji, A., and Ishimaru, M., Excitatory amino acids: implications for psychiatric disorders research, *Life Sci.*, 55, 1683, 1994.

19. Benveniste, M. and Mayer, M. L., Multiple effects of spermine on N-methyl-D-aspartic acid receptor responses of rat cultured hippocampal neurons, *J. Physiol.*, 464, 131, 1993.

20. Gu, Y. and Huang, L.-Y. M., Modulation of glycine affinity for NMDA receptors by extracellular Ca^{2+} in trigeminal neurons, *J. Neurosci.*, 14, 4561, 1994.

21. Sullivan, J. M., Traynelis, S. F., Chen, H.-S. V., Escobar, W., Heinemann, S. F., and Lipton, S. A., Identification of two cysteine residues that are required for redox modulation of the NMDA subtype of glutamate receptor, *Neuron*, 13, 929, 1994.

22. Nakanishi, S. and Masu, M., Molecular diversity and functions of glutamate receptors, *Annu. Rev. Biophys. Biomol. Struct.*, 23, 319, 1994.

23. Hollmann, M. and Heinemann, S., Cloned glutamate receptors, *Annu. Rev. Neurosci.*, 17, 31, 1994.

24. Hollmann, M., Maron, C., and Heinemann, S., N-glycosylation site tagging suggests a three transmembrane domain topology for the glutamate receptor GluR1, *Neuron*, 13, 1331, 1994.

25. Wo, Z. G. and Oswald, R. E., Transmembrane topology of two kainate receptor subunits revealed by N-glycosylation, *Proc. Natl. Acad. Sci. U.S.A.*, 91, 7154, 1994.

26. Kuryatov, A., Laube, B., Betz, H., and Kuhse, J., Mutational analysis of the glycine-binding site of the NMDA receptor: structural similarity with bacterial amino acid-binding proteins, *Neuron*, 12, 1291, 1994.

27. Wong, L. A. and Mayer, M. L., Differential modulation by cyclothiazide and concanavalin A of desensitization at native α-amino-3-hydroxy-5-methyl-4-isoxazolepropionic acid- and kainate-preferring glutamate receptors, *Mol. Pharmacol.*, 44, 504, 1993.

28. Partin, K. M., Patneau, D. K., and Mayer, M. L., Cyclothiazide differentially modulates desensitization of α-amino-3-hydroxy-5-methyl-4-isoxazolepropionic acid receptor splice variants, *Mol. Pharmacol.*, 46, 129, 1994.

29. Ozawa, S., Iino, M., and Tsuzuki, K., Two types of kainate response in cultured rat hippocampal neurons, *J. Neurophysiol.*, 66, 2, 1991.

30. Ozawa, S. and Iino, M., Two distinct types of AMPA responses in cultured rat hippocampal neurons, *Neurosci. Lett.*, 155, 187, 1993.

31. Yamada, K. A. and Tang, C.-M., Benzothiadiazides inhibit rapid glutamate receptor desensitization and enhance glutamatergic synaptic currents, *J. Neurosci.*, 13, 3904, 1993.

32. Rammes, G., Parsons, C., Müller, W., and Swandulla, D., Modulation of fast excitatory synaptic transmission by cyclothiazide and GYKI 52466 in the rat hippocampus, *Neurosci. Lett.*, 175, 21, 1994.

33. Tang, C.-M., Shi, Q.-Y., Katchman, A., and Lynch, G., Modulation of the time course of fast EPSCs and glutamate channel kinetics by aniracetam, *Science*, 254, 288, 1991.

34. Henley, J. M., Kainate-binding proteins: phylogeny, structures and possible functions, *Trends Pharmacol. Sci.*, 15, 182, 1994.

35. Burnashev, N., Khodorova, A., Jonas, P., et al., Calcium-permeable AMPA-kainate receptors in fusiform cerebellar glial cells, *Science*, 256, 1566, 1992.

36. Huettner, J. E., Glutamate receptor channels in rat DRG neurons: activation by kainate and quisqualate and blockade of desensitization by ConA., *Neuron*, 5, 255, 1990.

37. Palmer, A. J. and Lodge, D., Cyclothiazide reverses AMPA receptor antagonism of the 2,3-benzodiazepine, GYKI 53655, *Eur. J. Pharmacol. Mol. Pharmacol.*, 244, 193, 1993.

38. Patneau, D. K., Vyklicky, L., Jr., and Mayer, M. L., Hippocampal neurons exhibit cyclothiazide-sensitive rapidly desensitizing responses to kainate, *J. Neurosci.*, 13, 3496, 1993.

39. Verdoorn, T. A., Johansen, T. H., Drejer, J., and Nielsen, E. O., Selective block of recombinant gluR6 receptors by NS-102, a novel non-NMDA receptor antagonist, *Eur. J. Pharmacol. Mol. Pharmacol.*, 269, 43, 1994.

40. Eccles, J. C. and McGeer, P. L., Ionotropic and metabotropic neurotransmission, *Trends Neurosci.*, 2, 39, 1979.

41. Sladeczek, F., Pin, J.-P., Recasens, M., Bockaert, J., and Weiss, S., Glutamate stimulates inositol phosphate formation in striatal neurons, *Nature*, 317, 717, 1985.

42. Berridge, M. J., Inositol trisphosphate and diacylglycerol as second messengers, *Biochem. J.*, 220, 345, 1984.

43. Nicoletti, F., Iadarola, M. J., Wroblewski, J. T., and Costa, E., Excitatory amino acid recognition sites coupled with inositol phospholipid metabolism: developmental changes and interaction with α_1-adrenoreceptors, *Proc. Natl. Acad. Sci. U.S.A.*, 83, 1931, 1986.

44. Nicoletti, F., Meek, J. L., Iadarola, M. J., Chuang, D. M., Roth, B. L., and Costa, E., Coupling of inositol phospholipid metabolism with excitatory amino acid recognition sites, *J. Neurochem.*, 46, 40, 1986.

45. Godfrey, P. P., Wilkins, C. J., Tyler, W., and Watson, S. P., Stimulatory and inhibitory actions of excitatory amino acids on inositol phospholipid metabolism in rat cerebral cortex, *Br. J. Pharmacol.*, 95, 131, 1988.

46. Desai, M. A. and Conn, P. J., Selective activation of phosphoinositide hydrolysis by a rigid analogue of glutamate, *Neurosci. Lett.*, 109, 157, 1990.

47. Palmer, E., Monaghan, D. T., and Cotman, C. W., *trans*-ACPD, a selective agonist of the phospho-inositide-coupled excitatory amino acid receptor, *Eur. J. Pharmacol.*, 166, 585, 1989.

48. Manzoni, O., Prezeau, L., Rassendren, F. A., Sladeczek, F., Curry, K., and Bockaert, J., Both enantiomers of 1-aminocyclopentyl-1,3-dicarboxylate are full agonists of metabotropic glutamate receptors coupled to phospholipase C, *Mol. Pharmacol.*, 42, 322, 1992.

49. Birse, E. F., Eaton, S. A., Jane, D. E., Jones, P. L., Porter, R. H., Pook, P. C., Sunter, D. C., Udvarhelyi, P. M., Wharton, B., Roberts, P. J., et al., Phenylglycine derivatives as new pharmacological tools for investigating the role of metabotropic glutamate receptors in the central nervous system, *Neuroscience*, 52, 481, 1993.

50. Jane, D. E., Jones, P. L. St. J., Pook, P. C.-K., Salt, T. E., Sunter, D. C., and Watkins, J. C., Stereospecific antagonism by (+)-α-methyl-4-carboxyphenylglycine (MCPG) of (1S,3R)-ACPD-induced effects in neonatal rat motoneurons and rat thalamic neurons, *Neuropharmacology*, 32, 725, 1993.

51. Baskys, A., *Metabotropic Glutamate Receptors*, (Medical Intelligence Unit series), R. G. Landes Company, Austin, 1994, 23.

52. Baskys, A. and Malenka, R. C., Agonists at metabotropic glutamate receptors presynaptically inhibit EPSCs in neonatal rat hippocampus, *J. Physiol.*, 444, 687, 1991.

53. Palmer, E., Nangel-Taylor, K., Krause, J., Roxas, A., and Cotman, C., Changes in excitatory amino acid modulation of phosphoinositide metabolism during development, *Dev. Brain Res.*, 51, 132, 1990.

54. Muller, D., Oliver, M., and Lynch, G., Developmental changes in synaptic properties in hippocampus of neonatal rats, *Dev. Brain Res.*, 49, 105, 1989.

55. Choi, D. W. and Rothman, S. M., The role of glutamate neurotoxicity in hypoxic ischemic neuronal death, *Annu. Rev. Neurosci.*, 13, 171, 1990.

56. Barks, J. D. E. and Silverstein, F. S., Excitatory amino acids contribute to the pathogenesis of perinatal hypoxic-ischemic brain injury, *Brain Pathol.*, 2, 235, 1992.

57. Chiamulera, C., Albertini, P., Valerio, E., and Reggiani, A., Activation of metabotropic receptors has a neuroprotective effect in a rodent model of focal ischemia, *Eur. J. Pharmacol.*, 216, 335, 1992.

58. Opitz, T. and Reyman, K. G., (1S,3R)-ACPD protects synaptic transmission from hypoxia in hippocampal slices, *Neuropharmacology*, 32, 103, 1993.

59. Siliprandi, R., Lipartiti, M., Fadda, E., Sautter, J., and Manev, H., Activation of the glutamate metabotropic receptor protects retina against N-methyl-D-aspartate toxicity, *Eur. J. Pharmacol.*, 219, 173, 1992.

60. McDonald, J. W. and Schoepp, D. D., The metabotropic excitatory amino acid receptor agonist 1S,3R-ACPD selectively potentiates N-methyl-D-aspartate-induced brain injury, *Eur. J. Pharmacol.*, 215, 353, 1992.

61. Sacaan, A. I. and Schoepp, D. D., Activation of hippocampal metabotropic excitatory amino acid receptors leads to seizures and neuronal damage, *Neurosci. Lett.*, 139, 77, 1992.

62. Choi, D. W., Excitotoxic cell death, *J. Neurobiol.*, 23, 1261, 1992.

63. Tymianski, M., Charlton, M. P., Carlen, P. L., and Tator, C. H., Source specificity of early calcium neurotoxicity in cultured embryonic spinal neurons, *J. Neurosci.*, 13, 2085, 1993.

64. Jaffe, D. B. and Brown, T. H., Metabotropic glutamate receptor activation induces calcium waves within hippocampal dendrites, *J. Neurophysiol.*, 72, 471, 1994.

65. Siliprandi, R., Lipartiti, M., Fadda, E., Arban, R., Kozikowski, A. P., and Manev, H., A derivative of a rigid glutamate analog protects the retina from excitotoxicity, *Neuroreport*, 5, 1227, 1994.

66. Bruno, V., Copani, A., Battaglia, G., Raffaele, R., Shinozaki, H., and Nicoletti, F., Protective effect of the metabotropic glutamate receptor agonist, DCG-IV, against excitotoxic neuronal death, *Eur. J. Pharmacol.*, 256, 109, 1994.

67. Koh, J. Y., Palmer, E., and Cotman, C. W., Activation of the metabotropic glutamate receptor attenuates N-methyl-D-aspartate neurotoxicity in cortical cultures, *Proc. Natl. Acad. Sci. U.S.A.*, 88, 9431, 1991.

68. O'Connor, J. J., Rowan, M. J., and Anwyl, R., Long-lasting enhancement of NMDA receptor-mediated synaptic transmission by metabotropic glutamate receptor activation, *Nature*, 367, 557, 1994.

69. Martin, R. L., Lloyd, H. G. E., and Cowan, A. I., The early events of oxygen and glucose deprivation: setting the scene for neuronal death?, *Trends Neurosci.*, 17, 251, 1994.

70. Ronne-Engstrom, E., Hillered, L., Flink, R., Spannare, B., Ungerstedt, U., and Carlson, H., Intracerebral microdialysis of extracellular amino acids in the human epileptic focus, *J. Cereb. Blood Flow Metab.*, 12, 873, 1992.

71. During, M. J. and Spencer, D. D., Extracellular hippocampal glutamate and spontaneous seizure in the conscious human brain, *Lancet*, 341, 1607, 1993.

72. Garcia, D. E., Cavalie, A., and Lux, H. D., Enhancement of voltage-gated Ca currents induced by daily stimulation of hippocampal neurons with glutamate, *J. Neurosci.*, 14, 545, 1994.

73. Choi, D. W., Calcium: still center-stage in hypoxic-ischemic neuronal death, *Trends Neurosci.*, 18, 58, 1995.

74. Beal, M. F., Mechanisms of excitotoxicity in neurologic diseases, *FASEB J.*, 6, 3338, 1992.

75. Beal, M. F., Role of excitotoxicity in human neurological disease, *Curr. Opinion Neurobiol.*, 2, 657, 1992.

76. DiFiglia, M., Excitotoxic injury of the neostriatum: a model for Huntington's disease, *Trends Neurosci.*, 13, 286, 1990.

77. Albin, R. L. and Tagle, D. A., Genetics and molecular biology of Huntington's disease, *Trends Neurosci.*, 18, 11, 1995.

78. Martin, S. J., Green, D. R., and Cotter, T. G., Dicing with death: dissecting the components of the apoptosis machinery, *Trends Biochem. Sci.*, 19, 26, 1994.

Chapter 8

PEPTIDES

*Heidi H. Swanson**

CONTENTS

* This chapter was written under the auspices of the Department of Psychobiology at the University of Seville, where I hold the position of Visiting Professor.

I. NEUROPEPTIDES

A. Introduction

The primary concern of any living organism is survival, which requires adaptation to the environment to assure stability. There must be constant monitoring and analysis of sensory input, followed by adjustments to maintain steady state, as survival requires the organism to be attuned to the rewarding and aversive properties of learning experiences. The pursuit of happiness or well-being is the ultimate aim of sentient organisms. The nervous, endocrine, and immune systems interact to ensure appropriate systemic and behavioral adaptations. The more research is carried out, the more obvious is the overlap and functional interaction between these systems. Over the last few decades one of the most prominent advances has been the identification of at least 50 neurotransmitters, the majority of these being peptides. It now seems that the brain is also a complex endocrine organ that synthesizes and releases a variety of hormone-like peptides. The functional significance of many of these neuropeptides is still unknown, but many have been found to be involved in pain reception, appetite, sexual and maternal behavior, affiliative and aggressive social behavior, cognitive function and learning, as well as the control of moods such as fear, anger, depression, love, and elation. The richness and diversity of peptide neurobiology is currently creating a revolution in our understanding of the nervous system, (for reviews see References 1, 4, 7, 23).

Peptides are single, unbranched amino acid chains covalently joined by amide linkages called peptide bonds. They occur in all parts of the body, particularly in the gastrointestinal tract. Neuropeptides are peptides found in the brain, which are believed to have neuroregulatory functions. Many peptides produced within the brain are similar to those in peripheral organs. Peptides are synthesized within ribosomes using messenger RNA (mRNA) sequences for instruction, unlike the classic catecholamines: norepinephrine, dopamine, and serotonin, which are built from amino acid building blocks. Since neuropeptides are produced from an mRNA template, the considerable power of molecular biology can be used in their study. Synthetic pathways can be elucidated and previously unknown fragments can be demonstrated. Genes have been discovered that lead to synthesis of multiple families of peptides. New peptides can be identified before their functional significance is known.

A computer analysis of the enormous growth over the last three decades of research on the role of peptides in the brain[7] showed that, since 1965, over a quarter of a million research papers have been published on peptides that have since been classified as neuroactive, although only 8.3% of these studies focused specifically on their neuroactive properties. The peak and at present the trend in research have leveled. Molecular biological and immunological progress has generally been greater than clinical correlation and relevance. The role that neuropeptides play in human behavior or particular neurological diseases is still uncertain, and we can expect to see more research in this field as new methods for assaying brain levels and mapping pathways become available.[23,24]

The proliferation of recognized neuroactive substances — many with nonclassic effects on neurons — has extended the concept of chemical transmission of nerve impulses. The term *neuroregulators* is useful for referring to the full range of compounds involved in the regulation of communication between nerve cells. Neuroregulators can be viewed as any substance that directly contributes to the regulation of signal transmission from one nerve cell to another. They fall into two categories, neurotransmitters and neuromodulators.

Neuropeptides are best classified as neuroregulators so as to include those which may be considered to be full neurotransmitters as well as those which function more as neuromodulators. Neurotransmitters are substances that relay a signal from one nerve cell to another within an identified synapse. A neurotransmitter should: (1) be synthesized and transported within presynaptic neurons, (2) be released with changes in excitability of neuronal membranes, (3) be bound to specific receptor sites, and (4) cause changes in membrane excitability in the presence of agonists and antagonists when applied exogenously to single cells. Acetylcholine and gamma-aminobutyric acid (GABA) are prime examples of neurotransmitters. For peptides, these criteria have been fulfilled for somatostatin, substance P, vasopressin, the tropic hormone-releasing factors, and some others (see Table 1).

Several aspects of synthesis and degradation distinguish neuropeptides from such classic neurotransmitters as norepinephrine and acetylcholine. Instead of being produced at the synapse, peptides are usually synthesized in the perikaryon or neuronal dendrites, at a distance from the release site. In this aspect they resemble hormones; indeed, they may be secreted directly into the bloodstream and are then classified as neurohormones. On the other hand, peptides which are transported along nerve fibers and released at synapses are known as neuromodulators. Furthermore, in contrast to other classes of neurotransmitters, there are no presynaptic receptors for neuropeptides and thus no reuptake mechanism to inhibit synthesis through negative feedback. Instead, destruction occurs by proteolytic enzymes. In the cerebrospinal fluid (CSF), which does not have significant protease activity, peptides may be stable for hours or days. They can still be found in post-mortem tissues. Peptides have effects on postsynaptic membranes which may last for long periods rather than the fast action of classic transmitters. Another mechanism of action of neuropeptides is as tropic agents, stimulating neurite outgrowth during development and regeneration after nerve damage. The concentration of brain peptides is several orders of magnitude less than that of transmitters of the amino acid or monoamine group. Thus, whereas amino acids occur in microgram (10^{-6}) amounts per liter, monoamines occur in nanograms (10^{-9}) and peptides in picograms (10^{-12}).

Despite being unable to fulfill strict neurotransmitter criteria, some peptides may provide interneural chemical communication as neuromodulators or neurohormones. Neuromodulators are substances with neuroregulatory functions other than neurotransmission. Synaptic neuromodulators act in synapses by altering the action of a neurotransmitter or receptor. They may affect synthesis, release, or inactivation of neurotransmitters, as well as alter receptor interactions. In contrast, neurohormones change neuronal activity at greater distances and in several sites, sometimes through specific receptors. By involving a greater number of cells than the transsynaptic processes related to classic neurotransmission and by acting over a longer period of time, neuromodulators provide a mechanism for fine-tuning neural signals. However, they differ from the usual endocrine hormones by producing specific alteration of neuronal signals as such, rather than by acting through a general increase or decrease of cellular metabolism.

B. Distribution

Neuroactive peptides were originally found in the hypothalamus. The remarkable research story of this decade is how these neuropeptides, initially thought to be localized only in the hypothalamus, have been identified throughout the brain and spinal cord.[8,23] Immunological techniques have proved to be powerful tools for peptide localization. Some peptides are produced in neurons distributed throughout the central nervous system (CNS), whereas others have widespread distribution of cell processes despite localization of the majority of the cell bodies in particular areas of the hypothalamus. It should be emphasized that all the neuropeptides isolated so far are present in the hypothalamus where their function is the maintenance of homeostasis in response to changes in the environment. However, the discovery that these same neuroactive peptides are also found in other parts of the brain has led to a great deal of information that is now being reevaluated.

The classic concept of neuronal transmission stipulated that there should only be one transmitter per nerve ending. This concept has been overturned by the discovery that different neuropeptides may coexist in the same neuron. One or more neuropeptides may also be localized in presynaptic endings together with other transmitters such as norepinephrine, acetylcholine, GABA, dopamine, or serotonin.

C. Neuropeptides and Behavior

We fall in love, we get angry, we acquire knowledge. All these events are realized with the help of billions of brain cells, which work in groups to process incoming stimuli. Neuropeptides improve the coordination of brain cells by delivering messages between cells and from one group of neurons to another. They grease, as it were, the cogs of the gray matter which is the motor of our living and being. Neuropeptide-induced changes in behavior such as social and sexual behavior, learning and memory reflect their role in behavioral adaptation to environmental demands.

It is possible to make a serious conceptual error by assuming that a single peptide is responsible for a single behavior. Only a few years ago, every major mental disorder and many behaviors such as eating, sexual activity, and aggression were explained on the basis of increases or decreases in six compounds then identified in the brain. With the discovery of the first peptide neuroregulators, a few simplistic hypotheses about their effects on mental disorders were proffered. Earlier studies dealing with different neuroregulator systems such as biogenic amines have demonstrated the complexity of behavioral processes in relation to neuroregulators. Increasingly sophisticated knowledge about colocalization of neuropeptides in the same synapses and the mechanisms by which the neuroregulators can alter brain activity have necessitated a reformulation of neuronal function. The same substance may have many roles within the brain, and those roles may be exercised at many different brain sites. Furthermore, the same neuroregulator may affect a given behavior in different ways, depending on the particular site within the brain at which it is acting. It is important to realize that we are dealing with extremely complex models, in which neuronal action results from a series of neuroregulatory processes involving multiple neurotransmitters and neuromodulators, some of which may have synergistic or antagonistic actions.

In studying the control of behavior one must consider the interaction between several neuroregulatory systems simultaneously. Technical difficulties make it hard to examine two or more systems at the same time, and often because researchers tend to be familiar with one particular system, they have limited their studies to that system. The issue of multiple neuroregulatory systems and their interaction is particularly important in the investigation of severe mental disorders and possibilities for treatment.

Finally, it is necessary to remember that not only does the brain produce behavior, but behavior in turn influences the brain. The whole neuro–endocrine–immune system is interrelated, the responses being superimposed on a genetic base modulated by development and life experiences. Although enhanced understanding of brain neuropeptides should lead to improved methods of diagnosing and treating severe mental illness and addictive disorders, caution must be exercised in the prescription of drugs. Antisocial behavior and other disturbances resulting from poor upbringing and environmental stress which are not associated with organic dysfunction may be better treated through psychotherapy and improved living conditions, with drugs given as temporary palliatives. It has often been said "that it is easier to repair a sail in a calm than in a storm."

D. Evolution of Peptides

It appears that early in evolution a restricted set of biologically active peptides may have constrained the scope of their subsequent diversification. As organisms became more complex and new peptides were needed, new functions were assumed by existing molecules. Some of the messenger molecules in evolution became functionally organized as neurotransmitters and some as neurohormones. Studies of peptide structure across multiple species show that portions of the peptide sequence are maintained throughout evolution, and that the sequences so conserved show phylogenetic differences in functional activity. Invertebrate peptides (restricted to neurons) may be found in vertebrates within both glandular and neuronal elements. Conversely, some peptides found in invertebrates solely in endocrine tissue may be found only in neural tissue in vertebrates. In addition to evolutionary change in peptide sequence, changes in receptors, processing, and target cells may result in altered function.

E. Classification of Neuropeptides

Peptides may be grouped in families[6] (Table 1). Almost all the neuropeptides are derived from cleavage at particular amino acid sequences in a large precursor molecule or prohormone. They may contain as few as 3 or as many as 50 amino acids. Most are unable to cross the blood-brain barrier. The processing of the fragments cleaved from the precursor may differ from one tissue to another and allow different ratios of particular end products, depending on the cell involved. For example, the prohormone proopiomelanocortin (POMC), which has been extensively studied, splits into adrenocorticotropic hormones (ACTH) and beta-lipotropin (B-LPH) in the anterior pituitary, whereas in the brain it forms melanocyte-stimulating hormone (MSH) and β-endorphin. This example illustrates the close relationship between peptides and classic hormones (some of which, as well as being produced by peripheral endocrine glands, also exist in the brain and are then called neurohormones). Only the neurohypophyseal hormones oxytocin and vasopressin and the opioid peptides will be considered in detail in this chapter.

TABLE 1.

Classification of Peptides

Neurohypophyseal Peptides	*Miscellaneous*
Oxytocin	Substance P
Vasopressin	Cholecystokinin
	Bombesin
Opioid Peptides	Vasoactive intestinal peptide
	Neuropeptide Y
β-Endorphin	Galanin
Met-enkephalin	Neurotensin
Leu-enkephalin	Angiotensin II
Dynorphin A	Calcitonin
Dynorphin B	Delta sleep–inducing peptide
Pituitary Releasing Hormones	
Corticotropin-releasing factor	
Thyrotropin-releasing factor	
Gonadotropin-releasing factor	
Growth hormone-releasing factor	
Somatostatin	

II. NEUROHYPOPHYSEAL HORMONES

As early as the 1920s the neurohypophyseal hormones oxytocin and vasopressin were among the first of the neuroactive substances to be identified in neural tissue. These substances are hormones secreted within the brain, in the so-called magnocellular compartments of the supraoptic and paraventricular nuclei of the hypothalamus. The concept of neurosecretion, i.e., that certain neurons are capable of releasing their products into the bloodstream in order to reach distant target structures, was a revolutionary idea first proposed in the 1930s. These neurons project their axons into the neural lobe of the pituitary gland, where oxytocin and vasopressin are released into the bloodstream. These substances are true hormones whose release is controlled directly by reflex nervous stimuli. Thus, the release of oxytocin is activated through the stimulus of suckling and vasopressin through changes in osmotic pressure relayed via osmotic receptors (for reviews, see Reference 37).

Vasopressin and oxytocin fibers have been demonstrated by means of immunocytochemistry in a large number of brain regions, from the olfactory bulb down to the spinal cord.[32,33,35,39] Limbic system brain structures, especially, are heavily innervated by vasopressin-containing fibers. The origin of the extrahypothalamic vasopressin and oxytocin fibers was originally thought to be only from the paraventricular nucleus (PVN). It is now known that an important source of vasopressin neurons is the suprachiasmatic nucleus. In addition, vasopressin cell bodies are much more widespread over the brain than had previously been assumed. In contrast to vasopressin, practically all oxytocin fibers seem to be derived from the PVN. A feedback link has also been established, with oxytocin and vasopressin fibers innervating both the paraventricular and supraoptic nuclei. The coupling of central and peripheral actions of the neurohypophyseal peptides is illustrated by the effect of vasopressin on temperature control. Vasopressin exerts an antipyretic effect (lowering of body temperature) in the CNS, while in the periphery it prevents the loss of water via the kidney to prevent dehydration due to fever-induced perspiration.

The first suggestion that pituitary hormones could have direct effects on behavior was made in 1965, when evidence was presented that ablation of the posterior lobe of the pituitary interfered with memory and learning in rats, whereas injection of vasopressin into the cerebral ventricles restored this function. With the refinement of immunocytochemical techniques, it was shown that a direct transport of vasopressin and oxytocin within the brain via nerve fibers might be the route through which these peptides reach their site of action. Recently, evidence has accumulated that peptides acting as hormones peripherally may also act directly in the brain to produce related behavioral effects.[10,15] Oxytocin, for example, has long been known to cause milk ejection from the lactating breast and contraction of the uterus at parturition. It appears that oxytocin secreted within the brain has direct behavioral actions related to maternal responses and sexual bonding in both females and males. One must not forget, however, that behavior can be the indirect consequence of cognitive perception of changes in body physiology produced by peripheral endocrine action.

A. Oxytocin: The Hormone of Love

What is love? Love may be defined as social bonding, which includes attachment of children to their parents, pair bonding between sexual partners, nurturing activities of parents toward their children, as well as affection toward other family members and friends. Love is necessary both for survival of the individual and for perpetuation of the species and is therefore strongly reinforcing, i.e., pleasurable. It may be considered a prerequisite for happiness and mental health.

Although such a complex emotion is bound to be influenced by many factors, there is evidence that oxytocin may play a prime role in the formation of interpersonal bonds.[15,18,25,26] Oxytocin is secreted in the hypothalamus and transmitted directly to the neurohypophysis from where it is released into the bloodstream during interpersonal acts, such as coitus, birth, and lactation, which result in the formation of affiliative bonds. In humans, all three are usually followed by some caretaking behaviors. Oxytocin acts peripherally in the female causing contractions of the uterus during labor and coitus and milk ejection during suckling. In the male, it facilitates sperm transport and ejaculation. At the same time, oxytocin is also secreted into specific regions of the brain where it binds to receptors. Its general effect seems to be to increase arousal and heighten the emotional tone so that the individual becomes more receptive to the sensory stimuli which promote bonding and less receptive to outside interference. There is also some recent evidence that oxytocin directly causes positively reinforcing mood elevation and a decrease in anxiety. Furthermore, oxytocin seems to have some potential in attenuating addiction to morphine-related drugs and in modulating withdrawal symptoms after naloxone.

1. Lactation and Maternal Behavior

Oxytocin secretion can easily be conditioned and inhibited.[25,26] During lactation, it is the stimulation of suckling which causes milk letdown. However, the sound of the baby crying can stimulate milk production. Blood levels of oxytocin were shown to rise in the blood of nursing women before the baby was put to the breast. The mothers had become conditioned to react to their babies before sucking occurred. On the other hand, disturbances and stress can decrease lactation, both in amount produced at one feeding and in the duration that breast feeding is kept up.

Maternal behavior refers to interactions between mother and young, which increase the probability that her offspring will survive to maturity.[20] It is most highly developed in mammals where nutrition through suckling needs a behavioral mechanism to assure proximity between mother and young.[17,28,29] Normal female mammals experiencing a natural gestation and parturition exhibit maternal care immediately after delivery. In animal studies, it has been shown that oxytocin produced in the brain is directly involved in the onset of maternal behavior.[11,15] In rats and sheep, intracerebral infusion of an oxytocin antagonist which binds to the brain receptors prevents normal maternal behavior from developing.[11,30] During labor, oxytocin receptors are increased in brain regions known to be involved in maternal behavior.[21] The olfactory bulb seems to be important for maternal recognition, and it has been suggested that it leads to olfactory imprinting which, for instance, allows a ewe to recognize her own lambs.[19,27] Human mothers also respond preferentially to the scent (as well as to the sound) of their own baby.[25,26] Bonding with caregivers other than the biological mother, such as the father or adopted parents can occur later through contact with infants, and once started it is mutually reinforcing so that it may forge a bond which can be just as strong as that formed after biological birth.[16,17] It has been known for a nonlactating bitch to start producing milk after adopting orphaned pups. Interestingly, oxytocin levels have been shown to rise in fathers who interact closely with their babies. Conversely, despite popular belief, in the absence of social pressure, no bond is formed where there is no physical contact between mother and child. This fact is well known by dairy farmers who have no problems of attachment with cows if they remove their calf at birth, whereas separation after a few hours of contact produces distress in both cow and calf. Furthermore, the fetus and young infant have relatively high oxytocin levels as compared with adults. Do oxytocin rises at these crucial times have any psychological significance in sustained family love?

2. Courtship, Mating, and Social Bonding

Sex hormones produced in large quantities at puberty prepare the male and female to engage in reproductive activities.[3,9,13,18,31] The first step is finding a mate. Courtship involves approach and the initiation of social contact and is usually accompanied by intense emotional arousal. In animals, courtship occurs only in the presence of active sex hormones which fluctuate in cycles of various lengths, and females are only receptive at the time of ovulation. In humans, sex hormones play a permissive role, the actual level not being critical. It is believed that the importance of establishing a permanent partnership for the prolonged rearing of immature children has resulted in allowing sexual reinforcement to occur throughout both menstrual and seasonal cycles.

Little is known regarding the neurobiological substrate of the human sexual experience. Until recently, because oxytocin is secreted in large amounts during parturition and lactation, this hormone has mainly been considered to be a "female" hormone, and previously no function was known for oxytocin in males. Animal experiments suggest that oxytocin, modulated by steroid hormones, plays a role in sexual arousal, orgasm, and sexual satiety in both sexes, as well as in other aspects of sociosexual behavior. Correlational studies suggest similar mechanisms in humans.[12-14,25]

Sexual arousal can be elicited by a variety of sensory stimuli. Genital and breast stimulation, but also light touch, causes the release of oxytocin. Oxytocin release can be influenced by conscious factors and is easily conditioned to occur in the absence of direct physical stimulation. Emotionally loaded environmental stimuli are particularly potent in eliciting sexual arousal. Recent studies indicate that oxytocin does indeed increase during sexual arousal and shows a marked rise during the orgasmic phase of sexual behavior.[25] Oxytocin release during eating and grooming strongly suggests that it may play a psychological role in family love behavior.[2] Eating together is an important activity promoting social bonding during courtship, family life, and even in general communal interactions.

Oxytocin released centrally during early stages of sexual excitement could, through positive feedback, prime the nervous system to permit a larger, pulsatile release of oxytocin. Oxytocin has been implicated in facilitating penile erection. During orgasm, oxytocin causes contractions of the male and female reproductive tracts (thereby facilitating sperm transport) and precipitates the autonomic responses of increased heart rate, respiration, and blood pressure. The ability to achieve orgasm, or variability in the subjective experience of this phenomenon, could reflect differences in the release of oxytocin possibly conditioned by association with previous pleasant or unpleasant experiences. Finally, intracerebral oxytocin has been implicated in the phenomenon of sexual refractoriness or satiety as an aftermath to orgasm.

B. Vasopressin

Vasopressin derives its name from its peripheral action on the blood vessels resulting in increased blood pressure, particularly in response to hemorrhage. Later, it was also found to have potent antidiuretic effects in the kidney. The role of vasopressin in the brain is poorly understood. Given intracerebrally, it elevates blood pressure, heart rate, and body temperature and under certain circumstances appears to act as an analgesic.[32] The collective evidence suggests that vasopressin modulates pain sensitivity through non-opiate mechanisms. Vasopressin and corticotropin-releasing factors are stored within the same synaptic vesicles. Clearly, vasopressin when it is co-released with corticotropin-releasing factor may influence the stress response. Vasopressin exerts an important function in the control of circadian rhythms. Its putative effects on memory enhancing are controversial.

1. Learning and Memory

a. Background

The hormone of the posterior pituitary gland, arginine vasopressin, and some synthetic analogues and antagonists have been studied for nearly two decades with respect to their ability to modify memory and learning.[34-36] As the investigations have proceeded, controversies have arisen regarding experimental results themselves, as well as the interpretation of these results. In part, these difficulties occur because of the complexities inherent in defining and measuring learning and memory, particularly in animals. Thus, we have to distinguish between the acquisition of a response, consolidation of short-term to long-term memory, and retrieval of information. Additionally, it is difficult to dissociate the confounding influence of motivational or attentional factors from learning and memory processes. There are also technical problems relating to doses, route of administration, and different experimental conditions. Nevertheless, there is overwhelming evidence that vasopressin and related peptides influence memory and behavior. Although blood-borne vasopressin, at physiological titers, seems to be effectively excluded from the CNS by the blood-brain barrier, larger pharmacological doses may pass through the blood-brain barrier into the brain. The most convincing evidence comes from animal studies of intracerebral infusion of vasopressin and/or its antagonists.

Experiments in which an animal is required to learn to avoid an unpleasant experience by active or passive responses have generally supported a role for vasopressin in the formation of long-term memory. Studies using the Brattleboro strain of rats, which is genetically deficient in vasopressin and shows some memory impairment, suggest that the endogenous hormone is important in this process under normal circumstances.[37,38]

The investigations of peptide effects on avoidance behavior have been criticized because they are stressful, and since vasopressin is involved in the stress response, results may be confounded. Animals (and humans) can also be motivated to perform tasks by positive reinforcement, and such tasks are referred to as appetitively motivated. The performance of positively reinforced behavior may be more closely related to natural situations. If it is assumed that the rewards obtained are reinforcing, then it might be postulated that peptides that enhance learning or memory of these tasks also enhance the perception of reward. The weight of evidence, from both aversely and appetitively motivated behavior, favors the hypothesis that vasopressin enhances both learning and consolidation of long-term memory and its retrieval.

The major source of disagreement arises from the proposal that vasopressin does not directly affect any aspect of learning or memory per se, but instead increases the arousal or alertness of animals and thereby alters their behavior. This theory is given support by the observation that small and large doses of vasopressin may have opposite effects. A bell-shaped dose–response pattern is characteristic for drugs that increase arousal, since a small increase in arousal enhances motivation, whereas a large increase may distract the animal and disrupt performance. There is electrophysiological evidence that vasopressin increases activity in midbrain structures, possibly by activation of the catecholamine system. It is difficult to satisfactorily distinguish between effects on arousal or motivation and effects on memory.[32,34]

b.   Clinical Applications

Considering the increasing loss of memory that occurs during aging, or in Alzheimer's disease, it is not surprising that the idea of a drug that can improve or reinstate memory is appealing. There now exists a vasopressin spray for intranasal administration, providing direct access to the brain. In spite of the fact that clinical trials are difficult to evaluate, it does seem that vasopressin produces some improvement in memory, particularly in people without too serious deficits. It is not clear whether this is an intrinsic effect or because of enhancement of attention or motivation. Mild cognitive impairment was observed in humans with vasopressin deficiency, as in diabetes insipidus, and improvement was found with vasopressin treatment. Such results are consistent with a modulatory role of vasopressin on the neuronal systems involved in the formation and/or retrieval of memory. In patients with serious brain injury or deficits, such as Korsakoff's syndrome of alcoholics or Parkinson's disease, vasopressin was ineffective in improving memory. This is not surprising, because degeneration of brain structures may destroy the site of action for the peptides. Positive effects have been found more frequently in less severely traumatized patients and in normal volunteers. These findings suggest that future studies should concentrate on patients with mild memory deficits without pronounced anatomical destruction, such as elderly people with mild senile forgetfulness, or on patients with amnesia after concussion. Possible action through improvement in motivation suggests that depressive patients should also prove good candidates.[36,39,42,44]

2.   The Biological Clock

The rotation of the earth results in a periodically changing environment of light and dark. All animals show an adaptive organization of activity in rhythm with the daily and seasonal fluctuations in illumination and temperature. The mammalian suprachiasmatic nucleus (SCN), a small group of neurons located in the basal part of the anterior hypothalamus, is generally considered to be the major component of the circadian (day–night) timing system. In addition to its role as a circadian pacemaker, the SCN also seems to be involved in the seasonal timing of physiological and behavioral processes such as reproductive activity and hibernation.[40,41]

The most evident circadian rhythm is the daily cycle of sleep and wakefulness, accompanied by synchronized changes in body temperature, hormones, and behavior. The implication of the SCN as the primary pacemaker was discovered by the loss of diurnal rhythms following bilateral removal of this nucleus in experimental animals. A few years later it was demonstrated that the SCN contained a prominent population of vasopressin-containing cells. The other major peptides found in the SCN include vasoactive intestinal peptide, neurotensin, and neuropeptide Y. The SCN receives input from the visual pathway, which allows it to become entrained to the light–dark cycle. The target areas of the SCN fibers are most probably the sites where, by periodic release of peptidergic transmitters, the circadian signal is translated into hormonal and behavioral events.[5,43,45]

Recently, histological and morphometrical examination of human brains obtained at autopsy have shown that the vasopressin population in the SCN shows distinct diurnal and seasonal variations. Peak values of vasopressin-immunoreactive neurons (reflecting active synthesis) were observed in the early morning. In addition to the diurnal cycle, a marked seasonal rhythm was observed, with a maximum in early autumn, a lower plateau in winter, and a deep trough in spring and summer.

During normal aging, the number of vasopressin neurons decreases. In Alzheimer's disease, the decrease in cell number is even more pronounced and degenerative changes within these cells have been observed.[46] Furthermore, in elderly subjects no diurnal variation in vasopressin-immunoreactive neurons was detected, although the seasonal variation persisted. The circadian timing system is progressively disturbed with aging, both in humans and other mammals. In humans, age-related changes have been described for body temperature, hormonal rhythms, and the sleep–wake cycle. In fact, night wakefulness is one of the major reasons older people are placed in institutions. In Alzheimer's disease, these changes are exaggerated. This pathology might be the neural basis for the nightly restlessness observed in elderly people and, more particularly, in patients with Alzheimer's disease.

A desynchronization of circadian rhythms has been implicated in other conditions which interfere with human well-being, such as affective disorders, night-shift work, and jet lag. Therapies aimed at resynchronizing the rhythms were able to alleviate such conditions. In particular, exposure to bright light during regular periods of the day, increasing social and physical activity, curtailment of daytime naps, and ensuring complete darkness during the night have proved to be effective in improving both mood and cognitive performance. It might also be mentioned that the toxicity, effectiveness, and clearance of many pharmaceutical compounds depend on the timing of administration.

III. ENDOGENOUS OPIATES

A. Opioid Peptides

Morphine, an opium derivative, has long been known to alleviate pain. When receptors for morphine were found in the brain, a search began for an endogenous substance which would attach to these receptors because it seemed unlikely that there would be receptors for an external substance present in the brain, without a similar chemical being produced within the brain. Indeed, soon after this, the endorphins were identified. The identification of endogenous opiates is unusual because, for most other neuropeptides, the endogenous neurotransmitter was first identified and synthesis of external analogues followed (for review, see References 1, 4, 6, 54.)

The word "opioid" is now often used to refer to all drugs with morphinelike actions, rather than to only endogenous ones as originally intended. As a result, current use does not distinguish between *opioid* and *opiate*, so these terms will be used interchangeably. The action of morphine is selectively antagonized by naloxone. The opioid peptides are those whose receptor binding can be blocked by naloxone or other morphine antagonists.

Chemically, the opioid peptides are derived from three separate, multifunctional precursors. These precursors show a structural homology, suggesting a common ancestor gene. There are three families of opioid peptides: (1) the POMC family, containing β endorphin, which are best known and most similar to the opium drugs, (2) the enkephalins, and (3) dynorphin. Although they are chemically diverse, they share an initial sequence of four amino acids which give them their analgesic potency. The other end of the amino acid chain gives receptor specificity, which allows individual opioids to exert other characteristic effects at various brain sites. The β endorphins have been the most studied. They are mainly produced in the hypothalamus. It appears that POMC is produced in the cell bodies and broken down to its component peptides along the axon.

B. Opioid Receptors

Four main receptors have been identified, for which each class of opioids has selective affinity, although there is some cross-reactivity. Thus, β endorphin (part of the POMC family) shows affinity mainly for μ receptors, whereas enkephalins are selective for ∂ receptors, and dynorphins for κ receptors. In pharmacological assays, analgesia can be elicited by stimulating any of the receptors. The most powerful action is through the μ receptor, which is where morphine acts and which can be blocked by naloxone binding. Tolerance to both endogenous and exogenous opiates develops very easily. The primary action of opioids on pain-sensitive neurons is a decrease in electrical excitability and reduction in transmitter release. Opioid receptors are found in greatest concentration in the hypothalamus and limbic system, as well as in the spinal cord.

Although opiates have been mainly implicated in perception and response to pain, their functions and interactions are not at all simple. They appear to be involved in analgesia, memory, immune responses, feeding, blood pressure, tumor growth, and endocrine regulation, as well as social and sexual behavior. A major impetus for the study of neuropeptides in relation to behavior came as a direct consequence of the discovery of the endorphins. Following the identification and investigation of the behavioral effects of endogenous opiates, there was renewed interest in discovering new, previously unrecognized behaviorally active peptides.

C. Pain, Fear, and Stress

To feel acute pain is an absolutely essential condition for survival. Pain-initiated avoidance behavior protects the individual. Fast "warning" pain is not opiate responsive because there are no opiate receptors in this neurological path. As pain persists, its protective function diminishes, and the second wave of pain-receiving input via a different path, which contains opiate receptors, is opiate responsive and sensitive to pain modulation by endogenous peptides. There are problems in studying pain modulation because pain is very subjective and difficult to describe.[53,59]

Fear is a powerful analgesic. The opioid system is activated in response to fear — the major function of endorphins being the inhibition of pain-related behaviors during self-defense, such as nursing an injured limb. Fear-induced analgesia has been frequently observed, particularly in victims of attack, such as prey animals. Although it is difficult to quantify, there are numerous anecdotes of warriors (and sportsmen) continuing with their activities after experiencing horrific injury. Analgesia has been reported during violent physical exercise accompanied by a high emotional state. Opiates also have amnesic properties, which is useful, as there is no value in remembering the intensity of the pain. Furthermore, opioids seem to play a major role in voluntary suppression of the pain reaction via a state of altered consciousness, which may be induced by hypnosis or trances.[50,51,57,61]

For ethical reasons it is difficult to approach clinically relevant pain levels in either animals or man. A few clinical studies have attempted to measure the protective role of endogenous opiates. In one group of patients CSF levels were measured prior to abdominal surgery. After the operation the patients were allowed to control the level of analgesia by self-medication of opioids. It was found that patients with a similar disease history differed from three to five times in analgesia demand. The negative correlation between preoperative CSF opioid levels and postoperative demand suggests that the endogenous opioid protected against pain and that this ability differs between persons. Individual differences in pain tolerance may have a genetic as well as an emotional basis.[49]

The role of endogenous opiates in labor has also been studied.[47,52] The CSF levels of women in the early stages of labor were inversely correlated with their demand for epidural sedation, again suggesting that endogenous opiates played a protective role against excessive pain. There is evidence from animal experiments that vaginal distention during labor induces opioid secretion. Because opiates can interfere with memory retention, this may have adaptive value during parturition, as it seems useful to forget the pain associated with delivery. Too much opiate, either through exogenous morphine or excessive stress may, however, prolong labor through inhibition of oxytocin.

Long-term pain is maladaptive and will contribute to decreased function. If the cause of pain cannot be eliminated, such as in degenerative disorders or malignant tumors, it will obviously persist. Such pain is responsive to external opiate administration. The endogenous systems are operative although it is not easy to evaluate how important this function is. Some types of chronic pain known as neurogenic, e.g., the phantom limb syndrome, and neuralgias are associated with abnormally low levels of endogenous opiates. An interesting finding was that patients with so-called idiopathic or psychogenic pain, which does not seem to have an organic base but may be regarded as a somatization of mental distress, had higher than normal levels of endogenous opioids. Although idiopathic pain (meaning "of unknown origin") is often interpreted as "all in the mind," it is no less real to the patient. The high levels of endogenous opiates may reflect an attempt of the brain to alleviate the pain. In this instance, as indeed in many other situations, it is difficult to separate cause from effect.

The experience of acute pain triggers the drive to survive and activates a stress response to mobilize all body resources. Although activation of the opiate system during short-term stress which an individual is able to handle is adaptive and pleasurable, long-term stress is usually harmful. Noncoping individuals maintain a long-term stress response which is maladaptive. This condition may be induced by the perception of helplessness (lack of control) and hopelessness (no matter what I do, it turns out bad) and is seen in chronic unhappiness, depression, and anxiety. It should be noted that a situation may not in itself be dangerous or hopeless, but if it is perceived to be so by an individual, it will be self-fulfilling. Some stressors may lead to an increased propensity for self-administration of morphine, suggesting that stress potentiates the process of addiction.[62]

Finally, it has been shown in animal experiments that both prenatal and neonatal stress can modulate the opioid system which may affect later responses to stress. It seems likely that this may also occur in humans.

D. Reward Mechanisms

Of the many drugs that over the centuries have been used (or abused) for their pleasurable or hedonistic properties, the most potent has been opium. It is therefore not surprising that endogenous opiates have been implicated in the reward system of the brain.[60] Simply put, the subjective feelings of euphoria associated with reward for successful completion of a task are accompanied by an increase in brain endorphins. This system can be studied through intracranial self-stimulation. Since the early reports that animals will press a lever to electrically stimulate certain parts of the brain, it has been recognized that this technique can contribute to an understanding of how behavior is controlled by drugs. Many drugs abused for their pleasurable qualities, such as morphine, increase the responding rate and lower the threshold for self-stimulation. As might be expected, opiate antagonists have the opposite effect. Furthermore, the level of β endorphin in the hypothalamus increases in proportion to the rate of self-stimulation. Other neuroregulator systems, especially dopamine, also seem to be involved, possibly as secondary responses. These studies provide some insight about the role of endogenous opioids in reward-motivated behavior and learning. Endorphins are also involved in addiction to gambling and other compulsive behaviors.

A constant high level of brain opioids as, e.g., in opiate abuse, leads to apathy and inactivity and is therefore maladaptive. The rapid development of tolerance to endogenous opiates is designed to prevent this state. Unfortunately, the same mechanism quickly leads to addiction to opiates from external sources.

One of the strongest sources of happiness and excitement is the feeling of having been able to cope with a challenging task. This is a state that is strongly motivating and drives people to challenges and sensation seeking. A study of parachutists-in-training showed that coping with such a challenge produces a state of temporary activation accompanied by a production of endorphins. Similar results have been obtained more recently in association with the euphoria produced by bungee jumping.[62]

An interesting sidelight is that eating hot spicy food such as curries is addictive. The active pain-giving agent, capsaicin, gives an enormous boost of intensity to taste, while at the same time causing the release of endorphins. Endorphins can give a person a feeling of pleasure and well-being, so that you get a "buzz" from eating capsaicin. Furthermore, because the body becomes used to the exposure, it increases the level of endorphins it releases, and so eating spicy foods becomes addictive. Beware: the first bite of a mild curry leads to the vindaloo!

E. Social Attachment

There are always elements of risk and pain associated with any kind of social bonding: the love/hate phenomenon. Consider the following examples: (1) mother–baby:

mother suffers pain during labor but loves her child, (2) child–parent: a child clings to its parents even though they inflict pain as punishment, (3) pair-bond between parents must survive hardship "for better or for worse" in order to provide a home for the children, and (4) the family unit must bond to society despite the pain required for conformity to social norms, in order to procure the necessities of life and to provide a safe and stable environment. Brain opioid systems which mediate the perception of pain may have evolved to provide neurochemical mediation of social bonding.[48,56]

The analysis of the physiological factors underlying social attachment remains rudimentary, in comparison with the study of other motivated behaviors, such as eating, drinking, sex, and aggression.[56,63] Clearly, the need for social attachment is not secondary to more primary drives, as is evident from the pattern of emotional behavior most young animals, including children, exhibit when they are socially isolated. Even when all their bodily needs are assured, youngsters promptly start to emit distress vocalization when separated from their parents and/or siblings and continue to cry until replaced in their normal social environment. This reaction to separation is immediate, reflexlike, and consistent across species, and its expression requires no previous learning. It is well known that in Victorian times parents used to feed their children opium-laced candy to suppress crying and to produce a contented child. Similar results have been documented in experiments with puppies and monkeys, where morphine suppressed separation stress and naloxone increased contact seeking.

Social bonds may develop because of the positive reward associated with social interactions or because of the capacity of social interactions to reduce negative emotions such as separation distress. Thus, both the direct rewarding properties of opiates in the brain and, indirectly, their ability to assuage withdrawal distress lead to the reinforcement of social attachment. It is interesting that brain areas which control specific social behavior patterns (the hypothalamus and limbic system) are rich in opioids. Here, they may interact with the oxytocin system, which, as mentioned earlier, is also concerned with social bonding. At spinal and lower brain stem levels, opioid systems are ideally situated to receive and organize incoming sensory information from which the social bond must be constructed.[58]

Opioids also interact with other neurotransmitter and hormone systems. Thus, they are known to be powerful stimulants of prolactin and growth hormone secretion. Opiates released during suckling may help to strengthen the mother–infant bond, whereas psychosocial dwarfism, which results from emotionally induced deficiency in growth hormone secretion, might be partially attributed to inadequate opioid release by the social environment of the afflicted children. In women, amenorrhea during stress or undernutrition is thought to be partially mediated via endorphins depressing gonadotropin release.

The opioid system has also been implicated in feeding and drinking behavior. Particularly in humans, eating and drinking are social behaviors with strong affiliative components. One needs only to consider the importance of food during courtship and general social interaction. Generally, opiates facilitate feeding and may affect food preference and palatability. The strong emotional component behind such eating disorders as obesity, bulimia, and anorexia suggests an involvement of the opiate system, although a specific role has not been defined.

There are striking similarities between the dynamics of opiate addiction and affiliative behaviors. Both are characterized by strong emotional attachments and severe physiological and psychological distress after withdrawal of narcotics or the dissolution of personal relationships. The degree of overlap of withdrawal symptoms — depression, lacrimation, irritability, insomnia, and anorexia — supports the contention that both processes may arise from a common neurophysiological base. The idea that brain opioid systems mediate social emotions and attachment provides a culturally disturbing explanation for the powerful urges which underlie narcotic addiction. In the absence of positive reinforcement during childhood and with the pressures in society for living in isolation, a drug which replaces the need for social bonding may be very attractive. Furthermore, the comfort derived from the addiction may counteract any desire to give up the habit.

F.   Mental Illness and Mood

It seems reasonable that, if brain opioids control the intensity of social emotions, there should also exist emotional disorders which correspond to imbalances within these systems.[54-56,64] From the present conceptualization of opioid functions, an excess of opioids in the brain may prevent normal socialization. Withdrawal from social contact could be the outcome of high brain opioid activity; such children may appear to be quite content. In its extreme form, it may be manifested as childhood autism. Indeed, naltrexone has been shown to attenuate some of the manifestations of this illness.

Conversely, if a child were unable to feel opioid reinforcement because of constitutional underactivity of this system, such a child might look for closer social contact by disruptive behavior designed to attract attention. It was found that endorphin levels in the CSF of children with disruptive conduct were lower than normal. It is difficult to establish whether the differences in CSF opioids reflect a cause or an effect of the psychological disturbance.

There is evidence that the opiate system — being intimately connected with the modulation of fear and anxiety — may also play a role in phobic disorders. Phobic patients are fearful of imagined stimuli to which they presumably respond by release of endorphins. Patients with phobias have been systematically desensitized by being repeatedly presented with increasingly fearful stimuli. Blocking the endorphin response by naloxone prevented the desensitization regime from being effective.

IV. OTHER PEPTIDES

A.   Pituitary Releasing Hormones

The anterior pituitary has long been known to produce tropic hormones, e.g., ACTH, gonadotropic hormones (FSH and LH), thyrotropic hormone (TSH), which stimulate appropriate endocrine secretion by their peripheral target glands. The secretion of pituitary secretion is controlled by a negative feedback mechanism from the hypothalamus. So-called releasing factors are peptides secreted in the hypothalamus, each specific for a given tropic pituitary hormone. External stimuli impinging on the brain from the environment influence the secretion of the releasing factors.

Thus, corticotropin-releasing factor (CRF) produced in response to stress leads to the production of ACTH, whereas in animals with seasonal reproduction, variations in daylight influence the secretion of gonadotropin-releasing hormone (GnRH). GnRH may be an example of a neuropeptide which subserves mainly one important function, that of regulating reproduction. It stimulates release of pituitary gonadotropins which lead to follicle production in the female ovary and sperm production in the male testis. In female rats, its cerebral injection produces the lordosis posture despite ovariectomy and hypophysectomy, suggesting a direct central effect on sexual behavior as well as gonadotropin-releasing action.

Thyrotropin-releasing factor (TRF) is widely distributed throughout the brain and spinal cord, with 70% found outside the hypothalamus. TRF from the hypothalamus releases TSH, as well as prolactin, from the pituitary. Intracerebrally it directly affects oxygen consumption, respiratory rate, temperature, and epinephrine and glucose levels in the blood. These findings suggest that it has a fundamental role in stabilization of energy resources.

B. Miscellaneous Peptides

A particularly interesting aspect of peptide distribution is the sharing of certain species of peptide between brain and gastrointestinal tract. The frequently noted property of peptides to show parallel actions in the brain and the periphery subservient to the same general function is again illustrated here. Many of these peptides are concerned with digestion and absorption from the gut, while in the brain they are associated with various aspects of feeding and drinking. For example, cholecystokinin (CCK) is associated with satiety. Bombesin administered peripherally increases insulin and glucagon secretion by the pancreas, whereas centrally it increases sympathetic outflow with increased epinephrine, glucagon, and glucose production. Overall, bombesin and the opposing actions of CCK may be important in cardiovascular, thermal, and metabolic homeostasis during nutrient deprivation. The pancreatic hormone, neuropeptide Y, has been associated with obesity.

There is also a group of peptides related to the renin–angiotensin system in the kidney. These hormones act at the peripheral and central level in regulating blood pressure, salt and water balance, as well as the subjective feeling of thirst.

V. CONCLUSIONS

The last 10 years have shown a consolidation of knowledge and understanding of the basic mechanisms of neuropeptide action in the brain. Although progress in molecular biology and immunology has so far been greater than clinical correlation or relevance, we can expect much progress in the coming years through the collaboration of cross-disciplinary research teams. The diversity of functions influenced by neuropeptides suggests that there is almost certainly no simple relation between a particular peptide and a particular behavior. The general picture that emerges is of a complex system of neuroregulators which integrate the action of the CNS, the endocrine system, and the immune system for adaptation to an ever-changing internal and external environment, so as to produce a contented individual functioning at optimal capacity in society. Generally, there is a close functional relationship between peripheral and central effects of a given neuropeptide.

Enhanced understanding of brain neuropeptides should lead to better diagnosis and treatment of mental illness and drug addiction. A realization of how life experiences can affect the physiology of both brain and body emphasizes the importance of combining drug treatment with psychotherapy. Despite the multifaceted functions attributed to neuropeptides, specific receptors have been identified in systems of the brain known to be concerned with particular functions. The development of agonists and antagonists which bind to specific receptors has given us more-precise control over the action of pharmaceutical compounds. This is a rapidly advancing field.

An appreciation of the role of biological factors, such as oxytocin and endorphin acting in the brain during the formation and consolidation of social bonds, may provide insight into the reasons for the breakdown of personal relationships only too common in contemporary society and provide a rationale for treatment. Practical applications have emerged for the use of vasopressin in attention and memory deficits as well as disturbances in the sleep/wake cycle in old age and some psychotic conditions.

The treatment of pain is a major challenge to the medical profession. An understanding of how the perception of pain may be altered through endogenous opiates has been an important factor in the clinical alleviation of pain, phobias, guilt feelings, and anxiety. The role of endogenous opiates in promoting healthy and pleasurable states of mind can help us to understand, prevent, and treat addiction to opiate drugs. Research concerning the possible malfunction of the opiate system in childhood autism and disruptive behavior has suggested novel approaches to treatment.

ACKNOWLEDGMENTS

I should like to dedicate this chapter to Professor C. P. Leblond, Department of Anatomy, McGill University, Montreal, who introduced me to the joys of research. It is remarkable that Dr. Leblond wrote a paper on maternal behavior in 1940 (see Reference 22), which 50 years later, is still relevant to the present state of knowledge. For providing me with library and computer facilities to write this chapter, I should like to thank Professor D. F. Swaab and the staff of the Netherlands Institute for Brain Research in Amsterdam, with particular thanks to Dr. J. P. C. de Bruin for his encouragement and advice.

REFERENCES

1. Barchas, J. D., Neuropeptides in behavior and psychiatric syndromes: an overview, in *Neuropeptides in Neurologic and Psychiatric Disease*, Martin, J. B. and Barchas, J. D., Eds., Raven Press, New York, 1986, 287.
2. Bjorkstrand, E., Hulting, A. L., Meister, B., and Uvnas-Moberg, K., Effect of galanin on plasma levels of oxytocin and cholecystokinin, *Neuroreport*, 4, 10, 1993.
3. Fink, G., Rosie, R., Sheward, W. J., Thomson, E., and Wilson, H., Steroid control of central neuronal interactions and function, *J. Steroid Biochem. Mol. Biol.*, 40, 123, 1991.
4. Krieger, D. T., An overview of peptides, in *Neuropeptides in Neurologic and Psychiatric Disease*, Martin, J. B. and Barchas, J. D., Eds., Raven Press, New York, 1986, 1.
5. Leibowitz, S. F., Brain neuropeptide Y: an integrator of endocrine, metabolic and behavioral processes, *Brain Res. Bull.*, 27, 333, 1991.
6. Moore, M. R. and Black, P. M., Neuropeptides, *Neurosurg. Rev.*, 14, 97, 1991.
7. Myers, R. D., Review: neuroactive peptides: unique phases of research in mammalian brain over three decades, *Peptides*, 15, 367, 1994.
8. Swaab, D. F., Hofman, M. A., Lucasse, P. J., Purba, J. S., Raadsheer, F. C., and Van de Nes, J. A., Functional neuroanatomy and neuropathology of the human hypothalamus, *Anat. Embryol.*, 187, 317, 1993.
9. Van de Poll, N. E. and Van Goozen, S. H. M., Hypothalamic involvement in sexuality and hostility; comparative psychological aspects, *Progr. Brain Res.*, 93, 343, 1992.
10. Argiolas, A. and Gessa, G. L., Central functions of oxytocin, *Neurosci. Biobehav. Rev.*, 15, 217, 1991.
11. Broad, K. D., Kendrick, K. M., Sirinathsinghji, D. J., and Keverne, E. B., Changes in oxytocin immunoreactivity and in RNA expression in the sheep brain during pregnancy, parturition and lactation and in response to oestrogen and progesterone, *J. Neuroendocr.*, 5, 435, 1993.
12. Carter, C. S., Oxytocin and sexual behavior, *Neurosci. Biobehav. Rev.*, 16, 131, 1992.

13. Carter, C. S., Hormonal influences on human sexual behaviour, in *Behavioural Endocrinology*, Becker, J. B., Breedlove, S. M., and Crews, D., Eds., MIT Press, Cambridge, MA, 1992, 131.

14. Carter, C. S. and Getz, L. L., Monogamy and the prairie vole, *Sci. Am.*, June, 70, 1993.

15. Dreifuss, J. J., Tribollet, E., Dubois-Dauphin, M., and Raggenbass, M., Receptors and neural effects of oxytocin in the rodent hypothalamus and preoptic region, *Ciba Found. Symp.*, 168, 187, 1992.

16. Eisenberg, L., The biosocial context of parenting in human families, in *Mammalian Parenting*, Krasnegor, N. A. and Bridges, R. S., Eds., Oxford University Press, New York, 1990, 9.

17. Fleming, A. and Corter, C., Factors influencing maternal responsiveness in humans: usefulness of an animal model, *Psychoneuroendocrinology*, 13, 189, 1988.

18. Hatfield, E. and Rapson, R. L., Passionate love: new directions in research, *Adv. Personal Relationships*, 1, 109, 1987.

19. Keverne, D. B., Central mechanisms underlying the neural and neuroendocrine determinants of maternal behaviour, *Psychoneuroendocrinology*, 13, 127, 1988.

20. Krasnegor, N. A. and Bridges, R. S., Eds., *Mammalian Parenting: Biochemical, Neurobiological and Behavioural Determinants*, Oxford University Press, New York, 1990.

21. Ingram, C. D. and Wakerley, J. B., Post-partum increase in oxytocin-induced excitation of neurones in the bed nuclei of the stria terminalis in vitro, *Brain Res.*, 602, 325, 1993.

22. Leblond, C. P., Nervous and hormonal factors in the maternal behaviour of the mouse, *J. Genet. Psychol.*, 57, 327, 1940.

23. Legros, J. J., Neurohypophyseal peptides and psychopathology, *Horm. Res.*, 37 (Suppl. 3), 16, 1992.

24. Legros, J. J., Ansseau, M., and Timsit-Berthier, M., Neurohypophyseal peptides and psychatric diseases, *Regul. Pept.*, 45, 133, 1993.

25. Newton, N., The role of oxytocin reflexes in three interpersonal reproductive acts: coitus, birth and breast feeding, in *Clinical Psychoneuroendocrinology in Reproduction*, Carenza, L., Pancheri, R., and Zichella, L., Eds., Academic Press, London, 1978, 411.

26. Newton, N. and Modahl, C., Oxytocin — psychoactive hormone of love and breast feeding, in *The Free Woman: Women's Health in the 1990's*, Van Hall, E. V. and Everard, W., Eds., Parthenon, Carnforth, England, 1989, 75.

27. Poindreau, P., Levy, F., and Krehbiel, D., Genital, olfactory and endocrine interactions in the development of maternal behaviour in the parturient ewe, *Psychoneuroendocrinology*, 13, 99, 1988.

28. Sobrinho, L. G., The psychogenic effects of prolactin, *Acta Endocrinol. (Copenhagen)*, 129 (Suppl. 1), 38, 1993.

29. Stern, J. M., Licking, touching and suckling: constant stimulation and maternal psychobiology in rats and women, *Ann. N.Y. Acad. Sci.*, 474, 95, 1986.

30. Swanson, H. H. and van Leengoed, E., Oxytocin antagonist injected into the cerebral ventricle inhibits spontaneous postpartum maternal behavior in Wistar rats, in *The Free Woman: Women's Health in the 1990s*, Van Hall, E. V. and Everard, W., Parthenon, Eds., Carnforth, England, 1989, 326.

31. Winslow, J. T., Hastings, N. Carter, C. S., Harbaugh, C. R., and Insel, T. R., A role for central vasopressin in pair bonding in monogamous prairie voles, *Nature*, 365, 545, 1993.

32. Buijs, R. M., Vasopressin and oxytocin — their role in neurotransmission, *Pharm. Theor.*, 22, 127, 1983.

33. Buijs, R. M., The development of vasopressin and oxytocin systems in the brain, *Handbo. Chem. Neuroanat.* 10, 547, 1992.

34. Dantzer, R. and Bluthe, R. M., Vasopressin and behaviour: from memory to olfaction, *Regul. Pept.*, 45, 121, 1993.

35. De Wied, D., Diamant, M., and Fodor, M., Central nervous system effects of the neurohypophyseal hormones and related peptides, *Front. Neuroendocrinol.*, 14, 251, 1993.

36. Fehm-Wolfsdorf, G. and Born, J., Behavioural effects of neurohypophyseal peptides in healthy volunteers: 10 years of research, *Peptides*, 12, 1399, 1991.

37. Gash, D. M. and Boer, G. H., Eds., *Vasopressin: Principles and Properties*, Plenum Press, New York, 1987.

38. Gash, D. M., Herman, J. P., and Thomas, G. J., Vasopressin and animal behaviour, in *Vasopressin: Principles and Properties.*, Gash, D. M. and Boer, G. H., Eds., Plenum Press, New York, 1987, 517.

39. Hoffman, P. L., Central nervous system effects of neurohypophyseal peptides, *Peptides*, 8, 239, 1987.

40. Hofman, M. A. and Swaab, D. F., The human hypothalamus: comparative morphology and photoperiodic influences, *Progr. Brain Res.*, 93, 133, 1992.

41. Hofman, M. A. and Swaab, D. F., Diurnal and seasonal rhythms of neuronal activity in the suprachiasmatic nucleus of humans, *J. Biol. Rhythms*, 8, 283, 1993.

42. Jolles, J., Vasopressin and human behaviour, in *Vasopressin: Principles and Properties*, Gash, D. M. and Boer, G. H. Eds., Plenum Press, New York, 1987, 549.

43. Kalsbeek, A. and Buijs, R. M., Peptidergic transmitters of the suprachiasmatic nuclei and the control of circadian rhythmicity, *Progr. Brain Res.*, 92, 321, 1992.

44. Legros, J. J. and Ansseau, M., Neurohypophyseal peptides and psychopathology, *Progr. Brain Res.*, 93, 455, 1992.

45. Moore, R. Y., The organization of the human circadian timing system, *Progr. Brain Res.*, 93, 99, 1992.

46. Van Someren, E. J. W., Mirmiran, M., and Swaab, D. F., Non-pharmacological treatment of sleep and wake disturbances in aging and Alzheimer's disease: chronobiological perspectives, *Behav. Brain Res.*, 57, 235, 1993.

47. Dyer, R. G. and Bicknell, R. J., Eds., *Brain Opioid Systems in Reproduction*, Oxford University Press, Oxford, 1989.

48. Herbert, J., Specific roles for beta-endorphin in reproduction and sexual behaviour, in *Brain Opioid Systems in Reproduction*, Dyer, R. G. and Bicknell, R. J., Eds., Oxford University Press, Oxford, 1989, 167.

49. Hill, H. F. and Mather, L. E., Patient-controlled analgesia. Pharmacokinetic and therapeutic considerations, *Clin. Pharmacokinet.*, 24, 124, 1993.

50. Hunt, D. D., Adamson, R., Egan, K., and Carr, J. E., Opioids: mediators of fear and mania, *Biol. Psychiatr.*, 23, 426, 1988.

51. Kalin, N. H., The neurobiology of fear, *Sci. Am.*, May, 54, 1993.

52. Keverne, E. B. and Kendrick, K. M., Morphine and corticotropin-releasing factor potentiate maternal acceptance in multiparous ewes after vaginocervical stimulation, *Brain Res.*, 540, 55, 1991.

53. Levine, J. D., Fields, H. L., and Basbaum, A. I., Peptides and the primary afferent nociceptor, *J. Neurosci.*, 13, 2273, 1993.

54. Olson, G. A., Olson, R. D., and Kastin, A. J., Review: endogenous opiates: 1992, *Peptides*, 14, 1339, 1993.

55. Panksepp, J., A neurochemical theory of autism, *Trends Neurosci.*, 2, 174, 1979.

56. Panksepp, J., Herman, B. H., Vilberg, T., Bishop, P., and De Eskinazi, F. G., Endogenoous opioids and social behaviour, *Neurosci. Behav. Rev.*, 4, 473, 1978.

57. Rodgers, R. J. and Shepherd, J. K., Prevention of the analgesic consequences of social defeat in male mice by 5-HT-1A anxiolytics, buspirone, gepirone and ipsopirone, *Psychopharmacology*, 99, 374, 1989.

58. Seckl, J. and Lightman, S. L., Opioid peptides, oxytocin and human reproduction, in *Brain Opioid Systems in Reproduction*, Dyer, R. G. and Bickwell, R. J., Eds., Oxford University Press, Oxford, 1989, 309.

59. Terenius, L., Opioid peptides, pain and stress, *Progr. Brain Res.*, 92, 375, 1992.

60. Schaeffer, G. J., Opiate antagonists and rewarding brain stimulation, *Neurosci. Biobehav. Rev.*, 12, 1, 1988.

61. Schwarz, L. and Kindermann, W., Changes in beta-endorphin levels in response to aerobic and anaerobic exercise, *Sports Med.*, 13, 25, 1992.

62. Ursin, H., Psychobiology of stress and attachment: biobehavioural view, in *Health Promotion Research: Toward a New Social Epidemiology*, Badura, B. and Kickbusch, I., Eds., WHO Regional Publ., Europ. Series No. 37, Copenhagen, 1991, 173.

63. Von Holst, D., Psychosocial stress and its pathophysiological effects in tree shrews (*Tupaia belangeri*), in *Biological and Psychological Factors in Cardiovascular Disease*, Schmidt, T. H., Dembroski, T. M., and Blumchen, G., Eds., Springer-Verlag, Berlin, 1986, 476.

64. Wiegant, V. M., Ronken, E., Kovacks, G., and De Wied, D., Endorphins and schizophrenia, *Progr. Brain Res.*, 93, 433, 1992.

Part III

PSYCHOTROPIC DRUGS

Chapter **9**

ANTIDEPRESSANTS

Rifaat Kamil

CONTENTS

0-8493-8386-0/96/$0.00+$.50

I. INTRODUCTION

Depressive illness constitutes one of the most frequently seen mental disorders. The morbidity and mortality associated with alteration of mood is one of the major problems facing psychiatry, with the National Comorbidity Study showing a lifetime prevalence of major depression of 12.7% in men and 21.3% in women.[1] Along with the recognition of the high incidence and prevalence of mood disorders, there is sobering evidence that most mood disorders are chronic and require long-term management.[2] The landmark studies by Frank, Kupfer, and colleagues have shown high rates of relapse and recurrence when depression is inadequately treated.[3,4] Pharmacotherapy continues to be a mainstay treatment of depression.

The discovery of the antidepressant effect of medications was coincidental to their use for other disorders. Selikoff's[5] initial work showed that iproniazid (originally marketed for the treatment of tuberculosis) could elevate mood. The usefulness of this medication was confirmed in subsequent studies.[6,7] The use of iproniazid was discontinued after it was noted that hepatic necrosis was an unfortunate side effect. Imipramine, a derivative of chlorpromazine, was initially marketed as an antipsychotic agent. Subsequently, it was also shown to have antidepressant properties.[8]

Since the somewhat serendipitous but fortunate discovery of the initial antidepressants, many additional medications have been tested and approved for the treatment of depression. Although much work has been done on brain neuropeptide and second messenger systems, the principal groups of antidepressants available today are all presumed to exert their action via alteration of brain monoamine metabolism. Many of the new antidepressants, however, offer significant advantages with regard to reduced side effects and low toxicity in overdose. The advent of newer compounds has increased the options available to the clinician but also has created some uncertainty in establishing a clinically relevant algorithm for the treatment of depression.

II. CLASSIFICATION

There is no uniform classification system for currently available antidepressants. A system based either on chemical structure or biological action would be desirable. However, because of the pervasive nature of clinically established groupings, it is prudent to identify these compounds according to easily recognized classification schema. An overview of antidepressant compounds is provided in Table 1.

A. Monoamine Oxidase Inhibitors

The monoamine oxidase inhibitors (MAOIs) exert their primary action on the mitochondrial enzyme monoamine oxidase (MAO). This enzyme is responsible for the degradation of biogenic monoamines (norepinephrine, dopamine, serotonin, and melatonin). MAOIs block the action of this enzyme and thus increase the postsynaptic availability of these neurotransmitters.[9] Two subtypes of the MAO enzyme have been

TABLE 1.

Selected Antidepressant Drugs: Transmitter Effects

Drug	Neurotransmitter Enhancement or Reuptake Inhibition			Receptor Antagonism			
	NE	5–HT	DA	ACh	H_1	alpha$_1$	5-HT$_2$
MAOIs							
Irreversible MAOIs							
Phenelzine	+	+	+				
Isocarboxazid	+	+	+				
Tranylcypromine	+	+	+				
RIMAs							
Moclobemide	+	+	+				
Brofaramine	+	+	+				
TCAs							
Amitriptyline	+		+	+	+	+	
Imipramine	+		+	+	+	+	
Clomipramine	+		+	+	+	+	
Nortriptyline	+			+?	+?	+?	
Desipramine	+			+?	+?	+?	
Tetracyclic							
Maprotiline	+						
SSRIs							
Fluoxetine			+				
Fluvoxamine			+				
Paroxetine			+				
Sertraline			+				
Atypical							
Phenylpiperazines							
Nefazodone			+				+
Trazodone			+		+?		+
Phenylaminoketones							
Bupropion			+				
Phenylethylamine							
Venlafaxine	+	+	+				

[Handwritten margin note: Monoamines — Norepinephrine — Dopamine — Serotonin — melatonin]

Note: Data indicate qualitative effects on reuptake inhibition and receptor antagonism. NE, norepinephrine; 5-HT, serotonin; DA, dopamine; ACh, acetylcholine; H_1, histamine; alpha$_1$, alpha$_1$ adrenergic.

identified and are classified as MAO-A and MAO-B. The former catabolizes norepinephrine, serotonin, and dopamine, while the latter functions principally on phenylethylamine and dopamine (as well as exogenously occurring monoamines such as tyramine). MAOIs are classified as selective (in that they inhibit either MAO-A or MAO-B) or nonselective (exerting effects on both MAO-A and MAO-B). MAOI antidepressants are also classified as reversible or irreversible, based on their inhibition of monoamine oxidase.

1. Irreversible MAOIs

Four currently available medications are classified in this category. The hydrazines, which include isocarboxazid (Marplan) and phenelzine (Nardil), are chemically related to iproniazid. The nonhydrazines include tranylcypromine (Parnate) and pargyline (Eutonyl). Tranylcypromine has received some attention as it structurally resembles amphetamine and may be partially metabolized to this compound.[10] All the above compounds are nonselective with the exception of pargyline, which is relatively selective for MAO-B at lower doses.

The irreversible inhibition of MAO means that neurons must synthesize additional enzyme before biological activity is reestablished. The action of the enzyme is only apparent 10 to 14 days after the discontinuation of an irreversible MAOI. As well, irreversible MAOIs require the maintenance of dietary and drug monitoring compliance to prevent rapid and deleterious increases in monoamines (see Sections IV and V).

2. Reversible MAOIs

This recently introduced class of MAOIs includes moclobemide (Mannerix) and brofaramine (currently in clinical trials). Both these compounds reversibly inhibit MAO-A. Although they can produce elevations of norepinephrine and serotonin, they have reduced effect on dopamine levels and no effect on the degradation of exogenous amines such as tyramine. The clinical advantage of the reversible inhibitors of monoamine oxidase A (RIMAs) is that they do not require dietary compliance or a prolonged interval for MAO activity to be reestablished after discontinuation of the medication. They have thus improved safety in managing patients on MAOIs.

B. Tricyclic Antidepressants

Imipramine was the first of the tricyclic antidepressants (TCAs).[8] The nomenclature *TCA* for imipramine and related compounds has become universal even though some of the compounds are better classified structurally as heterocyclic or tetracyclic. The two original TCAs are tertiary amines, specifically imipramine (Tofranil and others) and amitriptyline (Elavil and others). These have demethylated secondary amine derivatives, desipramine (Norpramine) and nortriptyline (Aventyl and others). Additional tertiary compounds include clomipramine (Anafranil), doxepin (Adapin and others), and trimipramine (Surmontil). Maprotoline (Ludiomil) is a related tetracyclic medication. (See Table 1.)

All TCAs (and structurally related moieties) inhibit the presynaptic reuptake of the monoamine neurotransmitters norepinephrine and serotonin. They thus acutely enhance the availability of these monoamines. The relative effect on serotonin or norepinephrine reuptake inhibition varies from one TCA to another. Some TCAs affect serotonin and norepinephrine reuptake approximately equally (imipramine, amitriptyline), while others preferentially affect norepinephrine (desipramine, nortriptyline, protriptyline, and maprotiline) or serotonin (clomipramine). Regardless of varying action on norepinephrine or serotonin reuptake, most TCAs are equivalent in their antidepressant efficacy.

In addition to the blockade of neurotransmitter uptake, most TCAs have direct affinities for several different heterogeneous receptors. Their principal antagonistic effects are noted at the histamine$_1$, alpha$_1$-adrenergic, and muscarinic cholinergic receptors.[11,12] The tertiary amines exert particularly potent antihistaminic, alpha-adrenergic blocking, and anticholinergic effects. The receptor antagonism is primarily responsible for the side effects and toxicity of TCA compounds.

One TCA, amoxapine (Ascendin), has *in vitro* antagonistic effects on the dopamine D$_2$ receptor, accounting for its extrapyramidal side effects.[14]

C. Selective Serotonin Reuptake Inhibitors

Principle serotonin (5-hydroxytryptamine, 5-HT) receptors are subclassified as 5-HT$_1$ to 5-HT$_7$, with the 5-HT$_1$ receptor further known to have several subtypes (5-HT$_{1a-f}$).[14] Serotonin systems are discussed in detail in Chapter 4. A structurally heterogeneous group of compounds that have selective effect on the presynaptic reuptake of serotonin has recently become the most popular prescribed antidepressant. This group is classified as selective serotonin reuptake inhibitors (SSRIs).

Fluoxetine (Prozac) was the first marketed SSRI antidepressant. Other SSRIs are fluvoxamine (Luvox), sertraline (Zoloft), paroxetine (Paxil), and citalopram (not released). Most SSRIs have only modest clinically relevant effects on other brain receptors. As such, their clinical profile is primarily a reflection of their effect in acutely enhancing the synaptic availability of serotonin. These compounds have fewer antihistaminic, alpha-adrenergic, and anticholinergic effects compared with the TCAs. Many of the side effects associated with SSRIs are related to their enhancement of serotonin. In that they lack clinically significant effects on adrenergic and cholinergic receptors, these medications are free of the cardiac and anticholinergic side effects related to TCAs. They also have an extremely safe therapeutic/toxic ratio, perhaps partly accounting for their current popularity.

D. Novel/Atypical Antidepressants

This rather loose categorization includes antidepressant compounds that do not have the structural or functional similarities of the above classified compounds (MAOIs, TCAs, and SSRIs). However, as with all antidepressants, the primary mode of action is via monoamine transmitter reuptake inhibition.

The phenypiperazine compounds trazodone (Desyrel and others) and nefazodone (Serzone) preferentially inhibit the presynaptic reuptake of serotonin. Trazodone may achieve its antidepressant effect via antagonism of the 5-HT_2 receptor. Nefazodone is particularly interesting in that it not only displays antagonism at the 5-HT_2 receptor but has metabolites that have agonist properties at 5-HT_{1a} and 5-HT_{1c} receptor sites.[15] It also has a relative lack of antagonism at the histaminic, cholinergic, and alpha-adrenergic receptors, accounting for the paucity of side effects related to these receptors.[16]

Bupropion (Wellbutrin), a monocyclic phenylaminoketone, has relatively weak effects on the presynaptic reuptake of norepinephrine and serotonin.[17-19] It does have clinically significant reuptake inhibition of dopamine as well as direct agonist effects on dopamine receptors. Its mode of action is obscure, but it is presumed that its effect on dopamine underlies its effectiveness as an antidepressant and also is responsible for some of its side effects. Its low affinity for alpha-adrenergic and histaminic receptors accounts for a low incidence of side effects usually ascribed to receptor antagonism at these sites.

Venlafaxine (Effexor), a structurally novel phenylethylamine, strongly inhibits both norepinephrine and serotonin reuptake.[20,21] It may also exert clinically significant reuptake inhibition of dopamine and, therefore, may have the net bioamine effect similar to the MAOIs without incurring some of the consequences of MAO inhibition. In addition, its effect on neurotransmitter availability may be analogous to the TCAs which inhibit the reuptake of both serotonin and norepinephrine. However, since it lacks antagonism at the histaminic, alpha-adrenergic, and cholinergic receptor sites, it is also relatively free of side effects associated with blockade at these transmitter systems.[22]

III. MODE OF ACTION

As has been alluded to above, the principal mode of action of all available antidepressants is their alteration of the availability of brain monoamine neurotransmitters.

Antidepressants acutely enhance the availability of one or more monoamines either by blocking degradation (the MAOIs) or by preventing presynaptic reuptake (TCAs, SSRIs, and most novel agents). It is presumed that this initial enhancement of monoamine levels causes secondary receptor down regulation which coincides with elevation of mood. Most investigators have examined specific aspects of monoamine metabolism in depression, but it may be likely that a "final common pathway" may underlie the effects of varied antidepressants. The success of antidepressants which singly affect norepinephrine, serotonin, or dopamine has been seen as evidence that no single monoamine system is implicated in mood alteration. It is also established that alteration of the norepinephrine system has downstream effects on serotonin metabolism.[23,24] The complexity of interactions between all neurotransmitter systems and their effects on second messenger systems (particularly on adenylate cyclase, cyclic AMP, and the phosphoinositides) are just beginning to be elucidated.[23] The mode of action of antidepressants is complex and will likely continue to defy a simplistic or uniform explanation.

The biochemistry and role of brain monoamines has been reviewed in detail in Chapter 4 of this volume. A brief review of depression-related changes in monoamine metabolism is discussed with relevance to antidepressant mode of action.

A. Norepinephrine

The involvement of monoamines in the pathogenesis of depression was invoked as early as 1965 by Schildkraut.[25] He postulated that depression is related to a functional deficiency of norepinephrine. The early evidence for this from studies of plasma and urine metabolites of norepinephrine was confounded by the fact that most measures of norepinephrine and its metabolites were not measures of brain monoamine metabolism.[26,27] Subsequent measures of brain norepinephrine metabolism (in cerebrospinal fluid, CSF) have not been able to consistently confirm that depression is associated with a relative deficiency of norepinephrine or a decrease in norepinephrine turnover.[28] A major problem in interpreting these types of studies has been their inability to control significant factors such as age, sex, and assay compatibility. The NIMH Collaborative Study did, however, document that decreased norepinephrine metabolites are predictive of a response to specific antidepressants that preferentially enhance norepinephrine reuptake inhibition.[29] Moreover, brain receptor studies looking at the adrenergic system have noted that alpha$_2$-adrenergic and beta$_1$ receptor down regulation, as a response to antidepressant treatment, is related to therapeutic outcome and that the time course of receptor down regulation is temporally related to the therapeutic effect of antidepressant medications (usually 2 to 3 weeks).[30]

B. Serotonin

Coppen[31] suggested that a deficiency in brain serotonin is related to the clinical state of depression, and the serotonin system has come under intense scrutiny over the last two decades. Studies involving the CSF serotonin metabolite 5-hydroxyindoleacetic acid (5-HIAA) offer convincing evidence that brain serotonin metabolites are decreased in patients who attempt suicide.[32,33] Further, post-mortem studies have also confirmed that brain 5-HT receptors are increased in suicide victims.[34] The correlation between clinical response to an antidepressant and a reduction in 5-HT$_2$ receptor binding has been seen as a robust and consistent finding, so much so that reduced 5-HT$_2$ binding is currently taken as a benchmark in the identification of compounds with putative antidepressant properties at preclinical testing stages.[34]

Further intriguing studies implicating altered serotonin metabolism in depression have been derived from the imipramine-binding properties of platelets (with imipramine binding taken as a measure of serotonin receptors). Unfortunately, many of these studies have been difficult to interpret as imipramine has been shown to be somewhat nonspecific in its affinity for serotonin receptor sites.[35]

Clinical evidence for the importance of serotonin has been derived from studies by Delgado and others[36,37] investigating tryptophan depletion in patients receiving serotonin reuptake inhibiting antidepressants. Tryptophan is the amino acid precursor of serotonin in the rate-limiting step in its production, and its depletion has been shown to cause a reemergence of depression.[36-38]

C. Dopamine

Randrup et al.[39] were the first to formally propose the involvement of the dopamine system in depression. It has been postulated that dopaminergic systems underlie reward seeking and environmental sensitivity. Some early animal model evidence showing dopamine deficiency in depressive-analogue states was provided by Anisman et al.[40,41] Anisman's investigations of the model of learned helplessness (LH) showed that LH states in animals were correlated with dopamine depletion in the nucleus accumbens. It was further shown that pretreatment with dopamine-enhancing drugs could block LH development, whereas dopamine antagonists exacerbated LH. Analogous human dopamine depletion has been thought to underlie the anhedonia and psychomotor retardation seen in depression.[42]

Indirect effects of dopamine involvement in the alteration of mood have been derived from drug studies.[43] The mood-altering euphoriant effect of cocaine is believed to be mediated by its enhancement of the dopamine neuron firing rates. Further, the dopamine-depleting antihypertensive, alpha-methyl-dopa, has been known to induce depression, and antidepressants which preferentially inhibit dopamine presynaptic reuptake (bupropion, venlafaxine) are thought to have specific beneficial effects on psychomotor retardation.[19,20]

Finally, human studies by Roy et al.[44] on dopamine metabolism have shown that CSF dopamine is reduced in depression and that 24-h urine homovanillic acid (HVA, a dopamine metabolite) is reduced in suicidal depressed patients compared with nonsuicidal patients.

IV. CLINICAL USE

Although antidepressants have been used in a variety of psychiatric and non-psychiatric conditions, their primary indication is in the treatment of depressive disorders (particularly major depressive disorder as defined by DSM-IV).[45] In addition to the treatment of depression, antidepressants have also been used to treat posttraumatic stress disorder, anxiety disorders (obsessive-compulsive disorder, panic disorder, and specific phobias), chronic pain syndromes, eating disorders, and attention-deficit hyperactivity disorder.[46] The principal focus of this chapter will be to review the indications for using antidepressants in the treatment of major depressive disorder.

A decision to use a particular antidepressant is based on consideration of a patient's psychiatric history and symptom profile, evaluation of the side effects and safety of a medication, and finally, the putative neurotransmitter effect produced. A suggested algorithm for the treatment of depression is discussed after specific classes of medications are examined in detail. The use of augmentation therapy is also outlined.

A. Monoamine Oxidase Inhibitors

Traditionally, in North America MAOIs have been used as second- or third-line agents in patients who do not respond to TCAs or SSRIs.[47-49] More recently, considerable attention has focused on the utility of these agents in atypical depression.[50] This diagnostic entity has been defined by DSM-IV[45] as depression with the following prominent features:

1. Weight gain or increased appetite
2. Hypersomnia
3. "Leaden" paralysis
4. Interpersonal rejection sensitivity

Recent investigators have marshaled increasingly convincing evidence that MAOIs have particular efficacy in the treatment of atypical depression, even in circumstances where first-line treatment with TCAs has failed.[47,51]

1.   Irreversible MAOIs

The three irreversible MAOIs (phenelzine, isocarboxazid, and tranylcypromine) all have similar efficacy.[46] Special consideration must be given to the concept of drug "washout" with irreversible MAOIs as this depends not on the pharmacokinetics of the drug, but on the time taken to synthesize new mitochondrial monoamine oxidase (10 to 14 days). Data on phenelzine[52] shows that up to 2 weeks is required to achieve maximal inhibition of platelet monoamine oxidase and that a similar amount of time is required to recover enzyme activity after discontinuation of the medication. Robinson et al.[52] has established that 80% or greater inhibition of monoamine oxidase is correlated with a therapeutic antidepressant response. Clinically, routine determinations of monoamine oxidase activity are not possible, but the recommended doses of MAOIs that likely inhibit 80% of monoamine oxidase activity are given in Table 2.

MAOIs are relatively nonsedating medications and are typically administered during the first half of the day to prevent insomnia. Tranylcypromine has received particular attention in this regard because of its structural similarity to d-amphetamine.[10] Usually, the total dose may be divided into two and administered at breakfast and lunch. Treatment with phenelzine, the most extensively studied MAOI, is usually commenced at 15 mg twice a day with a final dose being in the range of 1 mg/kg body weight.[53,54]

The irreversible inhibition of monoamine oxidase has prompted many clinicians to regard MAOIs with extreme caution because of the potential of inducing severe hypertension. The unavailability of monoamine oxidase renders the patient vulnerable to situations which might result in increased sympathomimetic or serotonergic activity. Special dietary compliance is necessary for the period of administration of MAOIs and for a 2-week period following cessation of the drug. The necessity for this arises because of the presence of tyramine in some food products; intestinal MAO is inactivated by MAOIs, and dietary tyramine can act as a false neurotransmitter, as well as cause release of endogenously stored norepinephrine. This, in turn, can precipitate the "cheese" reaction, so named because of the relatively high concentrations of

TABLE 2.

Selected Antidepressants: Dosing Strategies

Drug	Maximal Dose (mg/day)	Starting Dose (mg/day)	Comments
Irreversible MAOIs			
Phenelzine	90	15 b.i.d	Require dietary
Tranylcypromine	60	10 b.i.d	compliance and 2-week washout
RIMAs			
Moclobemide	600–900	100 t.i.d	MAO-A inhibition reversed in 24 h
TCAs			
Amitriptyline	300	50	Tertiary amines
Imipramine	300	50	have potent H_1, Ch,
Clomipramine	300	50	alpha$_1$ blockade
Despramine	300	50	Secondary amines
Nortriptyline	125	25	have little receptor antagonism
SSRIs			
Fluoxetine	160	10–20	Long $t_{1/2}$
Fluvoxamine	200	50	
Sertraline	150	50	
Paroxetine	60	10–20	Shortest $t_{1/2}$ among SSRIs
Atypical			
Trazodone	600	50–100	
Nefazodone	600	100 b.i.d	
Venlafaxine	375	37.5 b.i.d	
Bupropion	450	75 t.i.d.	Doses over 450 mg/day associated with seizures

tyramine found in aged cheese. Patients receiving MAOIs are therefore asked to refrain from foods which may contain tyramine, and Jenike[55] has summarized key foods to avoid while on MAOI therapy. These include most cheeses (except soft cream cheeses), yeast-extract–containing foods (e.g., marmite, most commercial soup and gravy stocks), overripe or stewed fruits, broad beans, and aged meats. Previously suspect substances, such as wines (with the possible exception of Chianti), liquors, and chocolate, may be taken in small or moderate quantities without fear of precipitating a tyramine reaction.

In addition to foods, medications which have sympathomimetic or serotonergic activity cannot be used during MAOI treatment or in the 2-week period following cessation of therapy. TCAs and SSRIs must be stopped at least 10 days prior to commencing treatment with an MAOI, the exception to this being fluoxetine which requires a washout of 5 weeks because of the long half-life of its active metabolite, norfluoxetine.[56] Meperidine, a synthetic narcotic analgesic, is absolutely contraindicated during treatment with MAOIs as it may precipitate the serotonin syndrome (discussed below). Decongestants and cough medications, with the exception of Benadryl, must also be avoided during MAOI administration.

2. RIMAs

RIMAs represent an important step forward in the safe administration of MAOIs. Compared to classic MAOIs, the RIMAs, with their selective, reversible, and short-acting inhibition of MAO-A, have been shown to be safe with regard to interaction with sympathomimetic amines.[57-59] Much of the concern about dietary compliance is removed as the enzyme MAO-B is available to degrade tyramine. RIMAs have a large therapeutic/toxic ratio.

Moclobemide and brofaramine (not currently released) are two recently investigated RIMA antidepressants. Moclobemide, a benzamide derivative, has been the most extensively studied RIMA. In a recent review, Nair et al.[60] reported that it is rapidly absorbed after oral administration and MAO-A inhibition rises to 80% within 2 h, with the duration of MAO-A inhibition lasting 8 to 10 h. Activity of MAO-A is completely reestablished within 24 h of ingestion of the last dose.[61] The therapeutic advantage gained is the allowance of a quick switch to another antidepressant without the obligatory 10-day to 2-week washout required for irreversible MAOIs. Moclobemide was shown to have efficacy superior to placebo and equal to TCAs, SSRIs, and irreversible MAOIs in several studies.[47,57,62,63] However, in one multicenter trial, moclobemide was shown to have a weaker antidepressant effect than the tricyclic clomipramine.[64] Clinically, moclobemide was shown to be well tolerated and to have a side-effect profile superior to TCAs and irreversible MAOIs.[57,65] Its efficacy has also been documented in the treatment of atypical depression.[66] The usual starting dose is 300 mg/day. This may be increased up to 600 mg/day if needed, and divided doses are recommended based on the half-life of the drug.

Brofaramine is not currently released for clinical use. As with moclobemide, this RIMA antidepressant has not been noted to cause clinically significant tyramine reactions.[67] Two double-blind comparisons with tranylcypromine revealed brofaramine to be equally efficacious and better tolerated,[68,69] and it is likely that brofaramine will join moclobemide as a useful MAO-A–inhibiting antidepressant.

B. Tricyclic Antidepressants

The TCAs have been the most extensively studied medications used in the treatment of depression. Indeed, the antidepressant efficacy of the TCA imipramine is often used as a standard for comparison in controlled trials of new antidepressants. Until the introduction of SSRIs, TCAs were first-line agents in the pharmacotherapeutic management of unipolar mood disorders. Tertiary tricyclic compounds, especially imipramine and amitriptyline, however, are associated with potent antihistaminergic, anticholinergic, and alpha-adrenergic blocking effects. These medications also have cardiovascular effects similar to class I antiarrythymics, which renders them potentially cardiotoxic in doses only marginally more than their therapeutic daily dose (with a toxic dose being the equivalent of a 1-week cumulative dose).[70-72] As a result, current practice with TCAs is often limited to using secondary amine TCAs, such as desipramine or nortriptyline, which have clinically reduced receptor antagonist activity (see Table 1). It is recommended that to prevent potential cardiotoxicity in overdose, outpatient pharmacotherapy with TCAs should not be initiated with more than a 1-week supply prescribed at a time.

An important aspect of TCA use is the widespread clinical availability of plasma drug level assays. The TCA nortriptyline has been reported to have a "therapeutic window" of efficacy,[73] and this provides a useful rationale for the preferential use of this medication. A desired level of 50 ng/l to 150 ng/l is reported to be optimal for antidepressant effect, and this should be monitored 5 to 7 days after achieving stable doses of nortriptyline. Although plasma drug levels are available for most TCAs, the usefulness of these levels and their relationship to efficacy has been debated extensively. The APA Task Force on Laboratory Tests[73] recommends that clinical utility for TCA levels is limited and should be carried out under the following circumstances:

1. Patient nonresponse to an adequate trial of imipramine, desipramine, or nortriptyline;
2. Patient at high risk for toxicity requiring that the lowest and safest effective dose be used;
3. Patient requiring rapid increases of medication to arrive at optimal drug levels as quickly as possible;
4. Concern regarding noncompliance;
5. Potential for drug interactions with other concurrent drugs which might affect TCA levels; and
6. Documentation of TCA level to tailor future treatment.

Most TCA antidepressants have half-lives of about 1 day, thereby requiring once-daily dosing. As well, most TCAs have optimal dosing levels of 2.5 to 3 mg/kg body weight, the exceptions being nortriptyline and protriptyline, which are more potent and require a lower daily dose. The doses should be reduced for geriatric patients. Most tertiary TCAs are sedating because of antihistaminergic effects, and this may be usefully exploited by administration at night to improve sleep, while secondary amines can be administered during the day. Most recommendations suggest that initial doses commence at 25 to 50 mg/day; however, a recent study using nortriptyline in 26 depressed patients at starting doses of 75 to 125 mg/day found a more rapid response with significant mood improvement after 1 week of treatment.[74] The usefulness of plasma drug monitoring is highlighted here, where rapid early dosing may be guided by obtaining TCA levels 5 days after initiation of an aggressive dosing strategy.

C. Selective Serotonin Reuptake Inhibitors

Over the past 5 years, the SSRIs have rapidly assumed the role of first-line antidepressant agents in the management of depression. The basis for this is complex and reflects several reasons, including an increasing interest in the serotonergic system in the neurobiology of depression (outlined above), the extremely good safety profile of SSRIs in overdose, and the lower incidence of side effects compared with TCAs and classic MAOIs. Perhaps, increasing public awareness of depression and aggressive drug-marketing strategies also account for their current popularity.[75]

Four SSRIs are currently available and approved for clinical use in the treatment of depression. These are fluoxetine (Prozac), fluvoxamine (Luvox), paroxetine (Paxil), and sertraline (Zoloft). An SSRI still to be released, but showing good efficacy in clinical trials, is citalopram.[76,77] In numerous double-blind, placebo-controlled, and medication-controlled trials, all SSRIs have been shown to be efficacious in the treatment of major depression with treatment outcome superior to placebo and equal with TCAs and MAOIs.[78-85] Clinical trials comparing these medications to TCAs have consistently shown them to be better tolerated in double-blind studies. Data on suicide attempts have established that all SSRIs are remarkably safe in overdose and are almost devoid of any cardiotoxic effects.[71,72] In addition, in spite of their seemingly high prescription cost, a recent cost–benefit analysis comparing fluoxetine with imipramine showed that, when overall cost was evaluated in terms of efficacy, compliance, side effects, and safety profile, these medications are as cost effective as TCAs.

A recent safety concern was raised when an initial report linked the emergence of de novo suicidal or aggressive behavior in depressed patients taking fluoxetine,[87] and there have been several exhaustive reviews of the literature over the past few years which have analyzed the relationship between antidepressant use and new-onset suicidal ideation.[88-91] In large scale studies, there is no convincing evidence that use of SSRIs is related to an increase in suicidal ideation or attempts. Several investigators have, however, linked akathisia associated with SSRI use with possible exacerbation of depression and suicidal intent,[88] and caution is therefore advised in the use of SSRIs when akathisia is a prominent side effect. The recognition and treatment of SSRI-related akathisia and other side effects is discussed below (see Section V).

The clinically available SSRIs are markedly varied in their chemical structure but have somewhat similar neurotransmitter and clinical profiles. Among them, paroxetine is the most potent inhibitor of the 5-HT transporter, and fluoxetine has the least specificity. All, however, are highly selective for inhibition of reuptake of serotonin, as compared with their effect on norepinephrine reuptake. Selection of individual SSRI agents is based on considerations of diagnosis and side effects, as well as the use of concurrent medications. Investigators have determined that fluoxetine and paroxetine have potent inhibitory effects on cytochrome oxidase P-450, specifically the 2D6, 3A4, and possibly the 2C isoenzymes.[92] Consequently, the two mentioned SSRIs may increase the plasma levels of several concurrently used medications, including tricyclic antidepressants, neuroleptics, and such anticonvulsants as phenytoin.

Currently, SSRIs are widely used as first-line agents in all forms of major depression. The recommended starting and maximal doses of the SSRIs are given in Table 2. Although plasma levels are often used in clinical and research studies, there is no established validity of the use of SSRI plasma levels in clinical practice.

Fluoxetine, the first of the SSRIs, has been most extensively studied and is most remarkable for the very long half-life of its active metabolite norfluoxetine, i.e., up to 360 h.[83,85] Clinically, this may be exploited by allowing the interdosing interval to be increased to every few days. It is also possible to obtain a therapeutic response using small individual doses, and a unit dose of 20 mg/day can be administered without changing the dosage for age, sex, or body weight. This relative uniformity in dosing has facilitated the use of this medication in the treatment of uncomplicated major depression in primary-care settings. However, if side effects are encountered, there may be a prolonged interval of time before the drug is eliminated. Nuclear magnetic resonance studies using[19] fluorine have shown that both fluoxetine and norfluoxetine accumulate in brain at concentrations up to twice those found in plasma,[56] although the therapeutic and toxic effects of preferential accumulation of these SSRI compounds in brain tissue is unknown. Doses up to 80 mg/day have been used without major side effects. At the same time, there has been a report involving a study of four patients on 20 mg/day who did not respond to treatment at 20 mg/day until fluoxetine was withdrawn for 2 weeks and reinstituted at 20 mg every other day.[93] Clinically, a nonresponse to fluoxetine may possibly be secondary to serotonergic overstimulation, and consideration should be given to a dose reduction in circumstances where clinical improvement is absent in the face of high-dose treatment.

The long half-life of fluoxetine can be problematic as a washout period of 5 weeks is required prior to the institution of alternate pharmacotherapy. In addition, a routine change from fluoxetine to MAOIs as part of clinical management is also difficult since a rapid switch may precipitate the serotonin syndrome (discussed later in this chapter).[94]

Paroxetine has the shortest half-life, that is, approximately 24 h, with no identified active metabolites. As such, it lends itself well to situations where rapid drug clearance is important. Dose–response studies for paroxetine have established optimal benefit at 20 mg/day, with little benefit gained by exceeding this dose.[83] Paroxetine may also not be as likely to disturb sleep patterns and thus may be a relatively "sedating" SSRI.

The half-life of sertraline and its active metabolite is about 24 h. An important pharmacokinetic finding with this medication is the linear relationship between dose and plasma concentration. In contrast, for example, fluoxetine can inhibit its own metabolism and, therefore, cause its dose–concentration relationship to be nonlinear. As with paroxetine, sertraline is less likely to disturb sleep/wake cycles, and there may be some clinical utility in the preferential use of these medications when insomnia is a major part of the spectrum of depression.[95] The recommended starting dose of sertraline is 50 mg/day. The data on dose–response relationships shows a generally flat curve beyond this dose, and as such, a dose of 50 mg/day is often as effective as doses of 100, 150, or 200 mg.[95]

D. Novel/Atypical Antidepressants

Trazodone has been shown to be effective in the treatment of major depression in most controlled studies,[96,97] although many clinicians feel that this medication is clinically less effective than TCAs and SSRIs.[98] Doses from 300 to 600 mg/day are required for antidepressant action. More commonly, this medication has been used clinically in the relief of mild to moderate insomnia. Although not currently regarded as a first-line antidepressant medication, its use as an adjunct to other medications in relieving antidepressant-induced sleep disturbance has recently been advocated.[99] Care must be taken in the combined use of trazodone and SSRIs as the risk of the serotonin syndrome is theoretically increased with concurrent use of two medications manifesting serotonin-enhancing effects.

Nefazodone, a structural analog of trazodone, has reuptake inhibition effects on serotonin and, to a lesser degree, norepinephrine. However, its agonist properties at the $5-HT_{1a}$ and $5-HT_{1c}$ receptors and its antagonist effects at $5-HT_2$ and $5-HT_3$ receptors have given it some virtue in clinical applications.[16] In double-blind and controlled trials its antidepressant effect is equivalent to imipramine and greater than placebo.[16] In studies with anxious patients, its novel pharmacological and receptor properties have been thought to limit the akathisialike effect seen with other SSRIs.[16] As well, it is reported not to disturb sleep patterns and, in fact, to increase REM sleep time.[100] In this regard, it is distinct from most MAOIs and TCAs, which cause dramatic REM sleep suppression. It can be given at initial doses of 100 mg b.i.d. and increased up to 600 mg daily. Its pharmacological profile shows little cholinergic, histaminic, or adrenergic antagonism.[101] Currently, the lack of extensive studies limits this medication to second-line use in the management of depression.

Bupropion has received attention because of its action as a dopamine agonist and reuptake inhibitor. In controlled trials it has shown efficacy similar to TCAs and superior to placebo.[102] Special attention was generated after a report by Haykol and Akiskal[103] that bupropion was less likely than older antidepressants to induce mania in depressed bipolar patients. However, a recent review suggests that this medication may be as likely to precipitate mania as other antidepressants in this patient population.[104] Its dopamine-enhancing effect may make it specifically active against atypical depression in a manner similar to MAOI and RIMA antidepressants.[102]

Bupropion is moderately bound to plasma protein (82 to 88%) and has an elimination half-life of about 11 h.[105] The usual starting dose is 75 mg t.i.d., and this may be increased to a total daily dose of 450 mg. Initial reports suggested that a unit dose larger than 150 mg and a total daily dose larger than 450 mg were likely to be associated with an increase in the incidence of seizures.[106] A multicenter study examined this issue in 3341 patients and documented a seizure rate of 0.24% for all doses used and 0.36% when the dose was between 300 and 450 mg/day.[106] It seems that the association of this medication with seizures falls within accepted parameters for clinical use, as long as the dose does not exceed 450 mg/day. Evidence from studies of bupropion plasma levels indicates that the therapeutic effect of this medication is maximal at moderate plasma levels, i.e., 20 to 50 ng/ml, with either low (< 20-ng/ml) or high (>100-ng/ml) plasma levels being associated with a poor outcome.[107,108] The tentative establishment of this therapeutic window suggests that the efficacy of bupropion may be optimized with plasma level determination.

Venlafaxine has been shown to be superior to placebo and equal in efficacy to TCAs.[109,110] It may also have a role in the treatment of TCA-resistant patients, with more than a third of triple-drug-resistant patients showing a sustained response to venlafaxine in one study.[111] The usual starting dose is 37.5 mg b.i.d, with a maximal daily dose being 375 mg/day. The elimination half-life for the parent drug and its active metabolite (o-desmethyl venlafaxine) is about 10 h. In animal studies, it has been associated with rapid beta-adrenergic receptor down regulation, which would suggest that it may have an accelerated onset of action. Clinical studies suggest that this medication may have significant antidepressant activity within the first 2 weeks of treatment.[112] Additionally, in one trial venlafaxine was superior to fluoxetine in antidepressant efficacy at 4 and 6 weeks following initiation of treatment.[113] Finally, a rapid escalation to higher doses, i.e., 200 mg/day, has been also associated with improvement in depressive symptoms within the first 4 days of treatment.[109] The dose–response relationship for venlafaxine has been reported to be linear, suggesting that dose increments may be associated with increased clinical improvement.[109] If the initial reports of the rapid onset of action of venlafaxine are confirmed, this medication would be a welcome addition to the pharmacotherapeutic management of depression.

E. Augmentation Strategies

Most studies have shown that anywhere from 20 to 35% of patients fail to respond to adequate trials of antidepressants.[114-116] Accordingly, various augmentation strategies have been established to increase response rate, and the development and clinical use of augmentation strategies are important aspects in the use of antidepressant medications. The two most thoroughly investigated augmentation strategies are the addition of lithium or the thyroid hormone liothyronine (levorotatory isomer of triiodothyroinene, T_3).

The addition of lithium to concurrent antidepressant therapy was pioneered by deMontigny[117] in 1981. Since then, the clinical utility of this approach has been demonstrated with TCAs, MAOIs, SSRIs, and novel antidepressant agents in over 50 placebo-controlled studies.[118] The controlled trials have been metaanalyzed to suggest that the chances of improvement with lithium augmentation lie between 56 and 95%, with most investigators reporting the more conservative figure. The response can be rapid, often within 48 h, although there may also be a delay of up to 3 to 6 weeks.[119,120] Serum lithium levels do not seem to be highly correlated with clinical improvement, and responses have been documented with lithium levels being as low as 0.1 to 0.5 mEq/l.[121] It has been speculated that relatively low lithium levels can promote increased serotonin synthesis[121] and that this effect may "plateau" with levels higher than 0.4 to 0.6 mEq/l. Clinically, the addition of 300 to 600 mg/day of lithium to an antidepressant may be attempted for 2 weeks or more to document efficacy. There have been some reports of adverse consequences following the addition of lithium to SSRIs (suggestive of a central serotonin syndrome),[122,123] so caution should be exercised in using larger lithium doses with this group of antidepressants.

Perturbations in the hypothalamic–pituitary–thyroid axis in depression have been well documented. For example, a recent metaanalysis reported that up to 52% of patients with refractory depression evidence subclinical hypothyroidism, in contrast to thyroid abnormalities of 8 to 17% in nonselected groups of depressed patients.[124] It has been suggested that subclinical alterations in thyroid function might define a subcategory of depression, and several controlled studies have documented that the addition of low doses of triiodothyronine (T_3) can enhance the effects of antidepressants in patients with clinically normal thyroid function.[125] Joffe[126] has argued that this is a specific property of T_3 and not its precursor molecule levothyroxine (T_4). Clinically, augmentation is instituted with the addition of 25 to 50 µg of T_3 to the preexisting antidepressant, and a response may be seen in up to 60% of patients within 2 weeks.[126,127] There is some evidence to suggest that patients who do not respond to lithium augmentation may respond to the addition of T_3, as well as evidence that T_3-augmentation failures may respond to the addition of lithium.[120,127] The two augmentation strategies are therefore not considered interchangeable, and both should be tried in sequence. There is no rationale for continuing both augmentation approaches concurrently unless some benefit is gained by the first strategy and further antidepressant effect is still desirable.

Numerous additional augmentation strategies have been reported. These include the addition of the serotonin precursor amino acid l-tryptophan,[128] buspirone,[129] pindolol,[130] and trazodone.[131] For the most part these augmentation approaches have not been as rigorously defined as lithium and T_3, and as such, they should be reserved for severely treatment-resistant patients.

Combination treatments with antidepressants in refractory depression have also been suggested, and these include concurrent use of TCAs and SSRIs,[132] TCAs and MAOIs,[133] and multiple combinations.[134] Most of these combined treatment strategies are limited by side effects and, again, should be reserved for the most refractory patients.

Finally, the use of psychostimulants has also received attention in the treatment of depression.[135] Both d-amphetamine and methylphenidate have been used in medically ill patients and to augment conventional antidepressants. However, reports are anecdotal or based on open trials, and use should be limited to cases that are resistant to multiple drugs.

V. SIDE EFFECTS

All antidepressants have side effects, and the selection of a particular agent can often hinge on this component. Efforts have been made to link theoretically specific side effects with reuptake inhibition of particular neurotransmitters or receptor action, although the clinical situation is far less predictable. Individual patients may have varied side effects, ranging from severe to nonexistent on the same dose of medication. A general first principle of treatment of side effects should be dose reduction of the antidepressant. Often, however, medications need to be maintained at their optimal levels, and alternate approaches have to be used to treat or ameliorate side effects. The following is a brief review of common antidepressant-induced side effects and suggested treatments.

A. Monoamine Oxidase Inhibitors

I.   Irreversible MAOIs

Orthostatic hypotension is the most common side effect of classic MAOIs, with patients reporting dizziness, lightheadedness, and headaches in 50% of cases.[136] Dose reduction or change in MAOI may often be necessary. If dose reduction is not possible, a salt-heavy diet may be cautiously prescribed. Finally, the use of fludro-cortisone, a mineralocorticoid, has been recommended in doses of 0.1 mg once or twice daily.[136] Switching from one MAOI to another should be carried out with caution. A rapid switch from phenelzine to tranylcypromine is not advised as the stimulantlike activity of the latter medication may precipitate a hypertensive crisis.[137]

Hypertensive crises, although frequently feared, are infrequently encountered. Dietary controls have already been discussed above. Crises can be precipitated by the ingestion of tyramine-rich foods or use of sympathomimetics. The symptoms should be well explained to patients, they include severe headaches, visual blurring, nausea, and possible delirium. If blood pressure elevations are significant, they should be considered a medical emergency and treated with intravenous phentol-amine 5 mg (a short-acting alpha-adrenergic blocker). On an ambulatory basis, sub-lingual use of the calcium channel blocker nifedipine 10 mg has been found to be effective in reducing blood pressure.[138] It may be prudent clinical practice to ask patients to carry one or two capsules of nifedipine with them at all times.

Overall, with the exception of their effects on blood pressure, classic MAOIs have little effect on heart rate or myocardial contractility.[139]

The serotonin syndrome (discussed below) has been reported most often in conjunction with MAOI use. Usually it is associated with the combined use of MAOIs and serotonin-enhancing medications (l-tryptophan, lithium, meperidine).[94] Caution should be exercised in the use of these medications with MAOIs, and narcotics (other than codeine) should be scrupulously avoided.

Neurologically, MAOIs can cause a variety of side effects. These include tremor, myoclonus, peripheral neuropathy, carpal tunnel syndrome, and tinnitus. Hydrazine MAOIs like phenelzine have been associated with pyridoxine deficiency and any side effects suggestive of this should be treated with vitamin B_6 100 to 300 mg/day.

Sexual dysfunction is also associated with MAOIs.[140,141] Decreased libido and anorgasmia have both been reported. The serotonin antagonist cyproheptadine has been reported to reverse impaired sexual performance in doses of 2 to 4 mg taken 2 to 3 h before intercourse.[140,141]

Peripheral edema is also occasionally encountered, more frequently with hydra-zine MAOIs.[136] Dosage reduction and the use of support stockings and diuretics have all been prescribed as treatments.

2. RIMAs

RIMAs represent a significant advance over classic MAOIs with regard to pro-duction of side effects. There is no dietary compliance necessary with these medica-tions, and tyramine reactions are not significant. A double-blind, controlled study has shown that the RIMA moclobemide is as well tolerated as placebo and much better than TCAs.[63] Headache and insomnia are the most commonly reported side effects, with high doses being correlated with tremor, anxiety, and lightheadedness. Less commonly, nausea, dry mouth, and sweating are also reported.

It is important to be aware of potential drug interactions. Caution should be exercised in the concurrent use of moclobemide and SSRIs (especially fluoxetine).[142]

Over-the-counter decongestants containing ephedrine may have their sympathomimetic effects potentiated two- to fourfold.[142]

B. Tricyclic Antidepressants

TCAs have been associated with a myriad of side effects related to their receptor antagonism at cholinergic, alpha-adrenergic, and histaminergic sites. The tertiary TCAs are particularly likely to cause receptor antagonist–associated side effects.

Anticholinergic side effects produce blurred vision, dry mouth, constipation, urinary retention, and tachycardia.[143] Central anticholinergic effects (impaired concentration, confusion, and delirium) are possible and particularly seen in patients who may be already cognitively vulnerable. Desipramine and nortriptyline are TCAs less likely to cause anticholinergic side effects.

Cardiovascular effects can be varied. The most common, postural hypotension, is related to alpha-adrenergic blockade.[139] As was previously discussed, TCAs may act like class I antiarrythmics and may cause significant conduction abnormalities with increased dose (or in overdose).[139]

Sedation is commonly seen with tertiary TCAs and is likely related to antihistaminergic effects. Clinically, this is dealt with by changing to secondary desmethylated TCAs (e.g., desipramine) or by administration at night when the sedative effect may be exploited usefully.

Sexual function is less commonly impaired with TCA medications. However, their anticholinergic effects may impair orgasmic ability. Treatment with neostigmine 7.5 to 15 mg about 1 h prior to intercourse may reduce orgasmic dysfunction.[144] The use of the direct cholinergic agonist bethanechol has been reported to reverse impaired erectile function. Cyproheptadine has also been used to reverse orgasmic dysfunction.[145]

Abrupt withdrawal of TCAs may be associated with symptoms of cholinergic excess, including nausea, anorexia, diarrhea, increased salivation, and insomnia.[146] TCAs should be tapered over several weeks to prevent withdrawal symptoms.

C. Selective Serotonin Reuptake Inhibitors

SSRIs have a side effect profile markedly different from TCAs. This reflects the relative paucity of receptor antagonist effects. In particular, they have a very low incidence of anticholinergic effects and are remarkably free of cardiotoxicity.[71,83,84] However, their serotonin-enhancing action has been associated with many side effects.

Headache, insomnia, and nausea are the most frequently reported side effects. All SSRIs have been associated with these effects. Paroxetine and sertraline have been reported to cause less insomnia than fluoxetine.[83] Dose reduction is usually the first approach in treatment of these phenomena. Treatment of nausea may be attempted with antiemetic and gastric motility agents.[147]

SSRI-associated akathisia has received particular attention as it has been speculated that this may be related to the reports linking fluoxetine and suicidality.[88] As such, the early emergence of akathisia should be treated with great caution. Treatment of akathisia has been attempted with benzodiazepines and antihistaminic and anticholinergic agents.[88] As with other side effects, the principal treatment approach should be dose reduction or discontinuation of the offending SSRI.

Numerous other neurological symptoms have also been reported with SSRI use. These include movement disorders[88,148] and acute dystonia.[149]

Anorgasmia and reduced libido have been reported to be caused by SSRI use.[150-152] Treatment should be via dose reduction. Cyproheptadine and yohimbine have been used in the treatment of SSRI-related sexual dysfunction.[152,153]

Sundry side effects include yawning,[151] bruxism,[154] and hyponatremia.[155]

A serious complication of SSRIs is the possible induction of the serotonin syndrome. This is presumed to be because of overstimulation of 5-HT_1 receptors and is associated with changes in mental status, restlessness, myoclonus, hyperreflexia, diaphoresis, rigors, and tremor.[94] Prior to the introduction of SSRIs, the most frequent cause was interaction between MAOIs and serotonergic agents. However, SSRIs have been associated with this syndrome when they are used concurrently with other serotonin-enhancing medications (l-tryptophan, trazodone, lithium). SSRI/MAOI combinations have also been known to precipitate this syndrome and, as a rule, should be avoided or initiated with extreme caution.

A recent case series documented sudden deaths in medically ill patients on fluoxetine.[156] It has been speculated that this may be related to potential hyponatremic and/or vasoconstrictive effects of SSRIs.[156,157] Large-scale studies have not noted this effect, and it may reflect an extremely rare occurrence in patients who were medically unstable prior to antidepressant treatment.

D. Novel/Atypical Antidepressants

Trazodone has been most commonly associated with drowsiness, nausea, and headache. In spite of low affinity for cholinergic receptors, dry mouth and blurred vision are clinically reported with high doses. Rare cases of cardiac ventricular excitability have also been reported. In contrast to MAOIs, TCAs, and SSRIs, which usually inhibit sexual response, trazodone has been associated with increased libido and sexual arousal in women.[158] In men, priapism (involuntary, painful and prolonged erections) has also been reported.[159] Priapism is a medical emergency, and treatment of males with high doses of trazodone should be undertaken with caution.

Nefazodone has most commonly been associated with headache, lassitude, dry mouth, nausea, and constipation.[16] Unlike other serotonin-specific antidepressants, nefazodone has not been reported to subjectively interfere with sleep and may, in fact, increase REM sleep periods.[100] No cardiotoxic effects have been reported,[95,101] and currently there are no reports of sexual dysfunction associated with this medication. In particular, nefazodone has no effect on nocturnal penile tumescence.[160]

The effect of bupropion on dopamine accounts for some of its side effects. Its possible role in seizure induction has already been discussed. At high doses it has also been associated with the development of psychosis and delirium.[161] For the above reasons, it is prudent not to exceed a single dose of 150 mg and a daily total dose of 450 mg. The induction of parkinsonianlike side effects (particularly falling and retropulsion) has also been reported in conjunction with bupropion use.[162]

Venlafaxine may cause nausea, sweating, anxiety, tremor, and abnormal ejaculation as its more common side effects.[163] The incidence of side effects increases with increasing dose. There is considerable adaptation to side effects over a 6-week period. Treatment with venlafaxine has been associated with modest but sustained increases in blood pressure, with the incidence of increased diastolic blood pressure being 13% at doses over 300 mg/day. Clinically significant increases in diastolic blood pressure (diastolic greater than 90 mm/Hg or 10 mg above baseline) should be treated with dose reduction or discontinuation of venlafaxine.[163]

VI. TREATMENT ALGORITHM

The array of new antidepressants has increased therapeutic options available but has also caused confusion in the generation of a rational algorithm for the treatment of depression. Several general principles are common to all treatment approaches. First, all antidepressants have a time lag between initiation of therapy and production of beneficial effect. A therapeutic trial of an antidepressant should be between 6 and 8 weeks of treatment at optimal doses. A patient cannot be deemed nonresponsive to a particular treatment until this length of time has elapsed. Second, studies by Prien and Kupfer[164] have established the necessity for acute antidepressant treatment for up to 6 months to prevent relapse after recovery from an uncomplicated depression. Third, long-term follow-up studies of recurrent depression have established the efficacy of continued treatment of depression with adequate doses of antidepressants for up to 5 years.[3,4] There is no evidence to support the use of lower, "maintenance" doses of antidepressants.

The selection of an initial antidepressant should be based on patient profile, treatment efficacy, and side effects. The safety in overdose of antidepressants is also crucially important. Reviews of the safety of antidepressants have established the value of SSRIs, RIMAs, and most novel/atypical antidepressants over TCAs and classic MAOIs.[165-167]

The role of antidepressants in the induction of mania or hypomania has been much debated. MAOIs and TCAs have been reported to induce hypomania or rapid cycling in some studies.[168,169] It is not definitively known whether this represents an unmasking of an occult bipolar disorder or the *de novo* induction of cyclicity. As was mentioned above, bupropion was initially reported to be less likely than other antidepressants to induce hypomania.[103] If a history suggestive of subclinical bipolar disorder is obtained, it would be prudent to avoid antidepressants with long clinical half-lives (such as fluoxetine or the irreversible MAOIs) since they may not be able to be rapidly withdrawn.

The treatment of major depressive disorder with psychotic features has received special attention as it seems to respond poorly to antidepressants alone. Indeed, it is regarded as a distinct clinical syndrome separate from nonpsychotic depression.[170] Most investigators have shown greater efficacy with either electroconvulsive therapy or combination treatment with an antidepressant and a neuroleptic.[171-173] The use of amoxapine monotherapy has also been found efficacious, likely because of its *in vivo* metabolism to the neuroleptic loxapine.[174] The management of delusional depression should be qualitatively different from the treatment of nonpsychotic unipolar depression, with aggressive and early treatment using either combined antidepressant–neuroleptic treatment or an early switch to electroconvulsive therapy if pharmacotherapy fails.

It must be stressed that treatment algorithms are constantly changing as our knowledge of antidepressants increases and newer antidepressants are introduced. Therefore, flexibility must be exercised in making clinical decisions. On the basis of the information reviewed in this chapter, however, it is possible to make some recommendations on developing an algorithm for the treatment of nonpsychotic unipolar depression. This is schematically shown in Figure 1.

The efficacy, safety, and superior side effect profile of the SSRIs have established them as first-line antidepressant agents in most forms of depression. An SSRI with a short half-life should be selected if there is concern that the patient may not tolerate the medication or if there is an increased risk of the induction of hypomania. At the end of 6 to 8 weeks (or sooner if the patient cannot tolerate the medication), a clinical determination should be made of full, partial, or nonresponse. A partial or non-response

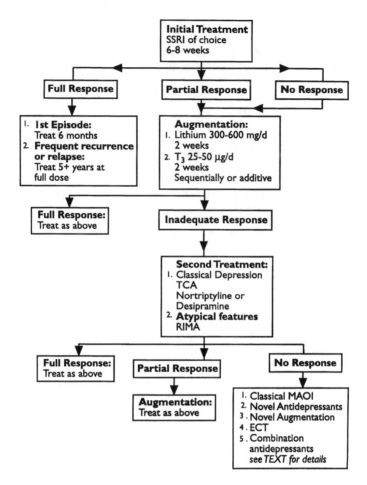

Figure 1
Algorithm for the treatment of nonpsychotic unipolar depression.

may be treated with a dose change, or augmentation with lithium or T_3. The value of trying an augmentation strategy at this juncture (for a partial or nonresponse) is that augmentation trials can be carried out for 2 weeks and may be more time effective than initiating another 8-week new drug trial. A nonresponse at the end of augmentation therapy may be a useful juncture to consider a second antidepressant.

If the depression has atypical features, consideration of RIMA or classic MAOI antidepressants would be appropriate (a 5-week washout of the SSRI would be required if fluoxetine was chosen as the first agent). If the depression does not meet criteria for atypical features, a TCA or novel/atypical antidepressant may be selected. The value of TCAs lies in their well-established efficacy as well as the evidence that substantial numbers of SSRI nonresponders are likely to respond to TCAs. In addition, most augmentation strategies have been demonstrated with TCAs, thereby increasing clinician confidence in using lithium or T_3 if augmentation is considered next. MAOIs have also been shown to be effective in patients who have failed TCA *and* lithium and T_3 augmentation approaches.[47] Their use in patients who have failed the first tier of antidepressant treatment would be highly recommended.

Currently the use of novel/atypical antidepressants is often as tertiary agents in the treatment of depression. This may change with their increasing clinical use and their established superiority to the TCAs with regard to side effects and safety in overdose. It is possible that venlafaxine, in particular, may become a more commonly used antidepressant based on its putative rapid onset of action.

Although a discussion of electroconvulsive therapy is beyond the scope of this chapter, its efficacy in the treatment of depression is unquestioned.[175] It should be considered in the following circumstances:

1. Severe and persistent suicidal ideation and attempts;
2. Intolerance of antidepressant medication;
3. Two or three failed pharmacological trials;
4. Severe physical debilitation which may not allow the time necessary for antidepressant medication to take effect;
5. Severe depression in first trimester of pregnancy; and
6. Depression with psychotic features.

VII. FUTURE DIRECTIONS

The established antidepressants all affect brain monoamine metabolism. Unfortunately, no medications have been shown to have robust antidepressant activity independent of their effects on amines. Conceptually, this has presented a barrier to the development of antidepressants which may have novel neurotransmitter effects. Several new compounds have been investigated, but definitive evidence of their clinical usefulness is lacking.

s-Adenosyl methionine (SAM), a naturally occurring compound, has been reported to have antidepressant activity.[176,177] The mode of application in most studies has been parenteral which may account for the paucity of data reporting clinical or investigational use.

The benzodiazepine alprazolam has been reported to have antidepressant activity in high doses.[178] However, most researchers in mood disorders have not been able to convincingly exclude the anxiolytic properties of alprazolam as accounting for its apparent efficacy in treating mood disorders. In addition, concerns about benzodiazepine dependency and withdrawal have mitigated against the wider use of alprazolam as an antidepressant.

Attention has also focused on the use of agents that act as gamma-aminobutyric acid (GABA) agonists. Case reports have identified progabide and fengabine (GABAmimetic agents) as having antidepressant activity.[179,180] Several animal-model studies also support the role of GABAmimetics in the treatment of depression.[181] Current development of GABAmimetic antidepressants has been limited by the central nervous toxicity shown by many of these compounds.

Indirect manipulation of adrenal and thyroid parameters has received some attention in the treatment of depression. An initial open study looked at suppression of endogenous steroids using aminoglutethamide, ketoconazole, or metyrapone.[182] Six of eight patients with treatment-resistant depression were reported to have sustained remissions for up to 5 months after treatment. The investigators suggested that readjustment of the hypothalamic-pituitary-adrenal (HPA) axis accounted for the remissions. Further attempts to alter the HPA axis have been documented with the use of the antihypertensive medication captopril (an angiotensin-converting–enzyme inhibitor, ACE-inhibitor). When used for control of blood pressure, this medication has been associated with mood-elevating effects.[183] The use of this medication has been limited to open trials, and the putative mechanism of action is thought to be via reduction in corticotropin production and secondary reduction in cortisol levels. Methimazole, a thyroid hormone antagonist, has been tried in one open trial for the management of treatment-resistant depression.[184] The results show only modest benefit from this approach.

l-Sulpiride, a substituted benzamide, has been usually associated with dose-related disinhibitory and neuroleptic activity. However, its use in depressed bipolar patients has shown that it had antidepressant activity equivalent to amitriptyline in a 4-week, double-blind, controlled trial.[185] Several older studies with unipolar patients have also shown efficacy similar to TCAs.[186,187] It is speculated that the antidepressant activity of *l*-sulpiride may be related to its antagonist action at low doses on D_2 presynaptic receptors, which may enhance dopamine transmission.

REFERENCES

1. Kessler, R. C., McGonagle, K. A., Zhao, S., Nelson, C. B., Hughes, M., Eshleman, S., Wittchen, H.-U., Kendler, K. S. Lifetime and 12-month prevalence of DSM-III-R psychiatric disorders in the United States. Arch Gen Psychiatry. 1994; 51:8-19.
2. Coryell, W., Akiskal, H. S., Leon, A. C., Winokur, G., Massen, J. D., Meuller, T. I., Keller, M. B. The time course of non-chronic major depressive disorder: uniformity across episodes and samples. Arch Gen Psychiatry. 1994; 51:405-410.
3. Frank, E., Kupfer, D. J., Perel, J. M., Cornes, C. L., Jarret, D., Mallinger, A. G., Thase, M. E., McEachran, A. B., Grochocinski, V. J. Three-year outcomes for maintenance therapies in recurrent depression. Arch Gen Psychiatry. 1990; 47:1093-1099.
4. Kupfer, D. J., Frank, E., Perel, J. M., Cornes, C. L., Mallinger, A. G., Thase, M. E., McEachran, A. B., Grochocinski, V. J. Five-year outcome for maintenance therapies in recurrent depression. Arch Gen Psychiatry. 1992; 49:769-773.
5. Selikoff, I. J., Robitzek, E. H., Ornstein, G. G. Toxicity of hydrazine derivatives of isonicotinic acid in the chemotherapy of tuberculosis. Q Bull Seaview Hosp.1952; 13:17-26.
6. Crane, G. E. Iproniazid (Marsilid) phosphate, a therapeutic agent for mental disorders and debilitating disease. J Psychiatr Res. 1957; 8:142-152.
7. Kline, N. S. Clinical experience with iproniazid (Marsilid). J Clin Exp Psychopathol. 1958; 19 (suppl 1):72-78.
8. Kuhn, R. The treatment of depressive states with G 22355 (imipramine hydrochloride). Am J Psychiatry. 1958; 115:459-464.
9. Klein, D. F., Davis, J. Diagnosis and Treatment of Psychiatric Disorders, ed. 1. Baltimore:Williams and Wilkins, 1968.
10. Quitkin, F., Rifkin, A., Klien, D. F. Monoamine oxidase inhibitors: a review of antidepressant effectiveness. Arch Gen Psychiatry. 1979; 35:749-760.
11. Cusack, B. M., Richelson, E. Antagonism by antidepressant and antihistaminics at the 5 cloned human muscarinic cholinergic receptors in CHO-KI cells. Faseb J. 1992; 6:1880-1892.
12. Rachelson, E., Nelson, A. Antagonism by antidepressants of neurotransmitters of normal human brain *in vitro*. J Pharmacol Exp Ther. 1984; 230:94-105.
13. Steele, T. E. Adverse effects suggesting amoxapine-induced dopamine blockade. Am J Psychiatry. 1982; 139:1500.
14. Harrington, M. A., Zhong, P., Garlow, S. J., Ciaranello, R. D. Molecular biology of serotonin receptors. J Clin Psychiatry. 1992; 53 (suppl 10):8-27.
15. Sharpley, A. L., Walsh, A. E. S., Cowen, P. J. Nefazodone — a novel antidepressant — may increase REM sleep. Biol Psychiatry, 1992; 31:1070-1073.
16. Fontaine, R. Novel serotonergic mechanisms and clinical experience with nefazodone. Clin Neuropharmacol. 1993; 16 (suppl 2):S45-S50.
17. Branconnier, R. J., Cole, C. O., Ghazvinian, S. Clinical pharmacology of bupropion and imipramine in elderly depressives. J Clin Psychiatry. 1983; 44:130-133.
18. Farid, F. F., Wenger, T. L., Tsai, S. Y. Use of bupropion in patients who exhibit orthostatic hypotension on tricyclic antidepressants. J Clin Psychiatry. 1983; 44:170-173.
19. Blackwell, B. Newer antidepressant drugs. In: Meltzer, H. Y., ed. Psychopharmacology: The Third Generation of Progress. New York: Raven Press; 1987: 1041-1050.

20. Howell, S. R., Husbands, G. E., Scatina, J. A., Sisenwine, S. F. Metabolic disposition of 14c-venlafaxine in mouse, cat, dog, rhesus monkey and man. Xenobiotica. 1993; 23:349-359.

21. Klamerus, K. J., Maloney, K., Rudolph, R. L., Sisenwine, S. F., Jusko, W. J., Chiang, S. T. Introduction of a composite parameter to the pharmacokinetics of venlafaxine and its active O-desmethyl metabolite. J Clin Pharmacology. 1992; 32:716-724.

22. Montgomery, S. A. Venlafaxine: a new dimension in antidepressant pharmacotherapy. J Clin Psychiatry. 1993; 54:119-126.

23. de Montigny, C., Aghajanian, G. K. Tricyclic antidepressants: long term treatment increases responsivity of rat forebrain neurons to serotonin. Science. 1978; 202:1303-1306.

24. Blier, P., de Montigny, C., Chaput, Y. Modification of the serotonin system by antidepressant treatments: implications for the therapeutic response in major depression. J Clin Psychopharmacol. 1987; 7:24S-35S.

25. Schildkraut, J. J. The catecholamine hypothesis of affective disorders: a review of supporting evidence. Am J Psychiatry. 1965; 122:509-522.

26. Maas, J. W., Fawcett, J. A., Dekirmenjian, H. 3-Methoxy-4-hydroxy phenol glycol (MHPG) excretion in depressive states: a pilot study. Arch Gen Psychiatry. 1968; 19:129.

27. Maas, J. W., Koslow, S. H., Davis, J., Katz, M., Frazer, A., Bowden, C. L., Berman, N., Gibbons, R., Strokes, P., Landis, H. Catecholamine metabolism and disposition in healthy and depressed subjects. Arch Gen Psychiatry. 1987; 44:337-344.

28. Roy, A., Pickar, D., Dejong, J., Karoum, F., Linnoila, M. Norepinephrine and its metabolites in cerebrospinal fluid, plasma and urine: relationship to hypothalamic-pituatary-adrenal axis function in depression. Arch Gen Psychiatry. 1988; 45:849-857.

29. Davis, J. M., Koslow, S. H., Gibbons, R. D., Maas, J. W., Bowden, C. L., Casper, R., Hanin, I., Javaid, J., Chang, S., Stokes, P. Cerebrospinal fluid and urinary biogenic amines in depressed patients and healthy controls. Arch Gen Psychiatry. 1988; 45:705-717.

30. Caldicott-Hazard, S., Morgan, D. G., DeLeon-Jones, F., Overstreet, D. H., Janowsky, D. Clinical and biochemical aspects of depressive disorders: II. transmitter/receptor theories. Synapse. 1991; 9:251-301.

31. Coppen, A. J., The biochemistry of affective disorders. Br J Psychiatry. 1967; 113:1237-1264.

32. Traskman-Bendz, L., Asperg, M., Schalling, D. Serotonergic function and suicidality in personality disorders. Ann N.Y. Acad Sci. 1986; 487:168-174.

33. Van Praag, H. M., Indoleamines in depression and suicide. Prog Brain Res. 1986; 65:59-71.

34. Mann, J. J., Stanley, M., McBride, P. A., McEwen, B. S. Increased serotonin$_2$ and beta-adrenergic receptor binding in the frontal cortices of suicide victims. Arch Gen Psychiatry. 1986; 43:954-959.

35. Marcusson, J., Fowler, C. J., Hall, H., Ross, S. B., Winbald, B. "Specific" binding of tritiated-imipramine to protease sensitive and protease resistant sites. J Neurochem. 1985; 44:705-711.

36. Delgado, P. L., Charney, D. S., Price, L. H., Aghajanian, G. K., Landis, H., Heninger, G. R. Serotonin function and the mechanism of antidepressant action. Arch Gen Psychiatry. 1990; 47:411-418.

37. Delgado, P. L., Price, L. H., Miller, H. L., Salomon, R. M., Aghajanian, G. K., Heninger, G. R., Charney, D. S. Serotonin and the neurobiology of depression: effects of tryptophan depletion in drug-free depressed patients. Arch Gen Psychiatry. 1994; 51:865-874.

38. Barr, L. C., Goodman, W. K., McDougle, C. J., Delgado, P. L., Heninger, G. R., Charney, D. S., Price, L. H. Tryptophan depletion in patients with obsessive compulsive disorder who respond to serotonin reuptake inhibitors. Arch Gen Psychiatry. 1994; 51:309-317.

39. Randrup, A., Munkvad, I., Fog, R. Mania, depression and brain dopamine. In: Essman, W. B., Vaizelli, L. (eds.), Current Developments in Psychopharmacology. New York: Spectrum, 1975; 207-229.

40. Anisman, P. H., Irwin, J., Skiar, L. S. Deficits of escape performance following catecholamine depletion: implications for behavioral deficits induced by uncontrollable stress. Psychopharmacology. 1979; 64:163-170.

41. Anisman, P. H., Remington, G., Skiar, L. S. Effect in inescapable shock on subsequent escape performance: catecholamine and cholinergic mediation of response initiation and maintenance. Psychopharmacology. 1979; 61:107-124.

42. Kapur, S., Mann, J. J. Role of the dopaminergic system in depression. Biol Psychiatry. 1992; 32:1-17.

43. DiChiara, G., Acquas, E., Carboni, E. Dopamine and drug-induced motivation. In Gessa, G. L., Serra, G (eds), Dopamine and Mental Depression. Oxford: Pergamon Press, 1990; 27-38.

44. Roy, A., Farouk, K., Pollack, S. Marked reduction in indexes of dopamine metabolism among patients who attempt suicide. Arch Gen Psychiatry. 1992; 49:447-450.

45. American Psychiatric Association: Diagnostic and Statistical Manual of Mental Disorders, 4th Ed. Washington, D.C.: American Psychiatric Association, 1994.

46. Goodman, W. K., Charney, D. S. Therapeutic applications and mechanisms of action of monoamine oxidase inhibitors and heterocyclic antidepressants. J Clin Psychiatry. 1985; 46 (suppl 10):6-22.

47. Thase, M. E., Frank, E., Mallinger, A. G., Hamer, T., Kupfer, D. J. Treatment of imipramine-resistant depression. III: Efficacy of monoamine oxidase inhibitors. J Clin Psychiatry. 1992; 53:5-11.

48. Shawcross, C., Tyrer, P. The place of monoamine oxidase inhibitors in the treatment of depression. In: Zohar, J., Belmaker, R. H. (eds). Treating Resistant Depression. New York: PMA Publishing, 1987; pp 113-129.

49. Devlin, M. J., Walsh, B. T. Use of monoamine oxidase inhibitors in refractory depression. In: Tasman, A., Goldfinger, S. M., Kaufmann, C. A. (eds). Review of Psychiatry, vol 9. Washington, D.C.: American Psychiatric Press. 1990; pp 74-90.

50. McGrath, P. J., Stewart, J. W., Harrison, W. M. The predictive value of symptoms of atypical depression for differential drug treatment outcome. J Clin Psychopharmacol. 1992; 12:197-202.

51. Quitkin, F. M., McGrath, P. J., Stewart, J. W., Harrison, W., Tricamo, E., Wagar, S. G., Ocepek-Welikson, K., Nunes, E., Rabkin, J. G., Klein, D. F. Atypical depression, panic attacks, and response to imipramine and phenelzine: a replication. Arch Gen Psychiatry. 1990; 47:935-941.

52. Robinson, D. S., Nies, A., Ravaris, L. Clinical psychopharmacology of phenelzine. Arch Gen Psychiatry. 1978; 35:629-638.

53. Neis, A. Differential response patterns to MAO inhibitors and tricyclics. J Clin Psychiatry. 1984; 45:70-77.

54. Richelson, W. Treatment of acute depression. Psych Clin North Am. 1993; 461-478.

55. Jenike, M. A. Affective illness in elderly patients, part 2. Psychiatr Times. 1987; 4:1-10.

56. Renshaw, P. F., Guimares, A. R., Fava, M., Rosenbaum, J. F., Pearlman, J. D., Flood, J. G., Puopolo, P. R., Clancy, K., Gonzalez, G. Accumulation of fluoxetine and norfluoxetine in human brain during therapeutic administration. Am J Psychiatry. 1992; 149:1592-1594.

57. Alevizos, B., Hatzimanolis, J., Markianos, M., Stefanis, C. N. Clinical, endocrine and neurochemical effects of moclobemide in depressed patients. Acta Psychiatr Scand. 1993; 87:285-290.

58. Korn, A., Gasic, S., Jung, M., Eichler, H. G., Raffesberg, W. Influence of moclobemide (ro 11-1163) on the peripheral adrenergic system: interaction with tyramine and tricyclic antidepressants. In: Tipton, K. F. (ed). Monoamine Oxidase: Prospects for Therapy with Reversible Inhibitors. London: Academic Press, 1984; 487-496.

59. Korn, A., DaPrada, M., Raffesberg, W., Allen, S., Gasic, S. Tyramine pressor effect in man: studies with moclobemide, a novel, reversible monoamine oxidase inhibitor. In: Youdim, M. B. H. (ed). The cheese effect and the reversible MAO-A inhibitors. J Neural Transm (suppl). 1988; 26:57-71.

60. Nair, N. P., Ahmed, S. K., Kin, M. H. Biochemistry and pharmacology of MAO-A agents: focus on moclobemide. J Psychiatr Neurosci. 1993; 18:214-225.

61. Holford, N. H., Guentert, T. W., Dingemanse, J., Banken, L. Monoamine oxidase-A: pharmacodynamics in humans of moclobemide, a reversible and selective inhibitor. Br J Clin Pharmacol. 1994; 37:433-439.

62. Freeman, S. Moclobemide: a drug profile. Lancet. 1993; 342:1528-1532.

63. UK Moclobemide Study Group. A multicenter comparative trial of moclobemide, imipramine and placebo in major depressive disorder. Int J Psychopharmacol. 1994; 9:109-113.

64. Danish University Antidepressant Group. Moclobemide: a reversible MAO-A inhibitor showing weaker antidepressant effect than clomipramine in a controlled multicenter study. J Aff Disorders. 1993; 28:105-116.

65. Heinze, G., Rossel, L., Gabelic, I., Galeano-Munoz, J., Stabl, M., Allen, S. R. Double-blind comparison of reversible monoamine oxidase-A inhibitors. Pharmacopsychiatry. 1993; 26:240-245.

66. Lonnqvist, J., Sihvo, S., Syvalahti, E., Kiviruusu, O. Moclobemide and fluoxetine in atypical depression: a double-blind trial. J Aff Disorders. 1994; 32:169-177.

67. Bieck, P. R., Antonin, K. H., Schmidt, E. Clinical pharmacology of reversible monoamine oxidase-A inhibitors. Clin Neuropharmacol. 1993; 16 (suppl 2):S34-41.

68. Volz, H.-P., Faltus, F., Magyar, I., Moller, H.-J. Brofaramine in treatment-resistant depressed patients — a comparative trial versus tranylcypromine. J Aff Disorders. 1994; 30:209-217.

69. Nolen, W. A., Haffmans, P. M., Bouvy, P. F., Duivenvoorden, H. J. Monoamine oxidase inhibitors in resistant major depression. A double-blind comparison of brofaramine and tranylcypromine in patients resistant to tricyclic antidepressants. J Aff Disorders. 1993; 28:105-116.

70. Glassman, A. H., Roose, S. P. Risks of antidepressants in the elderly: tricyclic antidepressants and arrythmia — revising risks. Gerontology. 1994; (suppl 1):15-20.

71. Laird, L. K., Lydiard, R. B., Morton, W. A., Steele, T. E., Kellner, C., Thompson, N. M., Ballenger, J. C. Cardiovascular effects of imipramine, fluvoxamine and placebo in depressed outpatients. J Clin Psychiatry. 1993; 54:224-228.

72. Rechlin, T. The effect of amitriptyline, doxepin, fluvoxamine and paroxetine on heart-rate variability. J Clin Psychopharmacol. 1994; 14:392-395.

73. Task Force on the Use of Laboratory Tests in Psychiatry, American Psychiatric Association. Am J Psychiatry. 1985; 142:155.

74. Warner, M. D., Griffin, M., Peabody, C. A. High initial dose nortriptyline in the treatment of depression. J Clin Psychiatry. 1993; 54:67-69.

75. Kramer, P. D. Listening to Prozac. New York: Viking Press. 1993.

76. Rosenberg, C., Damso, N., Fuglum, E., Jacobsen, L. V., Horsgard, S. Citalopram and imipramine in the treatment of depressive patients in general practice. A Nordic multicenter clinical study. Int Clin Psychopharmacol. 1994; 9 (suppl 1):41-48.

77. Montgomery, S. A., Pedersen, V., Tanghoj, P., Rasmussen, C., Rioux, P. The optimal dosing regimen for citalopram — a meta-analysis of nine placebo-controlled studies. Int Clin Psychopharmacol. 1994; 9 (suppl 1):35-40.

78. Claghorn, J.L., Kiev, A., Rickels, K., Smith, W. T., Dunbar, G. C. Paroxetine versus placebo: a double blind comparison in depressed patients. J Clin Psychiatry. 1992; 53:434-438.

79. Preskorn, S. Targeted pharmacotherapy in depression management: comparative pharmacokinetics of fluoxetine, paroxetine and sertraline. Int Clin Psychopharmacol. 1994; 9 (suppl 2):13-19.

80. DeWilde, J., Spiers, R., Mertens, C., Bartholome, F., Schotte, G., Leyman, S., A double-blind, comparative multicenter study comparing paroxetine with fluoxetine in depressed patients. Acta Psychiatr Scand. 1993; 87:141-145.

81. Dunbar, G. C., Claghorn, J. L., Kiev, A., Rickels, K., Smith, W. T. A comparison of paroxetine and placebo in depressed outpatients. Acta Psychiatr Scand. 1993; 87:302-305.

82. Neilsen, B. M., Behnke, K., Arup, P., Christiansen, P. E., Geisler, A., Ipsen, E., Maach-Moller, B., Ohrberg, S. C. A comparison of fluoxetine and imipramine in the treatment of outpatients with major depressive disorder. Acta Psychiatr Scand. 1993; 87:269-272.

83. Novel Selective Serotonin Reuptake Inhibitors, Part I. Communications presented at World Congress of Biological Psychiatry, Florence, Italy. June 9-14. Chairs: Waalinder, J., Feighner, J. P. J Clin Psychiatry. 1992; 53:107-112.

84. Nemeroff, C. B. The clinical pharmacology and use of paroxetine, a new selective serotonin reuptake inhibitor. Pharmacotherapy 1994; 14:127-138.

85. Alamura, A. C., Moro, A. R., Percudani, M. Clinical pharmacokinetics of fluoxetine. Clin Pharmacokin. 1994; 26:201-214.

86. LePen, C., Levy, E., Ravily, V., Beuzen, J. N., Meurgey, F. The cost of treatment dropout in depression. A cost-benefit analysis of fluoxetine vs. tricyclics. J Aff Disorders. 1994; 31:1-18.

87. Teicher, M. H., Glod, C., Cole, J. O. Emergence of intense suicidal preoccupation during fluoxetine treatment. Am J Psychiatry. 1990; 147:207-210.

88. Hamilton, M. S., Opler, L. A. Akathisia, suicidality and fluoxetine. J Clin Psychiatry. 1992; 53:401-406.

89. Mann, J. J., Kapur, S. The emergence of suicidal ideation and behavior during antidepressant pharmacotherapy. Arch Gen Psychiatry. 1991; 48:1027-1033.

90. Crundell, J. K. Fluoxetine and suicidal ideation — a review of the literature. Int J Neurosci. 1993; 68:73-84.

91. Moller, H. J., Stienmeyer, E. M. Are serotonergic reuptake inhibitors more potent in reducing suicidality? An empirical study on paroxetine. Eur Neuropsychopharmacol. 1994; 4:55-59.

92. Shader, R. I., Greenblat, D. J., von Moltke, L. L. Fluoxetine inhibition of phenytoin metabolism. J Clin Psychopharmacol. 1994; 14:375-376.

93. Cain, J. W. Poor response to fluoxetine: underlying depression, serotonergic overstimulation or a "therapeutic window"? J Clin Psychiatry. 1992; 53:272-277.

94. Sternbach, H. The serotonin syndrome. Am J Psychiatry. 1991; 148:705-713.

95. New directions in antidepressant therapy: a review of sertraline, a unique serotonin reuptake inhibitor. Proceedings of a clinical symposium, San Gimignano, Italy. June 12, 1991. Chair: Cole, J. O. J Clin Psychiatry. 1992; 53(9):333-340.

96. Workman, E. A., Short, D. D. Atypical antidepressants versus imipramine in the treatment of major depression: a meta-analysis. J Clin Psychiatry. 1993; 54:5-12.

97. Haria, M., Fitton, A., McTavish, D. Trazodone: a review of its pharmacology, therapeutic use in depression and therapeutic potential in other disorders. Drugs Aging. 1994; 4:331-355.

98. Preskorn, S. H., Burke, M. Somatic therapy for major depressive disorder: selection of an antidepressant. J Clin Psychiatry. 1992; 53 (suppl 9):5-18.

99. Nierenberg, A. A., Adler, L. A., Peselow, E., Zornberg, G., Rosenthal, M. Trazodone for antidepressant-associated insomnia. Am J Psychiatry. 1994; 151:1069-1072.

100. Armitage, R., Roffwarg, H. P., Cain, J., Trivedi, M., Rush, A. J. The effect of nefazodone on ultradian rhythms in EEG frequencies in major depression, abstract. Presented at the 7th annual meeting of the Association of Professional Sleep Societies, Los Angeles, June 1993.

101. Bruel, H. P., DeLeenheer, I., Coninx, L., Gammans, R. Comparison of the cardiovascular effects of nefazodone, imipramine and placebo in healthy elderly volunteers, abstract. Presented at the Affective Illness Conference, Switzerland, January 1993.

102. Zung, W. W. K. Review of placebo-controlled trials with bupropion. J Clin Psychiatry. 1983; 44:104-114.

103. Haykol, R. F., Akiskal, H. S. Bupropion: a promising approach to rapid cycling bipolar II patients. J Clin Psychiatry. 1991; 51 (suppl):17-19.

104. Masand, P., Stern, T. A. Bupropion and secondary mania. Is there a relationship? Ann Clin Psychiatry. 1993; 5:271-272.

105. Goodnick, P. J. Pharmacokinetics of second generation antidepressants: bupropion. Psychopharmacol Bull. 1991; 27:513-519.

106. Johnston, J. A., Lineberry, C. G., Ascher, J. A., Davidson, J., Khayrallah, M. A., Feighner, J. P., Stark, P. A 102-center prospective study of seizure in association with bupropion. J Clin Psychiatry. 1991; 52:450-456.

107. Preskorn, S. H. Should bupropion dosage be adjusted based upon therapeutic drug monitoring? Psychopharmacol Bull. 1991; 27:637-643.

108. Goodnick, P. J. Blood levels and acute response to bupropion. Am J Psychiatry. 1992; 149:399-400.

109. Venlafaxine: a new dimension in antidepressant pharmacotherapy. Clinical symposium presented at XVIIIth Collegium Internationale Neuro-Psychopharmacologicum (C.I.N.P.) Congress. June 28, 1992 at Nice, France. Chair: Montgomery, S. A., J Clin Psychiatry. 1993; 54:119-126.

110. Schweizer, E., Weise, C., Clary, C., Fox, I., Rickels, K. Placebo-controlled trial of venlafaxine for the treatment of major depression. J Clin Psychopharmacol. 1991; 14:170-179.

111. Nierenberg, A. A., Feighner, J. P., Rudolph, R., Cole, J. O., Sullivan, J. Venlafaxine for treatment resistant unipolar depression. J Clin Psychopharmacol. 1994; 14:419-423.

112. Khan, A., Fabre, L. F., Rudolph, R. Venlafaxine in depressed outpatients. Psychopharmacol Bull. 1991; 27:141-144.

113. Clerc, G. E., Ruimy, P., Verdeau-Pailles, J. The Venlafaxine French Inpatient Study Group. A double-blind comparison of venlafaxine and fluoxetine in patients hospitalized for major depression and melancholia. Int Clin Psychopharmacol. 1994; 9:139-143.

114. Klein, D. E., Gittelman, R., Quitkin, F. Diagnosis and drug treatment of psychiatric disorders: adults and children. 2nd ed. Baltimore: Williams and Wilkins, 1980.

115. Stark, P., Hardison, C. A review of multicenter controlled studies of fluoxetine vs. imipramine and placebo in outpatients with major depression. J Clin Psychiatry. 1985; 46:53-58.

116. Georgotas, A., McCue, D. S. Refractory depressions. In: Georgotas, A., Cancro, R. (eds). Depression and Mania: A Comprehensive Textbook. New York: Elsevier, 1988; 384-391.

117. deMontigny, C., Grunberg, F., Mayer, A., Dechenes, J. P. Lithium induces rapid relief of depression in tricyclic antidepressant drug non-responders. Br J Psychiatry. 1981; 138:252-256.

118. deMontigny, C. Lithium addition in treatment-resistant depression. Int Clin Psychopharmacol. 1994; 9 (suppl 2):31-35.

119. Price, L. H., Charney, D. S., Heninger, G. R. Variability of response to lithium augmentation in refractory depression. Am J Psychiatry. 1986; 143:1387-1392.

120. Thase, M. E., Kupfer, D. J., Frank, E., Jarret, E. B. Treatment of imipramine-resistant recurrent depression: II. An open trial of lithium augmentation. J Clin Psychiatry. 1989; 50:413-417.

121. Broderick, P., Lynch, V. Behavioral and biochemical changes induced by lithium and l-tryptophan in muridical rats. Neuropharmacology. 1982; 21:671-679.

122. Salama, A. A., Shafey, M. A case of severe lithium toxicity induced by combined fluoxetine and lithium carbonate. Am J Psychiatry. 1989; 146:278.

123. Noveske, F. G., Hahn, K. R., Flynn, R. J. Possible toxicity of combined fluoxetine and lithium. Am J Psychiatry. 1989; 146:1515.

124. Howland, R. H. Thyroid dysfunction in refractory depression: implications for pathophysiology and treatment. J Clin Psychiatry. 1993; 54:47-54.

125. Joffe, R. T., Post, R. M. Experimental treatment for affective illness. In: Brodie, H. K. H., Berger, P. A. (eds.) New York: Basic Books, 1985; 287-305.

126. Joffe, R. T. T_3 and lithium potentiation of tricyclic antidepressants. Am J Psychiatry. 1988; 145:1317-1318.

127. Joffe, R. T., Singer, W., Levitt, A. J., MacDonald, C. A placebo-controlled trial of lithium and tri-iodothyronine augmentation of tricyclic antidepressants in unipolar refractory depression. Arch Gen Psychiatry. 1993; 50:387-393.

128. Walinder, J., Skoff, A., Carlsson, A., Nagy, A., Ross, B.-E. Potentiation of the antidepressant action of clomipramine by tryptophan. Arch Gen Psychiatry. 1976; 33:1384-1389.

129. Joffe, R. T., Schuller, D. R. An open study of buspirone augmentation of serotonin reuptake inhibitors in refractory depression. J Clin Psychiatry. 1993; 54:269-271.

130. Artigas, F., Perez, V., Alverez, E. Pindolol induces a rapid improvement of depressed patients treated with serotonin reuptake inhibitors. Arch Gen Psychiatry. 1994; 51:248-251.

131. Nierenberg, A. A., Cole, J. O., Glass, L. Possible trazodone potentiation of fluoxetine: a case series. J Clin Psychiatry. 1992; 53:83-85.

132. Nelson, J. C., Mazure, C. M., Bowers, M. B., Jatlow, P. I. A preliminary, open study of the combination of fluoxetine and desipramine for rapid treatment of major depression. Arch Gen Psychiatry. 1991; 48:303-307.

133. Pare, C., Kline, N., Hallstrom, C., Cooper, T. Will amitriptyline prevent the "cheese" reaction of monoamine oxidase inhibitors? Lancet. 1982; 1:183-186.

134. Tanum, L. H. Combination treatment with antidepressants in refractory depression. Int J Psychopharmacol. 1994; 9 (suppl 2):37-40.

135. Satel, S. L., Nelson, J. C. Stimulants in the treatment of depression: a critical overview. J Clin Psychiatry. 1989; 50:241-249.

136. Rabkin, J. G., Quitkin, F. M., Harrison, W. Adverse reactions to monoamine oxidase inhibitors, Part I: Treatment correlates and clinical management. J Clin Psychopharmacol. 1985; 5:2-9.

137. Bazire, S. R. Sudden death associated with switching monoamine oxidase inhibitors. Drug Intell Clin Pharmacol. 1987; 48:249-250.

138. Clary, C., Schweizer, E. Treatment of MAOI hypertensive crises with sublingual nifedipine. J Clin Psychiatry. 1987; 48:249-250.

139. Goldman, L. S., Alexander, R. C., Luchins, D. J. Monoamine oxidase inhibitors and tricyclic antidepressants: comparison of their cardiovascular effects. J Clin Psychiatry. 1986; 47:225-229.

140. Rapp, M. S. Two cases of ejaculatory impairment related to phenelzine. Am J Psychiatry. 1979; 136:1200-1201.

141. Barton, J. L. Orgasmic inhibition by phenelzine. Am J Psychiatry. 1979; 136:1616-1617.

142. Dingemanse, J. An update of recent moclobemide interaction data. Int J Psychopharmacol. 1993; 7:167-180.

143. Pollack, M. H., Rosenbaum, J. F. Management of antidepressant induced side-effects: a practical guide for the clinician. J Clin Psychiatry. 1987; 48:3-8.

144. Kraupi-Taylor, F. Loss of libido in depression. Br Med J. 1972; 1:305.

145. Sovner, R. Treatment of tricyclic antidepressant-induced orgasmic inhibition with cyproheptadine. J Clin Psychopharmacol. 1984; 4:169.

146. Lawrence, J. M. Reactions of withdrawals of antidepressants, antiparkinsonian drugs and lithium. Psychosomatics. 1985; 26:869-875.

147. Wernicke, J. F. The side effect profile and safety of fluoxetine. J Clin Psychiatry. 1985; 46:59-67.

148. Settle, E. C. Akathisia and sertraline. J Clin Psychiatry. 1993; 54:321.

149. Black, B., Uhde, T. W. Acute dystonia and fluoxetine. J Clin Psychiatry. 1992; 53:327.

150. Patterson, W. M. Fluoxetine-induced sexual dysfunction. J Clin Psychiatry. 1993; 54:71.

151. Cohen, A. J. Fluoxetine-induced yawning and anorgasmia reversed by cyproheptadine. J Clin Psychiatry. 1992; 53:174.

152. Jacobsen, F. M. Fluoxetine-induced sexual dysfunction and open trial of yohimbine. J Clin Psychiatry. 1992; 53:119-122.

153. Kurt, U., Ozkardes, H., Altig, U., Germiyanoglu, C., Gurdal, M., Erol, D. The efficacy of anti-serotonergic agents in the treatment of erectile dysfunction. J Urol. 1994; 152:407-409.

154. Ellison, J. M., Stanzini, P. SSRI-associated nocturnal bruxism in four patients. J Clin Psychiatry. 1993; 54:432-444.

155. Pillans, P. I., Coulter, D. M. Fluoxetine and hyponatremia — a potential hazard in the elderly. N Z Med J. 1994; 107:85-86.

156. Spier, S. A., Frontera, M. A. Unexpected deaths in depressed medical patients treated with fluoxetine. J Clin Psychiatry. 1991; 52:377-382.

157. Fricchione, G. L., Woznicki, R. M., Klesmer, J., Vlay, S. C. Vasoconstrictive effects and SSRI's. J Clin Psychiatry. 1993; 54:71-72.

158. Gartrell, N. Increased libido in women receiving trazodone. Am J Psychiatry. 1986; 143:781-782.

159. Warner, M. D., Peabody, C. A., Whiteford, H. A. Trazodone and priapism. J Clin Psychiatry. 1987; 48:244-245.

160. Ware, J., Rose, V., McBrayer, R. The effects of nefazodone, trazodone, buspirone and placebo on sleep and sleep related penile erections (NTP) in normal subjects, abstract. Presented at the 5th Annual Meeting of the Association of Professional Sleep Societies, Snowbird, Utah. January 1991.

161. Ames, D., Wirshing, W. C., Szuba, M. P. Organic mental disorders associated with bupropion. J Clin Psychiatry. 1992; 53:53-55.

162. Szuba, M. P., Leuchter, A. F. Falling backward in two elderly patients taking bupropion. J Clin Psychiatry. 1992; 53:157-159.

163. Mendels, J., Johnston, R., Mattes, J., Riesenberg, R. Efficacy and safety of BID doses of venlafaxine in a dose-response study. Psychopharmacol Bull. 1993; 29:169-174.

164. Prien, R. F., Kupfer, D. J. Continuation therapy for major depression episodes: how long should it be maintained? Am J Psychiatry. 1986; 143:18-23.

165. Kapur, S., Mieczkowski, T., Mann, J. J. Antidepressant medications and the relative risk of suicide attempt and suicide. JAMA. 1992; 268:3441-3445.

166. Cassidy, S., Henry, J. Fatal toxicity of antidepressant drugs in overdose. Br Med J. 1987; 295:1021-1024.

167. Rotterstol, N. Norwegian data on death due to overdose of antidepressants. Acta Psychiatr Scand. 1989; 80 (suppl 354):61-68.

168. Wehr, T. A., Goodwin, F. K. Can antidepressants cause mania and worsen the course of affective illness? Am J Psychiatry. 1987; 144:1403-1411.

169. Wehr, T. A., Sack, D. A., Rosenthal, N. E. Rapid cycling affective disorder: contributing factors and treatment responses in 51 patients. Am J Psychiatry. 1988; 145:179-184.

170. Shatzberg, A. F., Rothschild, A. J. Psychotic (delusional) major depression: should it be included as a distinct syndrome in DSM-IV? Am J Psychiatry. 1992; 149:733-745.

171. Chan, C. H., Janicak, P. G., Davis, J. M. Response of psychotic and non-psychotic depression patients to tricyclic antidepressants. J Clin Psychiatry. 1987; 48:197-200.

172. Rothschild, A. J., Samson, J. A., Besette, M. P., Carter-Campbell, J. T. Efficacy of the combination of fluoxetine and perphenazine in the treatment of psychotic depression. J Clin Psychiatry. 1993; 54:338-342.

173. Spiker, D. G., Stein, J., Rich, C. L. Delusional depression and electroconvulsive therapy: one year later. Convulsive Ther. 1985; 3:167-172.

174. Anton, R. F., Burch, E. A. Amoxapine vs. amitriptyline combined with perphenazine in the treatment of psychotic depression. Am J Psychiatry. 1990; 147:1203-1208.

175. NIMH Consensus Development Conference on Electroconvulsive Therapy. JAMA. 1985; 245:2103-2108.

176. Berlanga, C., Ortega-Soto, H. A., Ontiveros, M. Efficacy of s-adenosyl-l-methionine in speeding the onset of action of imipramine. J Psychiatr Res. 1992; 44:257-262.

177. Fava, M., Rosenbaum, J. F., MacLaughlin, R. Neuroendocrine effects of s-adenosyl-l-methionine, a novel putative antidepressant. J Psychiatr Res. 1990; 42:177-184.

178. Rickels, K., Feighner, J. P., Smith, W. T. Alprazolam, amitriptyline, doxepin, and placebo in the treatment of depression. Arch Gen Psychiatry. 1985; 42:134-141.

179. Thaker, G. K., Moran, M., Tamminga, C. A. GABAmimetics: a new class of antidepressant agents? Arch Gen Psychiatry. 1990; 47:287-288.

180. Chabannes, J. P., Baro, P. Lambert, P. A., Decade, P., Musch, B. Antidepressant activity of fengabine (SL 79229): results from an open pilot study. In: Bartholini, G., Lloyd, K. G., Morselli, P. L. (eds). GABA and Mood Disorders: Experimental and Clinical Research. New York: Raven Press, 1986; 139-146.

181. Petty, F., Sherman, A. D. GABAergic modulation of learned helplessness. Pharmacol Bull. 1981; 15:567-570.

182. Murphy, B. E., Dhar, V., Ghaderian, A. M., Chouinard, G., Keller, R. Response to steroid suppression in major depression resistant to antidepressant therapy. J Clin Psychopharmacol. 1991; 11:121-126.

183. Germain, L. Chouinard, G. Captopril treatment of major depression with serial measurements of blood cortisol concentrations. Biol Psychiatry. 1988; 25:489-493.

184. Joffe, R. T., Singer, W., Levitt, A. J. Methimazole in treatment-resistant depression. Biol Psychiatry. 1992; 31:1235-1237.

185. Bocchetta, A., Bernardi, F., Burrai, C., Del Zompo, M. A. A double-blind study of l-sulpiride versus amitriptyline in lithium-maintained bipolar depressives. Acta Psychiatr Scand. 1993; 88:434-439.

186. Niskanen, P., Tamminen, T., Viukari, M. Sulpiride vs. amitriptyline in the treatment of depression. Curr Ther Res. 1975; 17:281-284.

187. Standish-Barry, H. M. A. S., Bouras, N., Bridges, P. K., Watson, J. P. A randomized double-blind comparative study of sulpiride and amitriptyline in affective disorder. Psychopharmacology. 1983; 81:258-260.

Chapter **10**

MOOD-STABILIZING AGENTS

Russell T. Joffe and L. Trevor Young

CONTENTS

I. INTRODUCTION

Mood stabilizers are a variety of different therapeutic agents which are used for the treatment of bipolar affective disorder. Until the introduction of lithium approximately 25 years ago, the treatment of bipolar illness was very inadequate. Lithium, the first treatment specifically indicated for the illness, had a dramatic impact on its course and prognosis.

Bipolar affective disorder, previously known as manic depressive illness, has a prevalence of about 1.5% of the general population.[1] It usually has an early age of onset, commonly in the late second or early third decade of life.[2] It is characterized by recurrent episodes of mania and depression. The likelihood of spontaneous cessation of episodes is extremely unlikely, and, in fact, the tendency is toward a greater frequency of longer episodes as the illness progresses over time.[3] Recently, Post[4] has hypothesized that the illness itself may increase vulnerability to subsequent episodes and to a chronic course. Although such a notion requires experimental confirmation, it is supported by several different pieces of clinical observation and data which provide compelling support for this sensitization hypothesis. If Post[4] is correct, it would suggest that the illness should be treated effectively as soon as possible so as to eliminate the current period of illness and thereby improve the long-term prognosis. Notwithstanding the issues raised by Post,[4] there is clear evidence that bipolar disorder causes an enormous burden of suffering for patients and their families, as well as having a significant impact on society.[2] Furthermore, it is associated with substantial mortality, particularly from suicide which is a consequence of both the depressed and the manic phase of the illness.[2] As a result of the considerable morbidity and mortality associated with the disorder, there has been a need to establish better and more-effective treatments.

Although lithium has had a substantial impact on the treatment of bipolar disorder, its limitations both with regard to efficacy and to side effects are well known. As a result, a substantial number of new treatment options have become available. These are listed in Table 1. In general, with the exception of the anticonvulsants, particularly carbamazepine and valproate, the efficacy of these treatments has not been rigorously evaluated. For most, there are few, if any, randomized, controlled trials comparing treatment either with placebo or with a standard treatment such as lithium. For the most part, the efficacy of these options has been documented in case reports or series and supported by physician satisfaction with general clinical use.

TABLE 1.

Mood Stabilizers

Lithium
Anticonvulsants
Carbamazepine
Valproate
Other
Calcium channel blockers
Thyroid hormone
Antipsychotics
Flupenthixol
Clozapine
Antidepressants
Bupropion
Electroconvulsive treatment
L-Tryptophan

Although the specific biological/biochemical defect(s) which are important in the pathophysiology of bipolar disorder has not yet been unequivocally identified, putative pathophysiological mechanisms have guided the selection of pharmaceutical agents in the treatment of this disorder. Moreover, investigations into the mechanisms of action of agents found to have mood-stabilizing properties have also informed our knowledge of the pathophysiology of this disorder.

In this chapter, we will briefly describe each of the current mood stabilizers available for use in bipolar disorder. We will briefly describe their documented efficacy in treating bipolar disorder and discuss their potential mechanisms of action. As a chapter like this, because of its brevity, cannot be comprehensive, we will rather focus on current issues related to each particular treatment option.

II. MOOD-STABILIZING AGENTS

A. Lithium

Of all the options (see Table 1), lithium has the most comprehensive data base on its efficacy, therapeutics, and side effect profile. Lithium has clearly been shown to be effective in the treatment of acute mania and in the prophylaxis of both manic and depressive episodes.[5] Its antidepressant properties remain a subject of debate, although it is generally agreed that they are much less well established and certainly appear to be inferior to antidepressants, particularly in unipolar depression.[2] Lithium is effective for both acute and prophylactic treatment in only about half of bipolar patients.[2] Furthermore, predictors of response to lithium treatment include the most typical forms of illness with less than four episodes per year and a family history both of affective illness and lithium response.[6] It follows, therefore, that atypical forms of bipolar disorder, such as rapid cycling and mixed states or dysphoric mania, are less likely to benefit from lithium treatment. As these comprise a substantial proportion of bipolar episodes, it would suggest that lithium has a more restricted therapeutic spectrum within the bipolar disorders than previously thought.[7,8]

The side effect profile of lithium is well established. It includes gastrointestinal symptoms, tremor, polyuria and thirst, various skin reactions, weight gain, and cognitive difficulties.[2] Collectively, the large side effect burden worsens noncompliance, although weight gain and, particularly, cognitive impairment appear to be the main contributors identified.[9] Long-term side effects are also well known. Approximately 10 to 15% of subjects on chronic lithium therapy will develop clinical hypothyroidism requiring replacement therapy with thyroid hormone.[10] The issue of whether or not lithium causes renal damage is still hotly debated, although, if it does, this is likely rare.[2]

Recent data suggest that the recommended therapeutic range for plasma lithium levels of 0.5 to 1.5 mmol/l requires reevaluation. First, Gelenberg et al.[11] showed that a group treated with standard lithium levels of 0.8 to 1.0 mmol/l had a significantly lower rate of relapse than a comparison group treated with low lithium levels of 0.4 to 0.6 mmol/l. Second, levels in excess of 1.0 mmol/l are usually associated with increased side effects and, therefore, reduce compliance without necessarily improving efficacy. These data imply that failed lithium trials with plasma levels in the range 0.5 to 0.7 mmol/l may not be considered optimum. Regardless of lithium level, Keller et al.[12] observed that the presence of subsyndromal symptoms of mania or depression substantially increased the risk of relapse. Therefore, one requires optimum lithium treatment with adequate lithium levels not allowing residual symptoms in order to improve prognosis. This, of course, is not always possible because of the high side effect burden associated with lithium therapy and resultant problems with noncompliance. Even if lithium treatment is optimized, there is still a substantial rate of nonresponse. In classic bipolar disorder, this may be as high as 30 to 40%, but in particular subtypes the nonresponse rate is much higher, about 60 to 80% in mixed states and rapid cyclers.[13]

Despite enormous effort, elucidation of the precise mechanism of lithium has not yet been accomplished. Since lithium has mood-stabilizing, antimanic, and antidepressant effects, it has been difficult to formulate a parsimonious explanation for all of these effects. Both animal and human studies suggest that the antidepressant action of lithium and its ability to augment partial responses to tricyclic antidepressants may be due to a net serotonergic-enhancing action which may be presynaptic.[14,15] Lithium has also been shown to down regulate β-adrenoceptor number in rodent brain, a property shared by the majority of other antidepressant treatments.[16] With respect to the antimanic effects of lithium, older literature suggested that it might block manifestations of dopaminergic supersensitivity and/or have a cholinergic-enhancing activity.[14,17]

The recent clarification of the molecular mechanisms that link receptors with cellular responses, i.e., the signal transduction pathways, has been very helpful in further clarification of the possible site of action of lithium. Within the β-adrenergic receptor–coupled signaling pathway, a well-replicated finding has been the ability of lithium to blunt noradrenergic-stimulated adenylyl cyclase activity.[18] The effect of lithium has been further localized to the level of the GTP-binding proteins (G proteins) which couple β-adrenergic (and other) receptors to the enzyme adenylyl cyclase.[19] More specifically, lithium has been shown to blunt agonist-stimulated GTP binding to G proteins,[20] to decrease the gene expression of specific G protein subunits,[21] and to decrease the coupling of G proteins to adenylyl cyclase.[22]

The second major signal transduction pathway, the polyphosphoinositide (PPI)-generated second messenger system, is also an important candidate site of action for lithium. The enzymes known as the inositol monophosphatases,[23] which metabolize PPI second messengers are inhibited in animal brain at clinically relevant concentrations which would be expected to blunt signaling through these second messengers. Furthermore, the translocation and subsequent blunting of protein kinase C activity from cytosol to the membrane is also enhanced after lithium treatment.[24,25] In general, the effects of lithium are to blunt the PPI-generated second messenger system which has been taken as evidence of a possible pathological increase in signaling through this system in bipolar disorder.

As the limitations of lithium have been confirmed, there has been a concerted effort to consider alternative treatments for bipolar illness. These will be considered in turn.

B. Anticonvulsants

Several of these compounds have shown efficacy in the treatment of bipolar disorder. Most of the anticonvulsants found to have efficacy are known to be useful for focal, particularly temporal lobe, epilepsy.[26] Furthermore, these same anticonvulsants are the most specific for inhibiting amygdala-kindled seizures in animal models.[26] These data have led to the suggestion that bipolar disorder may be a form of epilepsy, but there is no evidence for this. It is likely that the mood-stabilizing efficacy of each anticonvulsant has a unique mechanism of action unrelated to its effect on seizure disorder.

1. Carbamazepine

This anticonvulsant, with a tricyclic structure, has been used for the treatment of paroxysmal pain syndromes as well as epilepsy. Its efficacy in bipolar disorder is analogous to lithium in that it has well-recognized acute antimanic efficacy and is also effective in the prophylaxis of both manic and depressive episodes. Its antidepressant effect is less well established.[27]

Carbamazepine has been shown to have antimanic efficacy superior to placebo and comparable to both antipsychotics and lithium.[27] The time course of antimanic response to carbamazepine is also similar to lithium, occurring over about 10 days with the full effect noted in 3 weeks. Numerous open and controlled studies have documented the prophylactic efficacy of carbamazepine which has also been shown to be similar to lithium. There are only two controlled trials of the antidepressant effects of carbamazepine, one employing an A-B-A design, involving predominantly bipolar subjects,[28] and the other involving mostly subjects with unipolar depression.[29] Despite these differences, both trials showed carbamazepine to have modest to moderate clinical efficacy, having clinically substantial antidepressant efficacy in approximately one third of subjects. The conclusion of all these studies is that carbamazepine may be particularly effective in lithium-refractory patients, particularly rapid cyclers. These conclusions appear premature as most studies have included patients who have been selected for study because of lithium failure. There are no studies employing appropriate sample selection and directly comparing carbamazepine to lithium that shows carbamazepine to be superior to lithium in rapid cyclers. Predictors of carbamazepine response remain to be clarified but electroencephalogaphic abnormalities and the presence of psychosensory symptoms do not appear to be of utility in differentiating carbamazepine responders and nonresponders.[30] Paradoxically, despite the poorly documented antidepressant effect of carbamazepine, severity of depression appears to predict antidepressant response to this anticonvulsant.[30]

The mechanism of the mood-stabilizing action of carbamazepine is poorly understood. It has diverse effects on mumerous neurotransmitter and neuropeptide systems. These have been extensively reviewed elsewhere.[31] Although a clear understanding of the mechanism of action of carbamazepine is still lacking, much evidence suggests that it has markedly different effects from lithium. In common with valproate, carbamazepine has been shown to up regulate $GABA_b$ receptors and decrease GABA turnover.[32] The levels of and responses to serotonin (5-HT) may be enhanced by carbamazepine in rat brain,[33] which would also be consistent with lithium and other drugs with antidepressant effects that are thought to enhance decreased 5-HT neurotransmission which likely occurs in depression. Carbamazepine does not have as clear an effect on either G protein levels or the PPI-generated second messenger system in contrast to lithium.[31,34] Post et al.[31] have recently described several other novel mechanisms for this agent, such as enhancement of substance P levels or calcium channel blockade, which need further empirical investigation.

Carbamazepine is used in doses of 600 to 1600 mg/day for bipolar disorder. Plasma levels of the parent drug are not related to its psychotropic efficacy although there are preliminary data that mood-stabilizing response may be related to plasma and cerebrospinal fluid levels of its epoxide metabolite.[35] Although the drug is reasonably well tolerated, it has well-described side effects, some of which may be related to its tricyclic structure. In addition, allergic skin rashes are quite common and may limit treatment. Hematological monitoring is required because of the rare but fatal cases of bone marrow suppression which have to be distinguished from the common limited and benign reduction in all blood cell indices which occurs with treatment. A major limitation to carbamazepine therapy is that it is a potent inducer of hepatic microsomal enzymes so that drug interactions are common.[36] This is particularly so in bipolar disorder, where a substantial number of patients do not respond to monotherapy with any of the usual mood-stabilizing agents. Carbamazepine can be used in combination with lithium with resultant limited increase in central nervous system side effects, but drug interactions become more complex and clinically significant when it is used with other psychotropic agents and mood stabilizers.[36] Carbamazepine, combined with lithium, can be considered an option even in subjects who have failed to respond to the two agents administered as monotherapy.

2. Valproate

This is a simple chain fatty acid which is effective in various forms of epilepsy. During the last 30 years, there have been numerous case series suggesting that valproate, usually administered as divalproex sodium, may be effective as a mood stabilizer.[37] In recent years, there has been a concerted effort to document the efficacy of valproate in the treatment of acute mania and in the prophylaxis of bipolar disorder using randomized controlled trials. Two studies[38,39] clearly establish the efficacy of valproate as superior to placebo and comparable to lithium in acute mania. Data from large-scale controlled studies examining prophylaxis are still awaited, but the open studies and clinical experience indicate that the drug will be effective for this aspect of treatment. The antidepressant effects of valproate are uncertain, although preliminary data would suggest that this anticonvulsant does not have robust antidepressant properties.[37]

Valproate may be particularly effective in subtypes of bipolar disorder. Open studies suggest that it may be indicated in rapid cyclers and mixed states; these findings have been recently confirmed in a large, randomized controlled trial.[39]

Valproate is usually administered in doses of 500 to 2500 mg/day. Although there is no direct relationship between therapeutic response and plasma levels, there is convincing data that plasma levels above the lower limit of the therapeutic range for anticonvulsant effects are associated with a better therapeutic response in bipolar illness.[37]

Valproate differs in its side effects from both lithium and carbamazepine. These are compared and contrasted in Table 2. As can be seen from the table, some of the side effects of valproate are dose related and may be substantially reduced or eliminated by a dose reduction. In addition to the effects noted in the table, valproate may also lead to hematological and hepatic side effects. Suppression of hematological indices is a rare side effect.[37] Hepatic inflammation, which may be severe and potentially fatal, is extremely rare and is distinct from the benign elevation of hepatic transaminase enzymes frequently observed.[37] Risk factors for the severe hepatic side effects include age, most cases having been observed in children under the age of 2 years, previous history of liver disease, and use of multiple anticonvulsants.[40]

Even less is known about the mechanism of action of valproate than of lithium or carbamazepine. Valproate shares with these agents the ability to enhance brain 5-HT activity, possibly through increasing the availability in brain of the 5-HT precursor L-tryptophan.[41,42] In common with carbamazepine, valproate has been found to enhance central GABA activity and to up regulate $GABA_b$ receptors.[43,44] Clarification of the exact mechanism of action of valproate and whether or not it is distinct from other mood-stabilizing agents must await further investigation.

TABLE 2.

Comparison of Effects of Anticonvulsants Compared with Lithium

	Lithium	Carbamazepine	Valproate
Gastrointestinal	Common, blood level related	Less common	Common, reduced by enteric coating
Urinary	Polyuria, polydipsia, renal damage rare	Vasopressin activity	None
Hepatic	None	Enzyme increase, hepatitis rare	Enzyme increase, rare fatal hepatitis
Skin	Common	Common especially allergic	Dose-related alopecia
Hematological	Increase all indices	Decrease indices, rare marrow suppression	Rare marrow suppression
Body weight	Weight gain common	Rare	Common, dose-related
Central nervous system	Cognitive difficulties	Less cognitive problems, neurological symptoms at higher doses	Less cognitive problems
Thyroid	Decreased T_4, T_3; increased TSH; clinical hypothyroidism	Limited decrease in T_4, T_3	Limited decrease in T_4, T_3
Pregnancy	Possibly safe	Absolutely contraindicated	Absolutely contraindicated

3. Other Anticonvulsants

There are extremely limited data on the efficacy of other anticonvulsants in bipolar disorder. Early, uncontrolled case series suggested that diphenylhydantoin may have psychotropic efficacy in a mixed, poorly defined group of psychiatric patients.[45,46] There are, however, no randomized controlled studies to confirm these findings.

Barbiturates have not been systematically evaluated in bipolar disorder. Although clinical lore suggests that these compounds may aggravate depression, there are several open case reports suggesting that primidone may be effective in bipolar illness, both the manic and depressed phases.[47] Although these data have to be confirmed in controlled trials, they are of theoretical importance as they suggest that anticonvulsants which do not specifically target the temporal lobe may have mood-stabilizing effects. Similarly, there are case studies but limited systematic data to support the efficacy of benzodiazepines as mood stabilizers. Although there have been some studies suggesting that clonazepam may have specific antimanic efficacy, these studies have been methodologically flawed and have usually involved concomitant antipsychotic medication, making it very difficult to evaluate the clinical effects of clonazepam.[48] Furthermore, Bradwejn[49] showed that at comparable doses lorazepam had similar efficacy to clonazepam in acute manic subjects, thereby challenging the specificity of the antimanic effect of clonazepam. The antidepressant effect of clonazepam has not been established.

Acetazolamide has strong anticonvulsant properties, although it is not routinely used clinically as an anticonvulsant. There are several studies which suggest that this compound, although ineffective alone, may have robust therapeutic effects when combined with other mood stabilizers.[50]

In summary, the anticonvulsants, particularly carbamazepine and valproate, have a major role in the treatment of bipolar disorder. They are clearly the first alternative to lithium in classic bipolar disorder. In certain subtypes of illness, particularly mixed states and rapid cycling, and in special populations, such as adolescents, the neurologically impaired, and the elderly, they are increasingly first-line treatments. In this regard, valproate is emerging as the preferred alternative because of its favorable side effect profile, limited drug interactions, and the ever-enlarging data base on its efficacy. Moreover, as monotherapy is relatively uncommon in bipolar illness, the anticonvulsants are frequently used together with lithium or with each other to enhance mood-stabilizing effects.

C. Calcium Channel Blockers

The most commonly used calcium channel blocker is verapamil, although there is preliminary evidence that nimodipine may be effective in acute mania.[51] In acute mania, both crossover and parallel double-blind studies suggest that verapamil has comparable effects to lithium. Similarly, the prophylactic efficacy of this drug has also been shown although, as for the studies in acute mania, sample sizes are small in the few studies to date. In the prophylaxis studies, the efficacy of verapamil was partially attributed to the carryover effects of lithium, suggesting that verapamil may be better as an adjunct rather than as monotherapy in the maintenance phase.[51] The antidepressant efficacy of verapamil has not been demonstrated, and, in fact, preliminary evidence is that its antidepressant effects are inferior to standard antidepressants and comparable with placebo.[51] With respect to the mechanism of action of these agents, a consistent finding in patients with bipolar disorder is increased intracellular calcium levels and flux in platelets and leukocytes which normalize with treatment.[52,53] Such pathophysiological mechanisms in patients with bipolar disorder are very consistent with the role of calcium channel blockers in the treatment of bipolar disorder. Future studies with more neuronal specific calcium channel blockers or in combination with other mood stabilizers may help to establish their importance.

D. Thyroid Hormone

Thyroid hormones have been used to treat various types of affective disorder. In particular, Stancer and Persad[54] showed that high doses of T_4, so called hypermetabolic T_4, may be effective in the rapid-cycling form of bipolar illness. Although their findings were confirmed in case studies, there are still very little systematic data to confirm these observations. In a systematic trial involving 11 patients, Bauer and Whybrow[55] reported that high doses of T_4 reduced the frequency and amplitude of rapid cycling. However, they noted that T_4 was more likely to be effective if administered as an adjunct to another mood stabilizer rather than as monotherapy. Furthermore, their study[55] suggested that very high doses, 300 to 500 µg, may not be necessary and that doses of T_4 to maintain plasma T_4 levels just above the normal range may be sufficient for mood stabilization. The potential efficacy of T_4 in bipolar disorder does not necessarily imply that the disorder is associated with a deficiency of thyroid hormone. In fact, a specific thyroid axis abnormality has not been identified in bipolar illness, and there is no relationship between baseline plasma T_4 levels and subsequent response to T_4.[55]

E. Antipsychotics

The antipsychotics have well-established efficacy in the treatment of acute mania. However, there is limited evidence that they may have mood-stabilizing properties.

In an earlier study, it was suggested that flupenthixol may be particularly effective in rapid cyclers,[56] and recently there are emerging data that a novel antipsychotic, clozapine, may have mood-stabilizing effects in rapid-cycling bipolars.[57,58]

F.   Antidepressants

This class of drugs is frequently used during the depressed phase of the illness. However, antidepressants are not regarded as having mood-stabilizing properties and have been reported to induce mania or rapid cycling in bipolar subjects. They are, therefore, used with considerable caution and mostly in combination with a mood stabilizer. Recently, case studies suggest that the novel antidepressant bupropion may have both antidepressant and mood-stabilizing effects although this has not been confirmed in controlled studies.[59]

G. Electroconvulsive Treatment

Electroconvulsive treatment (ECT) has well-demonstrated efficacy in both acute depression and mania. It is feasible that long-term ECT may also offer efficacy in the maintenance treatment of this disorder.[60] Although case studies would support this notion, controlled studies have yet to be done. ECT may exert its mood effects, paradoxically, by its anticonvulsant properties on temporal limbic structures in the brain.[61]

H. L-Tryptophan

This amino acid is a precursor for serotonin, a deficiency of which has been postulated in mood disorders, in general, and bipolar disorder, in particular. As a single agent, the mood-stabilizing efficacy of L-tryptophan has not been convincingly demonstrated. In combination with lithium, there is some preliminary, clinical evidence that it may be effective in selected cases.[62]

III. CONCLUSIONS

There is an increasing number of mood-stabilizing compounds. Although lithium was the first and the best established, its limitations are being increasingly recognized, particularly with regard to its side effect profile and its use in anything other than absolutely typical bipolar disorder. The anticonvulsants, particularly valproate, are the next best established treatments, and not only do they present a viable alternative in those who are lithium nonresponders or lithium intolerant, but they are increasingly becoming first-line treatments in certain subtypes of bipolar illness and in special populations with the disorder. The other treatments reviewed are used in cases refractory to lithium and the anticonvulsants and are supported by extremely limited data documenting their efficacy. This disparate group of compounds, notwithstanding the limited data on efficacy, provide only limited information on the biological basis of bipolar illness, as their mechanism of action remains uncertain and is not necessarily the same as for the other, usually primary, indications for each compound.

REFERENCES

1. Regier, D. A., Farmer, N. E., Rae, D. S., Locke, B. Z., Keith, S. J., Judd, L. L. Comorbidity of mental disorders with alcohol and other drug abuse: results from the epidemiologic catchment (ECA) study. *JAMA*, 264, 2511, 1990.

2. Goodwin, F. K., Jamison, K. R. *Manic-Depressive Illness*, Oxford University Press, New York, 1990.

3. Roy-Byrne, B. P., Post, R. N., Uhde, T. W., Porai, D., Davis, D. The longitudinal course of a current affective illness: life chart data from research patients at the NIMH. *Acta Psychiatr. Scand.*, 71, 3S, 1985.

4. Post, R. M. Transduction of psychosocial stress into the neurobiology of recurrent affective disorder. *Am. J. Psychiatry*, 149, 999, 1992.

5. Goodwin, F. K., Murphy, D. L., Bunney, W. E. Lithium-carbonate treatment in depression and mania. *Arch Gen. Psychiatry*, 21, 486, 1969.

6. Dunner, D. L., Fieve, R. R. Clinical factors in lithium prophylaxis failure. *Arch. Gen. Psychiatry*, 30, 229, 1974.

7. McElroy, S. L., Keck, P. E., Pope, H. G. J., Hudson, J. I., Feedda, G. L., Swann, A. C. Clinical and research implications of the diagnosis of dysphoric or mixed mania or hypomania. *Am. J. Psychiatry*, 149, 1633, 1992.

8. Swann, A. C. Mixed or dysphoric manic states: psychopathology and treatment. In *Anticonvulsants in Mood Disorders*, Joffe, R. T. and Calabrese, J. R., Eds., Marcel Dekker, New York, 1994, Chapter 8.

9. Jamison, K. R., Akiskal, H. S. Medication compliance in patients with bipolar disorders. *Psychiatr. Clin. North Am.*, 6, 175, 1983.

10. Lazarus, J. H., McGregor, A. M., Ludgate, M., Drake, C., Creagh, F. M., Kingswood, C. J. Effects of lithium carbonate therapy on thyroid immune state in manic depressive patients: a prospective study. *J. Affective Disorders*, 11, 135, 1986.

11. Gelenberg, A. J., Kane, J. M., Keller, M. B., Lavori, P., Rosenbaum, J. F., Cole, K. Comparison of standard and low serum levels of lithium for maintenance treatment of bipolar disorder. *N. Engl. J. Med.*, 321, 1489, 1989.

12. Keller, M. B., Lavori, P. W., Kane, J. M., Gelenberg, A. J., Rosenbaum, J. F., Walzer, E. A., Baker, L. A. Subsyndromal symptoms of bipolar disorder. A comparison of standard and low serum levels of lithium. *Arch. Gen. Psychiatry*, 49, 371, 1992.

13. Calabrese, J. R., Delucchi, G. A. Spectrum of efficacy of Valproate in 55 rapid-cycling manic depressives. *Am. J. Psychiatry*, 147, 431, 1990.

14. Bunney, W. E., Garland-Bunney, B. L. Mechanisms of action of lithium in affective illness: basic and clinical implications. In *Psychopharmacology: The Third Generation of Progress*, Meltzer, H. Y., Ed., Raven Press, New York, 553.

15. Price, L. H., Charnery, D. S., Delgado, P. L., Miger, G. R. Lithium and serotonin function: implications for the serotonin hypothesis of depression. *Psychopharmacology*, 100, 3, 1990.

16. Treiser, S. L., Cascio, C. S., O'Donohue, T. L., Thoa, N. B., Jacobowitz, D. M., Keller, K. J. Lithium increases serotonin release and decreases serotonin receptors in the hippocampus. *Science*, 213, 1529, 1981.

17. Pert, A., Rosenblatt, J. E., Sivit, C., Pert, C. B., Bunney, W. E., Jr. Long-term treatment with lithium prevents the development of dopamine receptor supersensitivity. *Science*, 201, 171, 1978.

18. Ebstein, R. P., Hermoni, M., Belmaker, R. H. The effect of lithium on noradrenalin-induced cyclic AMP accumulation in rat brain: inhibition after chronic treatment and absence of supersensitivity. *J. Pharmacol. Exp. Ther.*, 213, 161, 1980.

19. Hudson, C. J., Young, L. T., Li, P. P., Warsh, J. J. CNS transmembrane signal transduction in the pathophysiology and pharmacology of affective disorders and schizophrenia. *Synapse*, 13, 278, 1993.

20. Avissar, S., Schrieber, G., Danon, A., Belmaker, R. Lithium inhibits adrenergic and cholinergic increases in GTP binding in rat cortex. *Nature*, 331, 440, 1988.

21. Li, P. P., Tam, Y. K., Sibony, D., Warsh, J. J. Lithium decreases G_{i-1} and G_{i-2} alpha-subunit mRNA levels in rat cortex. *Eur. J. Pharmacol. Mol. Pharmacol.*, 206, 165, 1991.

22. Hsiao, A. K., Manji, H. K., Chen, G., Bitran, J. A., Risby, E. D., Potter, W. Z. Lithium administration modulates platelet G_i in humans. *Life Sci.*, 50, 227, 1992.

23. Hallcher, L. M., Sherman, W. R. The effects of lithium ion and other agents on the activity of myo-inositol-1-phosphatase from bovine brain. *Gen. Biol. Chem.*, 255, 10896, 1980.

24. Lenox, R. H., Watson, D. G. Targets for lithium action in the brain: protein kinase C substrates and muscarinic receptor regulation. *Clin. Neuropharmacol.*, 15, S612A, 1992.

25. Li, P. P., Sibony, D., Green, M., Warsh, J. Lithium modulation of the phosphoinositide signalling system in rat cortex: selective effects on phorbol ester binding. *J. Neurochem.*, 61, 1722, 1993.

26. Post, R. M. Mechanisms of action of carbamazepine and related anticonvulsants in affective illness. In *Psychopharmacology: The Third Generation of Progress*, Meltzer, H. Y., Ed., Raven Press, New York, 1987, 567.

27. Martin, L. S., Bebchuk, J. M., Joffe, R. T. Clinical efficacy of carbamazepine. In *Anticonvulsants in Mood Disorders*, Joffe, R. T., Calabrese, J. R., Eds., Marcel Dekker, New York, 1994, 111.

28. Post, R. M., Uhde, T. W., Roy-Byrne, P. P., Joffe, R. T. Antidepressant effects of carbamazepine. *Am. J. Psychiatry*, 143, 29, 1986.

29. Small, J. G. Anticonvulsants in affective disorders. *Psychopharmacol. Bull.*, 26, 25, 1990.

30. Post, R. M. Effectiveness of carbamazepine in the treatment of bipolar affective disorder. In *Use of Anticonvulsants in Psychiatry: Recent Advances*, McElroy, S. L., Pope, H. G., Jr., Eds., Oxford Health Care, Clifton, NJ, 1988, 25.

31. Post, R. M., Weiss, S. R. B., Chuang, D.-M., Ketter, T. A. Mechanism of action of carbamazepine in seizure and affective disorders. In *Anticonvulsants in Mood Disorders*, Joffe, R. T., Calabrese, J. R., Eds., Marcel Dekker, New York, 1994, 43.

32. Lloyd, K. G., Thuret, E. W., Pilc, A. GABA and the mechanism of action of antidepressant drugs. In *GABA and Mood Disorders: Experimental and Clinical Research*, Bartholini, G., Lloyd, K. G., Morselli, P. L., Eds., LERS Monograph Series, Vol. 4, Raven Press, New York, 1986, 33.

33. Yan, Q. S., Mishra, P. K., Burger, R. L., Bettendorf, A. F., Jobe, P. C., Dailey, J. N. Evidence that carbamazepine and antiepilepsirine may produce a component of their anticonvulsant effects by activating serotonergic neurons in genetically epilepsy-prone rats. *J. Pharmacol. Exp. Ther.*, 261, 652, 1992.

34. Li, P. P., Young, L. T., Tam, Y. K., Sibony, D., Warsh, J. J. Effects of chronic lithium and carbamazepine treatment on G-protein subunit expression in rat cerebral cortex. *Biol. Psychiatry*, 34, 167, 1993.

35. Post, R. M., Uhde, T. W., Ballenger, J. C., Chatterji, D. C., Green, R. F., Bunney, W. E., Jr. Carbamazepine and its 10,11-epoxide metabolite in plasma and CSF. Relationship to antidepressant response. *Arch. Gen. Psychiatry*, 40, 673, 1983.

36. Ketter, T. A., Post, R. M. Clinical pharmacology and pharmacokinetics of carbamazepine. In *Anticonvulsants and Mood Disorders*, Joffe, R. T., Calabrese, J. R., Eds., Marcel Dekker, New York, 1994, 147.

37. Calbrese, J. R., Woyshville, M. J., Rapport, D. J. Clinical efficacy of valproate. In *Anticonvulsants in Mood Disorders*, Joffe, R. T., Calabrese, J. R., Eds., Marcel Dekker, New York, 1994, 131.

38. Pope, H. G., Jr., McElroy, S. L., Keck, P. E., Jr., Hudson, J. I. Valproate in the treatment of acute mania: a placebo-controlled study. *Arch. Gen. Psychiatry*, 48, 62, 1991.

39. Bowden, C. L., Brugger, A. M., Swann, A. C., Calabrese, J. R., Janicak, P. G., Petty, F., Dilsaver, S. C., Davis, J. M., Rush, A. J., Small, J. G., Garza-Trevino, E. S., Risch, S. C., Goodnick, P. J., Morris, D. D. Efficacy of divalproex versus lithium and placebo in the treatment of mania. *JAMA*, 271, 91A, 1994.

40. McElroy, S. L., Keck, P. E., Pope, H. G., Jr., Hudson, J. I. Valproate in primary psychiatric disorders: literature review and clinical experience in a private psychiatric hospital. In *Use of Anticonvulsants in Psychiatry: Recent Advances*, McElroy, S. L., Pope, H. G., Jr., Eds., Oxford Health Care, Clifton, NJ, 1988, 25.

41. Maes, M., Calabrese, J. R. Mechanism of action of valproate in affective disorders. In *Anticonvulsants in Mood Disorders*, Joffe, R. T., Calabrese, J. R., Eds., Marcel Dekker, New York, 1994, 93.

42. Hwang, E. C., van Woert, M. H. Effect of valproic acid on serotonin metabolism. *Neuropharmacology*, 18, 1093, 1979.

43. Chapman, A. G. Mechanisms of anticonvulsant action of valproate. *Prog. Neurobiol.*, 19, 315, 1982.

44. Motohashi, N. GABA receptor alterations after chronic lithium administration: comparison with carbamazepine and sodium valproate. *Prog. Neuropsychopharmacol. Biol. Psychiatry*, 16, 571, 1992.

45. Kalinowsky, L. B., Putnam, T. J. Attempts at treatment of schizophrenia and other non-epileptic psychoses with Dilantin. *Arch. Neurol. Psychiatry*, 49, 414, 1943.

46. Freyhan, F. A. Effectiveness of diphenylhydantoin in management of non-epileptic psychomotor excitement states. *Arch. Neurol. Psychiatry*, 53, 370, 1945.

47. Hayes, S. G. Barbiturate anticonvulsants in refractory affective disorders. *Ann. Clin. Psychiatry*, 5, 35, 1983.

48. Chouinard, G. Clonazepam in the treatment of psychiatric disorders. In *Use of Anticonvulsants in Psychiatry: Recent Advances*, McElroy, S. L., Pope, H. G., Jr., Eds. Oxford Health Care, Clifton, NJ, 1988, 43.

49. Bradwejn, J. Double-blind comparison of the effects of clonazepam and lorazepam in acute mania. *J. Clin. Psychopharmacol.*, 10, 403, 1990.

50. Hayes, S. G. Acetazolamide in primary psychiatric disorders. *Neuropsychopharmacology*, 9 (Suppl.), 131S, 1983.

51. Cooke, R. G., Young, L. T. Miscellaneous anticonvulsants in affective disorders. In *Anticonvulsants in Mood Disorders*, Joffe, R. T., Calabrese, J. R., Eds., Marcel Dekker, New York, 1994, 215.

52. Dubovsky, S. L., Franks, R. D. Intracellular calcium ions in affective disorders: a review and an hypothesis. *Biol. Psychiatry,* 18, 781, 1983.

53. Dubovsky, S. L., Christiano, J., Danielle, L. C., Franks, R. D., Murphy, J., Adler, L., Baker, N., Harris, R. A. Increased platelet intracellular calcium concentration in patients with bipolar affective disorders. *Arch. Gen. Psychiatry,* 46, 632, 1989.

54. Stancer, H. C., Persad, E. Treatment of intractable rapid-cycling manic-depressive disorder with levothyroxine. *Arch. Gen. Psychiatry,* 39, 311, 1982.

55. Bauer, M., Whybrow, P. C. Rapid cycling bipolar affective disorder. II. Treatment of refractory rapid cycling with high-dose levothyroxine. *Arch. Gen. Psychiatry,* 47, 435, 1990.

56. Kielholz, P., Trzani, S., Poldinger, W. The long-term treatment of periodical and cyclic depressions with flupenthixol decanoate. *Int. Pharmacopsychiatry,* 14, 305, 1979.

57. Banov, M. D., Zarate, C. A., Tohen, M., Scialabba, D., Wines, J. D., Kolbrener, M., Kim, J.-W., Cole, J. O. Clozapine therapy in refractory affective disorders: polarity predicts response in long-term follow up. *J. Clin. Psychiatry,* 55, 295, 1994.

58. Calabrese, J. R., Meltzer, H. Y., Markovitz, P. J. Clozapine prophylaxis in rapid cycling bipolar disorder. *J. Clin. Psychopharmacol.,* 11, 396, 1991.

59. Haykel, R., Akiskal, H. Bupropion as a promising approach to rapid cycling bipolar II patients. *J. Clin. Psychiatry,* 13, 256, 1990.

60. Monroe, R. R. Maintenance electroconvulsive therapy. *Psychiatr. Clin. North Am.,* 14, 947, 1991.

61. Swartz, C. M. Seizure benefit: grand mal or grand bene? *Neurol. Clin.,* 11, 151, 1991.

62. Baldessarini, R. J. Treatment of depression by altering monoamine metabolism: precursors and metabolic inhibitors. *Psychopharmacol. Bull.,* 20, 224, 1984.

Chapter 11

NEUROLEPTICS

Gary Remington

CONTENTS

0-8493-8386-0/96/$0.00+$.50

I. INTRODUCTION

Neuroleptics were first synthesized in 1883, although it was not until the early 1950s that this class of medications found a role in psychiatry.[1] At that time, chlorpromazine was being investigated for its anesthetic properties when it was noted that a marked calming effect occurred following its administration. Subsequent clinical trials established the antipsychotic properties of these types of compounds, and they rapidly became the cornerstone of treatment programs for schizophrenia.[2,3] Indeed, they represented the first effective treatment for schizophrenia and firmly established it as an illness with biological origins. Moreover, their success as antipsychotics led to a process of deinstitutionalization beginning in the 1960s which is still underway today.

Clinical experience since has tempered initial enthusiasm, perhaps as it became clear that these agents had limitations, both in terms of clinical efficacy and side effects.[4,5] The result has been an ongoing search for better drugs, with a number of new compounds either available or under investigation during recent years.

II. CLASSIFICATION

Neuroleptics may be categorized in a number of ways: chemical class or potency, formulation, and biochemical/clinical profile, i.e., typical vs. "atypical."

A. Chemical Class/Potency

It is estimated that there are currently about 50 different neuroleptics representing 12 different chemical groups.[6] Table 1, for example, outlines those agents which are available for clinical use in Canada.[7] While it is characteristic to classify specific agents by their particular chemical class, it is often common as well to categorize compounds by their potency, that is, their dopamine (DA) D_2 affinity. Using this type of classification, neuroleptics are then described as low-potency (e.g., chlorpromazine), mid-potency (e.g., loxapine), and high-potency (e.g., haloperidol), with the high potency compounds reflecting the greatest D_2-binding affinity (see Figure 1).[8] This attribute also puts the high-potency neuroleptics more at risk of causing extrapyramidal side effects (EPS). In contrast, low-potency compounds substantially affect a number of other neurotransmitters which, in turn, contribute to their particular side effect profile. In the case of chlorpromazine, for example, its marked cholinergic binding accounts for its increased risk of cardiovascular side effects, as compared with a diminished risk of EPS because of its relatively low D_2 affinity.

B. Formulation

Neuroleptics are available in various formulations, including oral, intramuscular, and depot, with choice of formulation influenced by the clinical picture. For example, preference may be given to an intramuscular form in acutely psychotic patients, particularly those who are agitated or violent, as this offers greater and more rapid bioavailability.[9] At the other end of the treatment continuum, depot neuroleptics were introduced in the 1960s as alternatives to oral neuroleptics for the long-term management of schizophrenia since compliance is a significant problem associated with ongoing treatment.[10] By decreasing noncompliance, depot neuroleptics can reduce relapse rates by approximately 15 to 20%.[11,12] In actual clinical practice, it is

TABLE 1.

Oral Neuroleptics by Chemical Class and
Comparative Potency in Chlorpromazine
(CPZ) Equivalents

Chemical Class	CPZ Equivalents
Phenothiazine	
Aliphatic	
Chlorpromazine	100
Methotrimeprazine	70
Triflupromazine	25
Piperidine	
Mesoridazine	75
Pericyazine	15
Thioridazine	100
Piperazine	
Acetophenazine	20
Fluphenazine	2
Perphenazine	10
Thioproperazine	5
Trifluoperazine	5
Thioxanthene	
Chlorprothixene	100
Flupenthixol	5
Thiothixene	3
Butyrophenone	
Haloperidol	2
Diphenylbutylpiperidine	
Pimozide	2
Dibenzoxazepine	
Loxapine	15
Dihydroindolone	
Molindone	10
Novel Neuroleptics	
Dibenzodiazepine	
Clozapine	?
Benzisoxazole	
Risperidone	?

not uncommon to utilize all formulations during different phases of an individual's illness.

C. Pharmacological/Clinical Profile

Despite the clinical success of chlorpromazine, how such medications worked was initially poorly understood,[6] and it took a number of years to identify DA, and in particular the D_2 receptor, as critical to the antipsychotic activity.[13] This knowledge translated into an extensive search for more selective D_2 antagonists, culminating in the development of such high-potency agents as haloperidol. These various drugs are generally referred to as conventional or typical neuroleptics, sharing in common an ability to improve positive symptoms, e.g., hallucinations, delusions, but being minimally effective in the treatment of negative symptomatology, e.g., avolition, apathy, and having a propensity for EPS.

Generally speaking, atypical neuroleptics demonstrate antipsychotic effects clinically without causing significant EPS.[14] As early as 1970, clozapine was noted to be a potent antipsychotic without significant EPS,[6] but shortly after release it was

Relative Binding Affinities for Typical Neuroleptics

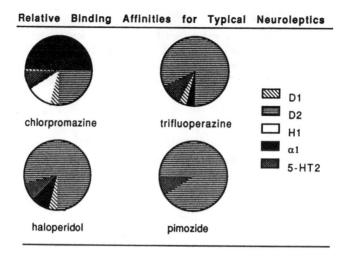

Figure 1
Relative binding affinities for typical neuroleptics. (Adapted from Hyttel, J., *Dyskinesia: Research and Treatment*, Casey, D. E., Chase, T. N., Christenssen, A. V., and Gerlach, J., Eds., Springer-Verlag, Berlin, 1985. With permission.)

withdrawn in most countries because of unexplained deaths, later associated with a risk for agranulocytosis.[15] It was reintroduced in the late 1980s and has come to represent the prototype of atypical neuroleptics,[16] with evidence that it is effective in treating negative, as well as positive, symptoms while at the same time having little, if any, risk of inducing acute EPS or tardive dyskinesia (TD).[17,18]

The precise definition of atypical neuroleptics remains somewhat ambiguous. Using animal models, induction of catalepsy and blockade of stereotyped behaviors in rodents following the administration of dopamimetic agents, e.g., apomorphine, amphetamine, characterize the action of typical neuroleptics and are thought to represent neuroleptic-induced EPS in humans. In contrast, clozapine fails to induce catalepsy or reverse the stereotypies.[16] Like typical neuroleptics, however, atypical neuroleptics may vary along a continuum in terms of side effects. Risperidone, for example, has been identified as atypical although it still has a risk for EPS, which appears dose-dependent.[19,20]

Along the same lines, lack of increased prolactin response in humans following neuroleptic administration has been identified as another atypical characteristic.[21] This is true for clozapine,[16] although it has not been demonstrated with other atypical compounds such as risperidone.[22]

There thus appears to be no unitary definition for an atypical neuroleptic, and even clozapine does not seem to meet all criteria absolutely.[23-26] The term "novel" is perhaps more to the point, as many of these newer compounds appear to meet each criterion to varying degrees.

III. MODE OF ACTION

A. Dopamine

The development of typical neuroleptics was predicated on the notion that schizophrenia reflects a state of hyperdopaminergic activity, and that neuroleptics manifest their antipsychotic action through an affinity for DA receptors and their blockade. A distinction between D_1 and D_2 receptor subtypes (i.e., the former iden-tifying receptors which stimulate adenyl cyclase compared with D_2 receptors which do not[27]) represented a further advance in the dopaminergic theory of schizophrenia. The development of more selective D_2 antagonists, and the finding of a correlation between the potency of neuroleptics as DA antagonists and their clinically effective doses, offered further support that the D_2 receptor plays a critical role in schizophre-nia.[28,29] Expanding this hypothesis, positron emission tomography (PET) data have indicated that patients on clinically effective doses of neuroleptics exhibit $78 \pm 6\%$ D_2 receptor occupancy.[30-33]

The exact involvement of the various DA receptor subtypes in schizophrenia has become increasingly complex. More recently, for example, an isoform of the D_2 receptor has been identified, i.e., D_4,[34] and it has been the subject of considerable attention with the finding that clozapine has a high affinity for this particular recep-tor.[16,34] Renewed interest in the possible role of D_1 receptors in psychosis has occurred with the recognition that clozapine is more potent in its blockade of central D_1 vs. D_2 receptors,[35] and it has been reported that D_1 and D_2 receptors may interact in a synergistic fashion.[36] Two isoforms of the D_1 receptor have been identified, specifi-cally the D_3 and D_5 receptors,[37,38] and the localization of D_3 receptors in the limbic region and the affinity of most antipsychotics, particularly the newer atypical com-pounds, for this receptor have been taken as evidence that the D_3 receptor may be integrally involved in psychosis.[37,39,40]

The notion that DA dysfunction in schizophrenia simply reflects a state of hy-perdopaminergic activity has been challenged recently. The DA system itself can be subdivided into a number of pathways: mesolimbic, nigrostriatal, tuberoinfundibu-lar, and mesocortical.[41] Mesolimbic dysfunction, represented by hyperdopaminergic activity, has been associated with psychosis and, in particular, positive symptoma-tology. The nigrostriatal pathway is linked with neuroleptic-induced movement disorders, i.e., EPS, while the tuberoinfundibular pathway has been associated with endocrine side effects, i.e., gynecomastia in males, amenorrhea in females.

There has been an increased focus recently on frontal lobe dysfunction in schizo-phrenia, with parallels drawn between damage to these areas and the deficit or negative symptoms of schizophrenia.[42] It has been postulated that mesocortical DA function may be underactive, secondary to a "lesion" which has occurred during neurodevelopment, adversely affecting DA afferents to the prefrontal cortex.[43] The effect of this lesion is one of impaired prefrontal cortical functioning, reflected in hypodopaminergic activity and negative symptoms. It is proposed that one of the tasks of the prefrontal cortex is modulation of mesolimbic activity and that dysfunction in the prefrontal cortical level results in impaired functioning of this feedback loop, i.e., decreased inhibition, leading to hyperdopaminergic activity at the mesolimbic level.

This has important implications regarding the pathophysiology of schizophrenia and the role for medications. For example, this would account for the lack of efficacy of conventional neuroleptics to treat negative symptomatology as, unlike positive symptoms, these are associated with hypo- vs. hyperdopaminergic activity. One could further speculate that changes in mesolimbic activity can be affected by agents acting directly on this system or indirectly through changes at the prefrontal cortex, while other neurotransmitters can influence DA functioning at any one of these levels.

B. Serotonin

Currently, serotonin (5-HT) is the other neurotransmitter receiving the most attention in terms of the pathophysiology of schizophrenia and the development of new antipsychotics. This is based on several lines of evidence. First, there is a substantial body of evidence to indicate that 5-HT fibers innervate those DA pathways thought to play a role in both positive (mesolimbic) and negative (mesocortical) symptoms, as well as neuroleptic-induced EPS (nigrostriatal).[44-47] For example, there is evidence of 5-HT inhibition of DA neurotransmission in the nigrostriatal pathway at both cell body and terminal regions, as well as reversal of neuroleptic-induced catalepsy through decreasing 5-HT neurotransmission,[48] suggesting that 5-HT–blocking agents may improve or diminish the risk of EPS. Similarly, 5-HT may play an inhibitory role on prefrontal DA activity, and 5-HT–blocking properties of certain novel antipsychotics can lead to an increased release of DA from prefrontal neurons and subsequent improvement in negative symptomatology.

That a combined profile of $5\text{-HT}_2/D_2$ antagonism may offer such compounds clinical benefits not seen with D_2 antagonism alone has been supported by recent reports involving clozapine and risperidone, both classified as novel antipsychotics and having such a profile. Each has been shown to have superior clinical efficacy in positive symptoms with patients refractory to conventional neuroleptics, as well as efficacy in the treatment of negative symptoms and a diminished risk of EPS.[17,19,20,49-51] Indeed, the unique clinical profile of these compounds has led to the development of numerous additional compounds with similar profiles (see Section VI).

IV. CLINICAL USE

While neuroleptics have been used in a variety of clinical diagnoses and conditions, e.g., Tourette's, impulsivity, insomnia, their primary indication is for the treatment of psychosis, usually in the context of schizophrenia. Accordingly, the guidelines here will focus on this illness, although the general principles apply to their use in other diagnoses, such as mania or depression with psychotic features.

A. Choice of Neuroleptic and Route of Administration

In terms of conventional neuroleptics, a superior response profile has not been demonstrated for specific conventional neuroleptics.[4,52] A particular agent may be chosen for a number of reasons, including side effects, history of response, and formulation. An individual or family history of previous positive response to a neuroleptic may guide one's choice.[4,53,54] In the patient who requires medication rapidly, parenteral administration offers distinct advantages. For example, parenteral haloperidol reaches peak levels in 20 to 40 minutes with 100% bioavailability, in contrast to oral treatment which achieves peak levels in 4 to 6 h with only 60% bioavailability.[9]

Interchanging neuroleptics during the course of therapy may increase the risk of relapse,[55] so an early decision that long-term therapy is going to include a depot neuroleptic favors the initial use of a drug which has this option available, e.g., haloperidol.

Depot neuroleptics may be considered in individuals where there appears to be problems with bioavailability, although this is thought to be relatively rare.[56] Noncompliance is an indication for depot neuroleptic use, as this is the major cause of relapse, and the use of depot medications can diminish relapse rates by up to 20%.[11,12] Depots have probably been underutilized,[12] and Johnson et al.[57] has suggested that depot therapy be considered for anyone with more than one episode.[58,59]

B. Acute Treatment

Monitoring the hospitalized, acutely psychotic patient off medication, or on the same dose prescribed prior to admission, should be considered initially because early and aggressive use of neuroleptics can obfuscate a proper evaluation.[60] Simply reinstating the medication as prescribed prior to hospitalization may be sufficient, given the frequency of noncompliance and its contribution to relapse rates.[61,62]

Early intervention with high-dose neuroleptic therapy does not produce more rapid recovery,[4,52,60] and it appears that doses beyond 10 to 12 mg of haloperidol equivalents show no improved benefit compared with lower doses.[6] Factors that may influence dosing in individual cases include age,[63,64] gender,[65] race,[66] and concomitant medications.[67,68] It is important to recognize that pharmacotherapy in the acute phase is often employed with two goals in mind: treatment of psychosis and behavioral control. Historically, neuroleptics have been used for both, perhaps contributing to the use of excessively high doses. This has been particularly true for the high-potency compounds, such as haloperidol, since their lack of cardiovascular complications permits disproportionately high doses.[69] It is now common practice, though, to use a benzodiazepine such as lorazepam in combination with lower doses of the neuroleptic, e.g., haloperidol 5 mg with lorazepam 2 mg, to achieve sedation without compromising antipsychotic effect.[70,71] The use of higher doses should stimulate a diagnostic reevaluation, as patients receiving higher doses in the acute setting may have a bipolar disorder.[72]

C. Maintenance Treatment

When parenteral administration is employed in the acute situation, the practice is to convert to oral medications within 24 to 48 h. A trial of a neuroleptic should be undertaken with the understanding that adequate doses need to be employed, and while it tends to take 10 to 14 days to begin to see clinically significant changes, improvement may continue for an additional 4 to 6 weeks or even longer.[73] The exception to this is clozapine, where there is evidence that an adequate trial should be in the range of 4 to 9 months.[74-76]

It is during this phase of treatment that one may consider a trial with a depot neuroleptic. Reasons for changing, as well as conversion formulae, have been detailed in several review articles focused on depot neuroleptics.[58,59]

In the longer term, the goal is to maintain an individual on the lowest therapeutic dose, and maintenance doses in the range of 6 to 12 mg haloperidol equivalents have been recommended.[60]

It is possible that in first episode patients even lower doses can be effective than those being recommended here for acute and maintenance treatment, but it remains for controlled investigations to clarify this and to determine whether this may also be the case for more-chronic patients if refractory and noncompliant patients are excluded.

The duration of treatment following a first episode of psychosis remains unclear, but continuation for at least 1 to 2 years has been recommended,[77,78] with maintenance doses decreased to approximately 50% of those used during the acute phase.[77] In the chronic population, intermittent, or targeted, pharmacotherapy has been suggested as an alternative to low-dose continuous medication, but the lack of benefits with respect to long-term side effects and the increased risk of relapse with intermittent treatment generally favor the use of a continuous low-dose strategy.[79]

D. Use of Novel Neuroleptics

In North America there are currently two novel compounds available for clinical use: clozapine and risperidone. Remoxipride, a substituted benzamide reflecting the search for highly selective D_2 antagonists,[80] was also released within the last several years here in Canada but subsequently withdrawn following scattered reports of aplastic anemia in association with its use.[81]

Because of the risk of agranulocytosis[82] and need for weekly blood monitoring, use of clozapine is confined to treatment-resistant patients or those with severe side effects related to conventional neuroleptic use. In refractory patients, clozapine has been shown to be superior in the treatment of both positive and negative symptoms.[48-51,74-76] Moreover, it has a very low risk of acute EPS and is the only neuroleptic to date which has clearly established itself as having little, if any, risk of inducing TD.[18]

Several published multicenter trials in North America also established that in chronic patients with schizophrenia risperidone is superior to haloperidol in terms of both positive and negative symptomatology.[19,20] A diminished risk of acute EPS was also noted in these trials, although it is dose-dependent and at higher doses, e.g., 16 mg daily, the EPS profile of risperidone paralleled haloperidol. Its risk for TD remains to be established, as long-term studies are required.

The issue of pharmacologic treatment of negative symptoms remains controversial. More recently, it has been common to distinguish primary and secondary negative symptoms, with primary symptoms more in keeping with deficit schizophrenia, whereas secondary negative symptoms may be related to EPS, environmental factors, and so on.[83,84] To date, no such studies have systematically evaluated this distinction with risperidone, while one study with clozapine has established that its improvement in negative symptoms is independent of its reduction in EPS.[50]

E. Refractory Schizophrenia

A systematic approach, as outlined in Table 2, should be employed when approaching patients who have been given the diagnosis of schizophrenia and who are not responding to treatment. The need to be aggressive must be emphasized, given evidence that persistence of symptoms and lack of effective treatment can be associated with poorer outcome.[94,95] At the same time, aggressive management is no longer synonymous with simply choosing a high-dose approach with neuroleptics.

V. SIDE EFFECTS

Side effects can seriously compromise compliance,[61,96] and it is therefore essential that clinicians be sensitive to the presence of side effects and address them as quickly as possible.

Although side effects are very individualized, they can be generally classified according to the receptor profile of a particular neuroleptic (Table 3).[97] As a rule, low-potency compounds are more prone to causing cardiovascular side effects, while the high-potency compounds are at greater risk for causing EPS. The shift over the years to the use of high-potency neuroleptics[69,98-100] has, in fact, established EPS as the most common side effect that clinicians must contend with.

TABLE 2.

Approach to Patients with Refractory Schizophrenia

Issue	Approach	References

A. Diagnosis	1. Rule out/treat:
	  • medical conditions
	  • substance abuse
	  • other psychiatric diagnoses
		80–90

B. Compliance	1. Address side effects
	  • decrease dose
	  • change neuroleptic
	  • side effect medications
	2. Prolactin and/or plasma neuroleptic level to evaluate compliance
	3. Trial of depot
		10–12, 56–59, 61, 91, 92, 111, 117

C. Neuroleptic trials	1. Identify target symptoms and outcome measures
	2. Trials of three neuroleptics
	  • at least two chemical classes
	  • may include risperidone here
	  • 6–8 weeks/trial
	  • up to 1000 CPZ equivalents
	3. Trial of clozapine
	  • trial of 4–9 months
	  • routine hematologic monitoring
		4, 17, 74–76, 92, 93, 186–189

D. Other	1. Adjunctive trial, guided by target symptoms
	  • positive: mood stabilizing drug (lithium, CBZ, valproate), benzodiazepines, ECT
	  • negative: L-dopa, benzodiazepines
	  • affective: lithium, CBZ, ECT, antidepressant
	  • impulsivity/aggression: CBZ, propanolol
	  • EEG abnormalities: CBZ, valproate
		92,93

Note: CPZ, chlorpromazine; CBZ, carbamazepine; ECT, electroconvulsive therapy.

TABLE 3.

Neuroleptic Side Effects According to Receptor Blockade

Receptor	Possible Clinical Consequence
Dopamine D_2	EPS Endocrine, e.g., galactorrhea
Histamine H_1	Sedation Weight gain Hypotension
alpha$_1$ adrenergic	Postural hypotension Reflex tachycardia
Muscarinic	Blurred vision Dry mouth Sinus tachycardia Constipation Urinary retention Memory dysfunction

Adapted from Richelson, E., *Psychiatr. Ann.*, 20: 640, 1990.

A. EPS (Dystonias, Akathisia, Parkinsonism, Dyskinesias)

It has been recommended that prophylactic use of antiparkinsonian medication be avoided.[101] At the same time, such an approach can significantly reduce the risk of acute dystonic reactions[102-110] and should be considered in patients with the following risk factors: young, male, high-potency neuroleptic, and history of significant EPS, particularly dystonic reactions.[111] Since approximately 85 to 95% of acute dystonic reactions occur in the first 3 to 4 days of treatment,[104,107] it may be useful in the population at risk to initiate a course of antiparkinsonian therapy during the first week, followed by a tapering over several days. If discontinuation cannot be achieved because of the coexistence of other EPS, ongoing use should be reevaluated at least every 2 to 3 months thereafter.

Akathisia may occur within hours of neuroleptic initiation[112] and is one of the most troublesome side effects, as it is common, distressing, and can often mimic anxiety or psychotic agitation,[113] which in turn can lead to inappropriate intervention with dose increments. Akathisia can be further subdivided into acute, chronic, tardive, withdrawal, and "pseudoakathisia."[114,115] Dose reduction or a switch to a lower potency compound are both treatment options. In terms of pharmacologic management, antiparkinsonian medications may be useful in some patients but, in general, prove less effective than the use of a benzodiazepine or β-blocker.[116]

Dose reduction or a switch to a lower potency neuroleptic should also be considered in patients with pseudoparkinsonian side effects (rigidity, akinesia, tremor). For details regarding decision making in the use of antiparkinsonian medications, the reader is referred to an algorithm outlining a systematic approach to the managment of EPS.[117]

Dyskinetic movements may be categorized as spontaneous, tardive (after at least 3 months of neuroleptic therapy), or withdrawal, and while certain risk factors remain relatively consistent, e.g., age, gender, it is still unclear who will develop TD.[118,119] The annual incidence is approximately 5% per year, at least for the first 4 to 5 years of treatment, with a mean prevalence of approximately 25%.[120] It has been suggested, however, that the risk may continue to rise with ongoing neuroleptic exposure.[12] Clinical variations have been described, e.g., tardive dystonias, tardive Tourette's,[119] although it is unclear whether or not these may be mediated by different pathophysiological mechanism.[121] Numerous treaments have been investigated, including tetrabenazine, vitamin E, benzodiazepines, and gamma-aminobutyric acid (GABA)mimetics, although none has proven to be consistently effective.[119] To date, clozapine remains the only neuroleptic clinically available which has been demonstrated to have little, if any, risk of causing TD,[122] as longer-term studies are required for risperidone. Interestingly, both clozapine and risperidone have been noted to demonstrate antidyskinetic properties in at least a subgroup of individuals with preexisting TD.[122-124]

B. Neuroleptic Malignant Syndrome

Neuroleptic malignant syndrome (NMS) has been comprehensively reviewed in a series of recently published articles.[125-130] While diagnostic criteria remain controversial,[131,132] the following features have been suggested: treatment with neuroleptics within 7 days prior to the onset of NMS (4 weeks for depot neuroleptics), hyperthermia (>38°C), muscle rigidity, and five of the following — altered mental status, tachycardia, diaphoresis or sialorrhea, dysarthria or dysphagia, tremor, incontinence, hypertension or hypotension, tachypnea or hypoxia, elevated creatinine phosphokinase levels or myoglobinuria, leukocytosis, and metabolic acidosis.[129] Proposed risk factors include previous NMS episodes; gender (male:female = 2:1); organic brain syndromes; mood disorders; agitation; dehydration; the potency, rate, route, and dosage of neuroleptic administration; recent history of alcohol or other substance abuse or dependence; electrolyte imbalance; thyrotoxicosis; and concurrent treatment with other psychotropic drugs, especially lithium.[125,126] Its incidence may vary from 0.07 to 1.5%,[133] with a mortality rate of approximately 10 to 20%.[128,134] The differential diagnosis is complex and extensive,[125,132] which likely contributes to the considerable variability in incidence and mortality figures.

Initial treatment consists of neuroleptic withdrawal and supportive medical therapy.[135] In addition, numerous additive treatments have been proposed, the most common being dantrolene, a peripheral muscle relaxant, and/or a dopamine agonist such as bromocriptine,[128,135-137] although the ability to change the course of the syndrome with such treatments has been challenged.[138] Other therapies that have been recommended include anticholinergics, benzodiazepines, curariform agents, calcium antagonists, vasodilators, and electroconvulsive therapy (ECT).[128,135-137]

Recurrence of NMS upon neuroleptic rechallenge is highly unpredictable, although as many as 87% of patients will be able to take neuroleptics again.[139] Recommendations regarding rechallenge include allowing a minimum of 1 to 2 weeks after recovery before neuroleptic reintroduction and choosing an agent of lower potency or one from another chemical class if a low-potency compound caused the original episode of NMS.[129] Using one of the treatment strategies prophylactically upon rechallenge is a further possibility, as is consideration of alternative therapies to neuroleptics if this is viable, e.g., ECT.[129] Introduction of a novel neuroleptic, such as clozapine, is an option, although reports of NMS have been reported with both clozapine[140-145] and risperidone.[146-148]

C. Novel Neuroleptics

As one might expect, the unique chemical profiles of the novel neuroleptics result in different side effect profiles. In the case of clozapine, its most troublesome side effect is the associated risk of agranulocytosis, estimated to occur in approximately 0.8% of patients during the first year of treatment.[149] Other more common side effects with clozapine include sedation, postural hypotension, tachycardia, hypersalivation, weight gain, constipation, increased liver enzymes, and electroencephalographic (EEG) alterations with an increased risk of seizures.[17,150,151]

In the two multicenter trials carried out in North America involving risperidone, the most common side effects identified were agitation, anxiety, EPS, headache, and insomnia.[19,20]

It is worth noting that both of these compounds have been associated with the induction of obsessive-compulsive symptoms, likely because of their 5-HT$_2$ antagonism.[152-156]

VI. FUTURE DIRECTIONS

As it became apparent that DA, and in particular the D$_2$ receptor, was integrally involved in schizophrenia, the search for neuroleptics turned to the development of highly selective D$_2$ antagonists, such as sulpiride and remoxipride. More recently, clozapine has forced a reevaluation of the search for "pure" D$_2$ agents, given its more prominent effect on the D_1 receptor and its relative lack of D_2 binding. In addition, its D_4 affinity and 5-HT$_2$ antagonism have suggested that either of these mechanisms may be important to its unique profile.[16]

The greatest focus currently in the development of novel compounds is on agents with combined DA/5-HT activity, specifically the D_2 and 5-HT$_2$ receptors, and there are now a number of such compounds under investigation, including savoxepine, seroquel, sertindole, and olanzapine.[47]

Various other receptors and neurotransmitters are also the subject of ongoing investigation. As mentioned, clozapine has raised the issue of D_4 involvement,[16,157-160] and there is considerable interest in the role of the D_3 receptor,[36,38,39,160,161] although in both cases there are not yet selective antagonists.[162] Selective D_1 antagonists have been associated with diminished EPS, but it remains to be established if these agents can demonstrate the degree of antipsychotic efficacy seen with D_2-blocking compounds.[163,164]

In terms of serotonin, 5-HT$_3$ antagonists such as ondansetron and zotepine have been investigated, but the results to date have not been particularly convincing.[165,166] There has been interest in the possible benefits of 5-HT$_{1A}$ agonists, buspirone being one such example, but again results have not been especially promising.[167] The notion of adding a selective 5-HT$_2$ antagonist, such as ritanserin, to standard neuroleptics, with their existing D_2 blockade, has been evaluated, and while preliminary reports indicated a reduction in EPS and negative symptoms,[168,169] studies verifying these findings have been lacking.

There has been ongoing attention to the possible role of norepinephrine,[170-172] acetylcholine,[173,174] GABA,[175] and the opioids.[176-179] In addition, there has been a growing interest in the excitatory amino acids such as glutamate, with evidence that phencyclidine (PCP) or "angel dust" can induce psychotic symptoms, possibly by interference with glutamatergic transmission at the level of the N-methyl-D-asparate (NMDA) receptor.[175,180-183] Further interest has turned to the possible role of second messengers, as these postreceptor molecular signaling processes may also be involved in the pathophysiology of schizophrenia.[184,185]

At a clinical level, there has been a shift in expectations and goals associated with novel antipsychotics. Improved effect in the treatment of positive symptoms remains a priority, particularly in light of the finding that as many as 15 to 30% of schizophrenics fail to respond adequately to conventional neuroleptics.[186] In addition, however, there is the expectation that the newer agents will be associated with fewer side effects, particularly EPS, and there has been increased emphasis on incorporating quality-of-life issues when evaluating new treatments.[187-189] It is now assumed that novel antipsychotics will be superior to standard agents in terms of improved negative symptomatology, but it remains to be firmly established whether this improvement is in primary or secondary negative symptoms. Finally, more recent evidence that novel compounds might alter neuropsychological decline suggests that considerable focus will turn to evaluating this potential benefit in the next years.[190-194]

For clinicians, this is an exciting and challenging era for the treatment of schizophrenia. For the first time in years, they have new pharmacological tools to add to their treatment armamentarium, tools which appear to have considerable benefits compared with the older drugs which were the mainstay of treatment for so many years. At the same time, to use these new medications successfully entails a more critical clinical evaluation of symptomatology, proactive treatment interventions with continuous reevaluation, and an understanding of the complex developments which reflect our current understanding of the pathophysiology of schizophrenia and its rational pharmacotherapy. For patients and their families, it offers them new hope in their struggle with a serious, and often extremely debilitating, illness.

REFERENCES

1. Jarvik, M.E., Drugs used in the treatment of psychiatric disorders, in *The Pharmacological Basis of Therapeutics*, 4th ed., Goodman, L.S. and Gilman, A., Eds. Collier-Macmillan, London, 1970, 151.

2. Deniker, P., From chlorpromazine to tardive dyskinesia (brief history of the neuroleptics). *Psychiatr J Univ Ottawa*, 14: 253, 1989.

3. Frankenburg, F.R., History of the development of antipsychotic medication. *Psychiatr Clin North Am*, 17: 531, 1994.

4. Kane, J.M. and Marder, S.R., Psychopharmacologic treatment of schizophrenia. *Schizophr Bull*, 19: 287, 1993.

5. Bristow, M.F. and Hirsch, S.R., Pitfalls and problems of the long term use of neuroleptic drugs in schizophrenia. *Drug Saf*, 8: 136, 1993.

6. Deniker, P., The neuroleptics: a historical survey. *Acta Psychiatr Scand*, 82 (suppl 358): 83, 1990.

7. Bezchlibnyl-Butler, K.Z., Jeffries, J.J., and Martin, B.A., *Clinical Handbook of Psychotropic Drugs*, 4th ed., Hogrefe & Huber, Toronto, 1994, 32.

8. Hyttel, J., Receptor binding profiles of neuroleptics, in *Dyskinesia: Research and Treatment*, Casey, D.E., Chase, T.N., Christenssen, A.V., and Gerlach, J., Eds., Springer, Berlin, 1985, 9.

9. Settle, E.C., Haloperidol: a quarter century of experience. *J Clin Psychiatry*, 44: 440, 1983.

10. Curry, S.H., Commentary: the strategy and value of neuroleptic drug monitoring. *J Clin Psychopharmacol*, 5: 263, 1985.

11. Davis, J.M. and Andriukaitis, S., The natural course of schizophrenia and effective drug treatment. *J Clin Psychopharmacol*, 6: 2S, 1986.

12. Glazer, W.M. and Kane, J.M., Depot neuroleptic therapy: an underutilized option. *J Clin Psychiatry*, 53: 426, 1992.

13. Seeman, P., Dopamine receptor sequences: therapeutic levels of neuroleptics occupy D_2 receptors, clozapine occupies D_4. *Neuropsychopharmacology*, 7: 261, 1992.

14. Wetzel, H., Wiedemann, K., Holsboer, F., and Benkert, O., Savoxepine: invalidation of an "atypical" neuroleptic response pattern predicted by animal models in an open clinical trial with schizophrenic patients. *Psychopharmacology*, 103: 280, 1991.

15. Naber, D. and Hippius, H., The European experience with use of clozapine. *Hosp Community Psychiatry*, 41: 886, 1990.

16. Coward, D.M., General pharmacology of clozapine. *Br J Psychiatry*, 160 (suppl 17): 5, 1992.

17. Kane, J., Honigfeld, G., Singer, J., Meltzer, H., and the Clozaril Collabarative Study Group, Clozapine for the treatment-resistant schizophrenic: a double-blind comparison with chlorpromazine. *Arch Gen Psychiatry*, 45: 789, 1988.

18. Casey, D.E., Clozapine: neuroleptic-induced EPS and tardive dyskinesia. *Psychopharmacology*, 99: S49, 1989.

19. Chouinard, G., Jones, B., Remington, G., Bloom, D., Addington, D., Macewan, G.W., Arnott, W., and Beaclair, L.A., Canadian multicenter placebo-controlled study of fixed doses of risperidone and haloperidol in the treatment of chronic schizophrenic patients. *J Clin Psychopharmacol*, 13: 25, 1993.

20. Marder, S.R. and Meibach, R.C., Risperidone in the treatment of schizophrenia. *Am J Psychiatry*, 151: 825, 1994.

21. Meltzer, H.Y., Clinical studies on the mechanism of action of clozapine: the dopamine-serotonin hypothesis of schizophrenia. *Psychopharmacology*, 99: S18, 1989.

22. Claus, A., Bollen, J., de Cuyper, H., Eneman, M., Malfroid, M., Peuskens, J., and Heylen, S., Risperidone vs. haloperidol in the treatment of chronic schizophrenic inpatients: a multicentre double blind comparative study. *Acta Psychiatr Scand*, 85: 295, 1992.

23. Doepp, S. and Buddeburg, C., Extrapyramidale Symptome unter Clozapin. *Nervenarzt*, 46: 589, 1975.

24. de Leon, J., Moral, L., and Camunas, C., Clozapine and jaw dyskinesia. *J Clin Psychiatry*, 52: 494, 1991.

25. Kane, J.M., Lieberman, J., Pollack, S., and Safferman, A., Clozapine and tardive dyskinesia: prospective data. *Schizophr Res*, 4: 364, 1991.

26. Cohen, B.M., Keck, P.E., Satlin, A., and Cole, J.O., Prevalence and severity of akathisia in patients on clozapine. *Biol Psychiatry*, 29: 1215, 1991.

27. Kebabian, J.W., Petzold, G.L., and Greengard, P., Dopamine-sensitive adenylate cyclase in caudate nucleus of rat brain, and its similarity to the "dopamine receptor." *Proc Natl Acad Sci U.S.A.*, 69: 2145, 1972.

28. Seeman, P., Lee, T., Chau-Wong, M., and Wong, K., Antipsychotic drug doses and neuroleptic/dopamine receptors. *Nature*, 261: 717, 1976.

29. Seeman, P., Antischizophrenic drugs — membrane receptor sites of action. *Biochem Pharmacol*, 26: 1741, 1977.

30. Brucke, T., Roth, J., Podreka, L., Strobl, R., Wenger, S., and Asenbaum, S., Striatal dopamine D_2-receptor blockade by typical and atypical neuroleptics. *Lancet,* 339: 497, 1992.

31. Farde, L., Nordstrom, A.-L., Wiesel, F.-A., Pauli, S., Halldin, C., and Sedvall, G., PET-analysis of D_1- and D_2-dopamine receptor occupancy in patients treated with classic neuroleptics and clozapine — relation to extrapyramidal side effects. *Arch Gen Psychiatry,* 49: 538, 1992.

32. Pilowsky, L.S., Costa, D.C., Ell, P.J., Murray, R.M., Verhoeff, N.L.P.G., and Kerwin, R.W., Clozapine, single photon emission tomography and the D_2 receptor blockade hypothesis. *Lancet,* 340: 199, 1992.

33. Pilowsky, L.S., Costa, D.C., Ell, P.J., Verhoeff, N.P.L.G., Murray, R.M., and Kerwin, R.W., Antipsychotic medication, D_2 dopamine receptor blockade and clinical response: a 123I-IBZM SPET (single photon emission tomography study). *Psychol Med,* 23: 791, 1993.

34. Van Tol, H.M.M., Bunzow, J.R., Guan, H.-C., Sunahara, R.K., Seeman, P., Niznik, H.B., and Civelli, O., Cloning of the gene for a dopamine D_4 receptor with high affinity for the antipsychotic clozapine. *Nature,* 350: 610, 1991.

35. Fitton, A. and Heel, R.C., Clozapine: a review of its pharmacological properties, and therapeutic use in schizophrenia. *Drugs,* 5: 722, 1990.

36. Walters, J.R., Bergstrom, D.A., Carlson, J.H., Chase, T.N., and Braun, A.R., D_1 receptor activation required for postsynaptic expression of D_2 agonist effects. *Am Assoc Adv Sci,* 236: 719, 1987.

37. Sokoloff, P., Giros, B., Martres, M.-P., Bouthenet, M.-L., and Schwartz, J.-C., Molecular cloning and characterization of a novel dopamine receptor (D_3) as a target for neuroleptics. *Nature,* 347: 146, 1990.

38. Sunahara, R.K., Guan, H.-C., O'Dowd, B.F., Seeman, P., Laurier, L.G., Ng, G., George, S.R., Torchia, J., Van Tol, H.H.M., and Niznik, H.B., Cloning of the gene for a human dopamine D5 receptor with higher affinity for dopamine than D1. *Nature,* 350: 614, 1991.

39. Sokoloff, P., Andrieux, M., Besancon, R., Pilon, C., Martres, M.-P., Giros, B., and Schwartz, J.-C., Pharmacology of human dopamine D_3 receptor expressed in a mammalian cell line: comparison with D2 receptor. *Eur J Pharmacol,* 225: 331, 1992.

40. Sokoloff, P., Martres, M.-P., Giros, B., Bouthenet, M.-L., and Schwartz, J.-C., The third dopamine receptor (D_3) as a novel target for antipsychotics. *Biochem Pharmacol,* 43: 659, 1992.

41. Pickar, D., Neuroleptics, dopamine, and schizophrenia. *Psychiatr Clin North Am,* 9: 35, 1986.

42. Damasio, A., The frontal lobes, in *Clinical Neuropsychology,* Heilman, K.M. and Valenstein, E., Eds., Oxford University Press, New York, 1979, 360.

43. Weinberger, D.R., Implications of normal brain development for the pathogenesis of schizophrenia. *Arch Gen Psychiatry,* 44: 660, 1987.

44. Meltzer, H.Y., The mechanism of action of novel antipsychotic drugs. *Schizophr Bull,* 17: 263, 1991.

45. Pancheri, P., Neuroleptics and 5-HT receptors: a working hypothesis for antipsychotic effect. *N Trends Exp Clin Psychiatr,* 7: 141, 1991.

46. Huttunen, M., The evolution of the serotonin-dopamine antagonist concept. *J Clin Psychopharmacol,* 15 (suppl 1): 4S, 1995.

47. Remington, G., Dopaminergic and serotonergic mechanisms in the action of standard and atypical neuroleptics, in *Contemporary Issues in the Treatment of Schizophrenia,* Shriqui, C.L. and Nasrallah, H.A., Eds., American Psychiatric Press, Washington, D.C., 1995, 295.

48. Meltzer, H.Y. and Nash, J.F., VII. Effects of antipsychotic drugs on serotonin receptors. *Pharmacol Rev,* 43: 587, 1991.

49. Meltzer, H.Y., Dimensions of outcome with clozapine. *Br J Psychiatry,* 160 (suppl 17): 46, 1992.

50. Miller, D.D., Perry, P.J., Cadoret, R.J., and Andreasen, N.C., Clozapine's effect on negative symptoms in treatment-refractory schizophrenia. *Compr Psychiatry,* 35: 8, 1994.

51. Brier, A., Buchanan, R.W., Kirkpatrick, B., Davis, O.R., Irish, D., Summerfelt, A., and Carpenter, W.T., Jr., Effects of clozapine on positive and negative symptoms in outpatients with schizophrenia. *Am J Psychiatry,* 151: 20, 1994.

52. Marder, S.R, Wirshing, W.C., and Van Putten, T., Drug treatment of schizophrenia: overview of recent research. *Schizophr Res,* 4: 81, 1991.

53. Seeman, M.V., Pharmacologic features and effects of neuroleptics. *Can Med Assoc J,* 125: 821, 1981.

54. Carpenter, W.T., Jr. and Keith, S.J., Integrating treatments in schizophrenia. *Psychiatr Clin North Am,* 9: 153, 1986.

55. Gardos, G., Are antipsychotics interchangeable? *J Nerv Ment Dis,* 159: 343, 1974.

56. Van Putten, T., Marder, S.R., Wirshing, W.C., Aravagiri, M., and Chabert, N., Neuroleptic plasma levels. *Schizophr Bull,* 17: 197, 1991.

57. Johnson, D.A., Kane, J.M., and Simpson, G.I., The use of depot neuroleptics: clinical experience in the United States and the United Kingdom, in *International Seminars on Depot and Oral Neuroleptics,* E.R. Squibb & Sons, Los Angeles, 1984, 5.

58. Remington, G.J., and Adams, M.E., Depot neuroleptics, in *Schizophrenia: Exploring the Spectrum of Psychosis,* Ancill, R.J., Holliday, S., and Higenbottom, J., Eds., John Wiley & Sons, Chichester, 1994, 171.

59. Remington, G.J. and Adams, M.E., Depot neuroleptic therapy: clinical considerations. *Can J Psychiatry,* 40 (suppl 1): S5, 1995.

60. Baldessarini, R.J., Cohen, B.M., and Teicher, M.H., Significance of neuroleptic dose and plasma level in the pharmacological treatment of psychoses. *Arch Gen Psychiatry,* 45: 79, 1988.

61. Van Putten, T., Why do schizophrenic patients refuse to take their drugs? *Arch Gen Psychiatry,* 31: 67, 1974.

62. Fuller Torrey, E.F., Management of chronic schizophrenic outpatients. *Psychiatr Clin North Am,* 9: 143, 1986.

63. Rosen, J., Bohon, S., and Gershon, S., Antipsychotics in the elderly. *Acta Psychiatr Scand,* 82 (suppl 358): 170, 1990.

64. Sakauye, K., Psychotic disorders: guidelines and problems with antipsychotic medications in the elderly. *Psychiatr Ann,* 20: 456, 1990.

65. Goldstein, J.M. and Tsuang, M.T., Gender and schizophrenia: an introduction and synthesis of findings. *Schizophr Bull,* 16: 179, 1990.

66. Bond, W.S., Ethnicity and psychotropic drugs. *Clin Pharmacol,* 10: 467, 1991.

67. Callahan, A.M., Fava, M., and Rosenbaum, J.F., Drug interactions in psychopharmacology. *Psychiatr Clin North Am,* 16: 647, 1993.

68. Goff, D.C. and Baldessarini, R.J., Drug interactions with antipsychotic agents. *J Clin Psychopharmacol,* 13: 57, 1993.

69. Baldessarini, R.J., Katz, B., and Cotton, P., Dissimilar dosaging with high-potency and low-potency neuroleptics. *Am J Psychiatry,* 141: 748, 1984.

70. Easton, M.S. and Janicak, P.G., The use of benzodiazepines in psychotic disorders: a review of the literature. *Psychiatr Ann,* 20: 535, 1990.

71. Salzman, C., Solomon, D., Miyawaki, E., Glassman, R., Rood, L., Flowers, E., and Thayer, S., Parenteral lorazepam versus parenteral haloperidol for the control of psychotic disruptive behavior. *J Clin Psychiatry,* 52: 177, 1991.

72. Remington, G., Pollock, B., Voineskos, G., Reed, K., and Coulter, K., Acutely psychotic patients receiving high-dose haloperidol therapy. *J Clin Psychopharmacol,* 13: 41, 1993.

73. Kane, J.M., Psychopharmacologic treatment issues. *Psychiatr Med,* 8: 111, 1990.

74. Meltzer, H.Y., Duration of a clozapine trial in neuroleptic-resistant schizophrenia. *Arch Gen Psychiatry,* 46: 672, 1989.

75. Brier, A., Buchanan, R.W., Irish, D., and Carpenter, W.T., Jr., Clozapine treatment of outpatients with schizophrenia II. Outcomes and long-term response patterns. *Hosp Community Psychiatry,* 44: 1145, 1993.

76. Kuoppasalmi, K., Rimon, R., Naukkarinen, H., Lang, S., Sandqvist, A., and Leinonen, E., The use of clozapine in treatment-refractory schizophrenia. *Schizophr Res* 10: 29, 1993.

77. Lieberman, J.A., Management of first-episode psychosis. *Curr App Psychoses,* 2: 8, 1993.

78. Zarate, C.A., Jr. and Cole, J.O., An algorithm for the pharmacological treatment of schizophrenia. *Psychiatr Ann,* 24: 333, 1994.

79. Jolley, A.G. and Hirsch, S.R., Continuous versus intermittent neuroleptic therapy in schizophrenia. *Drug Saf,* 8: 331, 1993.

80. Kohler, C., Hall, H., Magnusson, O., Lewander, T., and Gustaffson, K., Biochemical pharmacology of the atypical neuroleptic remoxipride. *Acta Psychiatr Scand,* 82 (suppl 358): 27, 1990.

81. Kerwin, R., Adverse reaction reporting and new antipsychotics. *Lancet,* 342: 1440, 1993.

82. Krupp, P. and Barnes, P., Clozapine-associated agranulocytosis. *Br J Psychiatry,* 160 (suppl 17): 38, 1992.

83. Carpenter, W.T., Jr., Heinrichs, D.W., and Wagman, A.M.I., Deficit and nondeficit forms of schizophrenia: the concept. *Am J Psychiatry,* 145: 578, 1988.

84. Carpenter, W.T., Jr., Serotonin-dopamine antagonists and treatment of negative symptoms. *J Clin Psychopharmacol,* 15 (suppl 1): 30S, 1995.

85. Galanter, M., Castanada, R., and Ferman, J., Substance abuse among general psychiatric patients: place of presentation, diagnosis, and treatment. *Am J Drug Alcohol Abuse,* 14: 211, 1988.

86. Munk-Jorgensen, P., The schizophrenia diagnosis in Denmark: a register-based investigation. *Acta Psychiatr Scand,* 72: 266, 1985.

87. Wilson, W.H., Reassessment of state hospital patients diagnosed with schizophrenia. *J Neuropsychiatr Clin Neurosci,* 1: 394, 1989.

88. Smith, G.N., Macewan, G.W., Ancill, R.J., Honer, W.G., and Ehmann, T.S., Diagnostic confusion in treatment-refractory psychotic patients. *J Clin Psychiatry,* 53: 197, 1992.

89. Lenz, G., Simhandt, C., and Thau, K., Temporal stability and predictive validity of diagnostic criteria for schizophrenia. *Schizophr Res,* 6: 170, 1992.

90. Fennig, S., Kovasznay, B., Rich, C., Ram, R., Pato, C., Miller, R., and Rubinstein, J., Six-month stability of psychiatric diagnoses in first-admission patients with psychosis. *Am J Psychiatry,* 151: 1200, 1994.

91. Hoge, S.K., Appelbaum, P.S., Lawlor, T., Beck, J.C., Litman, R., Greer, A., and Gutheil, T.G., A prospective, multicenter study of patients' refusal of antipsychotic medication. *Arch Gen Psychiatry,* 47: 949, 1990.

92. Meltzer, H.Y., Treatment of the non-responsive schizophrenic patient. *Schizophr Bull,* 18: 515, 1992.

93. Christison, G.W., Kirch, D.G., and Wyatt, R.J., When symptoms persist: choosing among alternative somatic treatments in schizophrenia. *Schizophr Bull,* 17: 217, 1991.

94. Wyatt, R.J., Neuroleptics and the natural course of schizophrenia. *Schizophr Bull,* 2: 325, 1991.

95. Wyatt, R.J., Risks of withdrawing antipsychotic medications. *Schizophr Bull,* 52: 205, 1995.

96. Perenyi, A., Goswami, U., Frecska, E., Majlath, E., Barcs, G., and Kassay-Farkas, A., A pilot study of the role of prophylactic antiparkinson treatment during neuroleptic therapy. *Pharmacopsychiatry,* 22: 108, 1989.

97. Richelson, E., Psychopharmacology of schizophrenia: past, present, and future. *Psychiatr Ann,* 20: 640, 1990.

98. Knudsen, P., Chemotherapy with neuroleptics. *Acta Psychiatr Scand,* 322 (suppl): 51, 1985.

99. Zito, J.M., Craig, T.J., Wanderling, J., and Siegel, C., Pharmaco-epidemiology in 136 hospitalized schizophrenic patients. *Am J Psychiatry,* 144: 778, 1987.

100. Remington, G.J., Prendergast, P., and Bezchlibnyk-Butler, K.Z., Dosaging patterns in schizophrenia with depot, oral and combined neuroleptic therapy. *Can J Psychiatry,* 38: 159, 1993.

101. World Health Organization, Prophylactic use of anticholinergics in patients on long-term neuroleptic treatment. *Br J Psychiatry,* 156: 412, 1990.

102. Stern, T.A. and Anderson, W.H., Benztropine prophylaxis of dystonic reactions. *Psychopharmacology,* 61: 261, 1979.

103. Keepers, G.A., Clappison, V.J., and Casey, D.E., Initial anticholinergic prophylaxis for neuroleptic-induced extrapyramidal syndromes. *Arch Gen Psychiatry,* 40: 1113, 1983.

104. Sramek, J.J., Simpson, G.M., Morrison, R.L., and Heiser, J.F., Anticholinergic agents for prophylaxis of neuroleptic-induced dystonic reactions: a prospective study. *J Clin Psychopharmacol,* 47: 305, 1986.

105. Winslow, R.S., Stillner, V., Coons, D.J., and Robinson, M.W., Prevention of acute dystonic reactions in patients beginning high-potency neuroleptics. *Am J Psychiatry,* 143: 706, 1986.

106. Boyer, W.F., Bakalar, N.H., and Lake, C.R., Anticholinergic prophylaxis of acute haloperidol-induced acute dystonic reactions. *J Clin Psychopharmacol,* 7: 164, 1987.

107. Keepers, G.A. and Casey, D.E., Prediction of neuroleptic-induced dystonia. *J Clin Psychopharmacol,* 7: 342, 1987.

108. Arana, G.W., Goff, D.C., Baldessarini, R.J., and Keepers, G.A., Efficacy of anticholinergic prophylaxis for neuroleptic-induced acute dystonia. *Am J Psychiatry,* 145: 993, 1988.

109. Lavin, M.R. and Rifkin, A., Prophylactic antiparkinsonian drug use: I. Initial prophylaxis and prevention of extrapyramidal side effects. *J Clin Pharmacol,* 31: 763, 1991.

110. Goff, D.C., Arana, G.W., Greenblatt, D.J., Dupont, R., Ornsteen, M., Harmatz, J.S., and Shader, R.I., The effect of benztropine on haloperidol-induced dystonia, clinical efficacy and pharmacokinetics: a prospective, double-blind trial. *J Clin Psychopharmacol,* 11: 106, 1991.

111. Remington, G., Pharmacotherapy of schizophrenia. *Can J Psychiatry,* 34: 211, 1989.

112. Barnes, T.R.E., The present status of tardive dyskinesia and akathisia in the treatment of schizophrenia. *Psychiatr Dev,* 4: 301, 1987.

113. Barnes, T.R.E., Braude, W.M., and Hill, D.J., Acute akathisia after oral droperidol and metoclopramide preoperative medication. *Lancet,* 2: 48, 1982.

114. Adler, L.A., Angrist, B., Reiter, S., and Rotrosen, J., Neuroleptic-induced akathisia: a review. *Psychopharmacology,* 97: 1, 1989.

115. Lang, A.E., Withdrawal akathisia: case reports and a proposed classification of chronic akathisia. *Mov Disord,* 9: 188, 1994.

116. Fleischhacker, W.W., Roth, S.D., and Kane, J.M., The pharmacologic treatment of neuroleptic-induced akathisia. *J Clin Psychopharmacol,* 10: 12, 1990.

117. Bezchlibnyk-Butler, K.Z. and Remington, G.J., Antiparkinsonian drugs in the treatment of neuroleptic-induced extrapyramidal symptoms. *Can J Psychiatry,* 39: 74, 1994.

118. Casey, D.E., Neuroleptic-induced extrapyramidal syndromes and tardive dyskinesia. *Schizophr Res,* 4: 109, 1991.

119. Gershanik, O.S., Drug-induced movement disorders. *Curr Opin Neurol Neurosurg,* 6: 369, 1993.

120. Jeste, D.V. and Caligiuri, M.P., Tardive dyskinesia. *Schizophr Bull,* 19: 303, 1993.

121. Wojcik, J.D., Falk, W.E., Fink, J.S., Cole, J.O., and Gelenberg, A.J., A review of 32 cases of tardive dystonia. *Am J Psychiatry,* 148: 1055, 1991.

122. Gerlach, J. and Hansen, L., Clozapine and D_1/D_2 antagonism in extrapyramidal function. *Br J Psychiatry,* 160 (suppl 17): 34, 1992.

123. Chouinard, G., Effects of risperidone in tardive dyskinesia: an analysis of the Canadian Multicenter Risperidone Study. *J Clin Psychopharmacol,* 15 (suppl 1): 36S, 1995.

124. Remington, G.J., Clinical considerations in the use of risperidone. *Can J Psychiatry,* 38 (suppl 3): S96, 1993.
125. Caroff, S.N., Mann, S.C., Lazarus, A., Sullivan, K., and Macfadden, W., Neuroleptic malignant syndrome: diagnostic issues. *Psychiatr Ann,* 21: 130, 1991.
126. Keck, P.E., Jr., McElroy, S.L., and Pope, H.G., Jr., Epidemiology of neuroleptic malignant syndrome. *Psychiatr Ann,* 21: 148, 1991.
127. Addonzio, G., The pharmacologic basis of neuroleptic malignant syndrome. *Psychiatr Ann,* 21: 152, 1991.
128. Sakkas, P., Davis, J.M., Hua, J., and Wang, Z., Pharmacotherapy of neuroleptic malignant syndrome. *Psychiatr Ann,* 21: 157, 1991.
129. Lazarus, A., Caroff, S.N., and Mann, S.C., Beyond NMS: after the acute episode. *Psychiatr Ann,* 21: 165, 1991.
130. Mann, S.C., Caroff, S.N., and Lazarus, A., Pathogenesis of neuroleptic malignant syndrome. *Psychiatr Ann,* 21: 175, 1991.
131. Gurrera, R.J., Chang, S.S., and Romero, J.A., A comparison of diagnostic criteria for neuroleptic malignant syndrome. *J Clin Psychiatry,* 53: 56, 1992.
132. Michalon, M. and Watler, C., Neuroleptic malignant syndrome — a misnomer: boundaries with other similar syndromes. *Ann RCPSC,* 28: 150, 1995.
133. Gelenberg, A.J., Bellinghausen, B., Wojcik, J.D., Falk, W.E., and Sachs, G.A., A prospective survey of neuroleptic malignant syndrome in a short-term psychiatric hospital. *Am J Psychiatry,* 145: 517, 1988.
134. Shalev, A., Hermesh, H., and Minitz, H., Mortality from neuroleptic malignant syndrome. *J Clin Psychiatry,* 50: 18, 1989.
135. Dickey, W., The neuroleptic malignant syndrome. *Prog Neurobiol,* 36: 425, 1991.
136. Harpe, C. and Stoudemire, A., Aetiology and treatment of neuroleptic malignant syndrome. *Med Toxicol,* 2: 166, 1987.
137. Ebadi, M., Pfeiffer, R.F., and Murrin, L.C., Pathogenesis and treatment of neuroleptic malignant syndrome. *Gen Pharmacol,* 21: 367, 1990.
138. Rosebush, P.I., Stewart, T., and Mazurek, M.F., The treatment of neuroleptic malignant syndrome: are dantrolene and bromocriptine useful adjuncts to supportive care? *Br J Psychiatry,* 159: 709, 1991.
139. Rosebush, P.I., Stewart, T.D., and Gelenberg, A.J., Twenty neuroleptic rechallenges after neuroleptic malignant syndrome in 15 patients. *J Clin Psychiatry,* 50: 295, 1989.
140. Pope, H.G., Cole, J.O., Choras, P.T., and Tulwiller, C.E., Apparent neuroleptic malignant syndrome with clozapine and lithium. *J Nerv Ment Dis,* 174: 493, 1986.
141. Muller, T., Becker, T., and Fritze, J., Neuroleptic malignant syndrome after clozapine plus carbamazepine. *Lancet,* 2: 1500, 1988.
142. DasGupta, K. and Young, A., Clozapine-induced neuroleptic malignant syndrome. *J Clin Psychiatry,* 52: 105, 1991.
143. Anderson, E.S. and Powers, P.S., Neuroleptic malignant syndrome associated with clozapine use. *J Clin Psychiatry,* 52: 102, 1991.
144. Miller, D.D., Sharafuddin, M.J.A., and Kathol, R.G., A case of clozapine-induced neuroleptic malignant syndrome. *J Clin Psychiatry,* 52: 99, 1991.
145. Redding, S., Minnema, A.M., and Tandon, R., Neuroleptic malignant syndrome and clozapine. *Ann Clin Psychiatry,* 5: 25, 1993.
146. Lee, H., Ryan, J., Mullett, G., and Lawlor, B.A., Neuroleptic malignant syndrome associated with the use of risperidone, an atypical antipsychotic agent. *Hum Psychopharmacol,* 9: 303, 1994.
147. Raitasuo, V., Vataja, R., and Elomaa, E., Risperidone-induced neuroleptic malignant syndrome in young patient. *Lancet,* 344: 1705, 1994.
148. Webster, P. and Wijeratne, C., Risperidone-induced neuroleptic malignant syndrome. *Lancet,* 344: 1228, 1994.
149. Alvir, J.M.J., Lieberman, J.A., Safferman, A.Z., Schwimmer, J.L., and Schaaf, J.A., Clozapine-induced agranulocytosis. Incidence and risk factors in the United States. *N Engl J Med,* 329: 162, 1993.
150. Naber, D., Holzbach, R., Perro, C., and Hippius, H., Clinical management of clozapine patients in relation to efficacy and side effects. *Br J Psychiatry,* 160 (suppl 17): 54, 1992.
151. Fleischhacker, W.W., Clozapine — beneficial and untoward effects: an update. *Pharmacopsychiatry,* 25: 74, 1992.
152. Patil, V.J., Development of transient obsessive-compulsive symptoms during treatment with clozapine. *Am J Psychiatry,* 149: 272, 1992.
153. Baker, R.W., Chengappa, K.N.R., Baird, J.W., Steingard, S., Christ, M.A.G., and Schooler, N.R., Emergence of obsessive-compulsive symptoms during treatment with clozapine. *J Clin Psychiatry,* 53: 439, 1992.
154. Patel, B. and Tandon, R., Development of obsessive-compulsive symptoms during clozapine treatment. *Am J Psychiatry,* 150: 836, 1993.

155. Steingard, S., Chengappa, K.N.R., Baker, R.W., and Schooler, N.R., Clozapine, obsessive symptoms, and serotonergic mechanisms. *Am J Psychiatry,* 150: 1435, 1993.

156. Remington, G. and Adams, M., Risperidone and obsessive-compulsive symptoms. *J Clin Psychopharmacol,* 14: 358, 1994.

157. Seeman, P., Dopamine receptor sequences: therapeutic levels of neuroleptics occupy D_2 receptors, clozapine occupies D_4. *Neuropsychopharmacology,* 7: 261, 1992.

158. Seeman, P. and Van Tol, H.H.M., Dopamine receptor pharmacology. *Curr Opin Neurol Neurosurg,* 6: 602, 1993.

159. Seeman, P., Guan, H.-C., and Van Tol, H.H.M., Dopamine D_4 receptors elevated in schizophrenia. *Nature,* 365: 441, 1993.

160. Meador-Woodruff, J.H., Update on dopamine receptors. *Ann Clin Psychiatry,* 6: 79, 1994.

161. Healy, D. D_1 and D_2 and D_3. *Br J Psychiatry,* 159: 319, 1991.

162. Lieberman, J.A., Understanding the mechanism of action of atypical antipsychotic drugs. *Br J Psychiatry,* 193 (suppl 22): 7, 1993.

163. Waddington, J.L. and Daly, S.A., The status of "second generation" selective D_1 dopamine receptor antagonists as putative atypical antipsychotic agents, in *Novel Antipsychotic Drugs,* Meltzer, H.Y., Ed., Raven Press, New York, 1992, 109.

164. Gerlach, J. and Casey, D.E., Drug treatment of schizophrenia: myths and realities. *Curr Opin Psychiatr,* 7: 65, 1994.

165. DeVeaugh-Geiss, J., McBain, S., Cooksey, P.G., and Bell, J.M., The effects of a novel 5-HT_3 antagonist, Ondansetron, in schizophrenia: results from uncontrolled trials, in *Novel Antipsychotic Drugs,* Meltzer, H.Y., Ed., Raven Press, New York, 1992, 225.

166. Newcomer, J.W., Faustman, W.O., Zipursky, R.B., and Csernansky, J.G., Zacopride in schizophrenia: a single-blind serotonin type 3 antagonist trial. *Arch Gen Psychiatry,* 49: 751, 1992.

167. Hollister, L.E., New psychotherapeutic drugs. *J Clin Psychopharmacol,* 14: 50, 1994.

168. Bersani, G., Grispini, A., Marini, S., Pasini, A., Valducci, M., and Ciana, N., 5-HT_2 antagonist ritanserin in neuroleptic-induced parkinsonism: a double-blind comparison with orphenadrine and placebo. *Clin Neuropharmacol,* 13: 500, 1990.

169. Duinkerke, S.J., Botter, P.A., Jansen, A.A.I., Van Dongen, P.A.M., Van Haaften, A.J., Boom, A.J., Van Laarhoven, J.H.M., and Busard, H.L.S.M., Ritanserin, a selective 5-$HT_2/1_C$ antagonist, and negative symptoms in schizophrenia. *Br J Psychiatry,* 163: 451, 1993.

170. Van Kammen, D.P. and Antelman, S., Impaired noradrenergic transmission in schizophrenia. *Life Sci,* 34: 1403, 1984.

171. Van Kammen, D.P., The biochemical basis of relapse and drug response in schizophrenia: review and hypothesis. *Psychol Med,* 21: 881, 1991.

172. Baldessarini, R.J., Huston-Lyons, D., Campbell, A., Marsh, E., and Cohen, B.M., Do central antiadrenergic actions contribute to the atypical properties of clozapine? *Br J Psychiatry,* 160 (suppl 17): 12, 1992.

173. Tandon, R., Shipley, J.E., Greden, J.F., Mann, N.A., Eisner, W.H., and Goodson, J., Muscarinic cholinergic hyperactivity in schizophrenia: relationship to positive and negative symptoms. *Schizophr Res,* 4: 23, 1991.

174. O'Keane, V., Abel, K., and Murray, R.M., Growth hormone responses to pyridostigmine in schizophrenia: evidence for cholinergic dysfunction. *Biol Psychiatry,* 36: 582, 1994.

175. Squires, R.F. and Saederup, E., A review of evidence for GABAergic predominance/glutamatergic deficit as a common etiological factor in both schizophrenia and affective psychoses: more support for a continuum hypothesis of "functional" psychosis. *Neurochem Res,* 16: 1099, 1991.

176. Schmauss, C. and Emrich, H.M., Dopamine and the action of opiates: a reevaluation of the dopamine hypothesis with special consideration of the role of endogenous opioids in the pathogenesis of schizophrenia. *Biol Psychiatry,* 20: 1211, 1985.

177. Borison, R.L., Diamond, B.I., and Dren, A.T., Does sigma receptor antagonism predict clinical antipsychotic efficacy? *Psychopharmacol Bull,* 27: 103, 1991.

178. Borison, R.L., Pathiraja, A.P., and Diamond, B.I., Clinical efficacy of sigma antagonists in schizophrenia, in *Novel Antipsychotic Drugs,* Meltzer, H.Y., Ed., Raven Press, New York, 1992, 203.

179. Taylor, D.P. and Schlemmer, R.F., Jr., Sigma "antagonists": potential antipsychotics?, in *Novel Antipsychotic Drugs,* Meltzer, H.Y., Ed., Raven Press, New York, 1992, 189.

180. Deutsch, S.I., Mastropaolo, J., Schwartz, B.L., Rosse, R.B., and Morihisa, J.M., A "glutamatergic hypothesis" of schizophrenia: rationale for pharmacotherapy with glycine. *Clin Neuropharmacol,* 12: 1, 1989.

181. Javitt, D.C. and Zukin, S.R., Recent advances in the phencyclidine model of schizophrenia. *Am J Psychiatry,* 148: 1301, 1991.

182. Csernansky, J.G., Murphy, G.M., and Faustman, W.O., Limbic/mesolimbic connections and the pathogenesis of schizophrenia. *Biol Psychiatry,* 30: 383, 1991.

183. Olney, J.W., Glutamatergic mechanisms in neuropsychiatry, in *Novel Antipsychotic Drugs*, Meltzer, H.Y., Ed., Raven Press, New York, 1992, 155.

184. Manji, H.K., G proteins: implications for psychiatry. *Am J Psychiatry*, 149: 746, 1992.

185. Hudson, C.J., Young, L.T., Li, P.P., and Warsh, J.J., CNS signal transduction in the pathophysiology and pharmacotherapy of affective disorders and schizophrenia. *Synapse*, 13: 278, 1993.

186. Meltzer, H.Y., New drugs for the treatment of schizophrenia. *Psychiatr Clin North Am*, 16: 365, 1993.

187. Collins, E.J., Hogan, T.P., and Desai, H., Measurement of therapeutic response in schizophrenia. *Schizophr Res*, 5: 249, 1991.

188. Awad, A.G., Quality of life of schizophrenic patients on medications and implications for new drug trials. *Hosp Community Psychiatry*, 43: 262, 1992.

189. Selai, C.E. and Trimble, M.R. The role of quality of life measures in psychopharmacology. *Hum Psychopharmacol*, 9: 211, 1994.

190. Strauss, W.H., Kleiser, E., and Luethcke, H., Dyscognitive syndromes in neuroleptic therapy. *Pharmacopsychiatry*, 21: 298, 1988.

191. Kenny, J.T., Friedman, L., Ubogy, D., and Meltzer, H.Y., Differential effect of clozapine on psychopathology and cognitive dysfunction in treatment-resistant schizophrenia. *Schizophr Res*, 4: 386, 1991.

192. Kenny, J.T. and Meltzer, H.Y., Effect of atypical and typical antipsychotic drugs on neuropsychological functions in early-stage schizophrenic patients. *Schizophr Res*, 6: 162, 1992.

193. Hagger, C., Buckley, P., Kenny, J.T., Friedman, L., Ubogy, D., and Meltzer, H.Y., Improvement in cognitive functions and psychiatric symptoms in treatment-refractory schizophrenic patients receiving clozapine. *Biol Psychiatry*, 34: 702, 1993.

194. Williams, R., Baillie, P., Dickson, R.A., and Dalby, J.T., Cognitive and behavioral efficacy of clozapine in clinical trials. *Can J Psychiatry*, 38: 522, 1993.

Chapter 12

BENZODIAZEPINES: MECHANISMS OF ACTION AND CLINICAL INDICATIONS

John Nelson and Guy Chouinard

CONTENTS

Benzodiazepines are today among the most widely prescribed drugs of any class of medication.[1] About one out of ten Canadians reports using a benzodiazepine at least once a year, and 10% of these have used benzodiazepines for a year or more.[2,3] In Canada, benzodiazepine use is most common in women and in persons over the age of 50.[4] In the U.S., by one estimate, 33% of daily users of benzodiazepines are at least 55 years of age.[5] Between 1980 and 1989, 265 million prescriptions were issued for benzodiazepines in Great Britain.[6] Anxiety and sleep disorders were the first indications for which benzodiazepines were found effective when they were introduced to clinical medicine in the early 1960s. Since then, a role has been established for benzodiazepines in the therapy of other psychiatric disorders (panic, mania, and psychotic agitation). The widespread use of benzodiazepines as hypnotics and anxiolytics has been encouraged, not only by the fact that they are prescribed for a variety of symptoms and conditions, but also because they are used to treat different populations in terms of age and sex for both short- and long-term periods.[7]

Benzodiazepines are unique among the psychotropic drugs for the scrutiny they have received in lay and medical circles since the early 1980s with respect to their clinical benefits and side effects.[8] Although anxiety and insomnia, on a short-term basis, are part of normal life for millions of people, contemporary cultures are ambivalent in attitude toward mental health and drug taking. As the social relations of health care have changed, so too have public attitudes toward the widespread use of benzodiazepines.[9] The recent reappraisal of benzodiazepines comes from several quarters and addresses the issue of acceptable risks vs. the therapeutic benefits of long-term use of benzodiazepines in different indications.[1] Patients, consumer groups, and the media have challenged what they perceive as the overprescription of these drugs, the high risk of "addiction," and their use in long-term therapy without regular reevaluation by a physician or alternatives to medication, such as counseling or psychotherapy. Psychiatrists have raised concerns that the newer, short half-life, high-potency compounds, such as alprazolam, triazolam, and lorazepam, may have the potential to produce serious side effects and to induce drug dependence. The 1990 Task Force Report of the American Psychiatric Association on benzodiazepine dependence, toxicity, and abuse concluded that "general concerns about the overprescribing and misuse of medication often do not hold up when they are translated into specific research questions and confronted by relevant data."[1] At the same time, the report calls for an examination of survey data as they become available on the newer, widely prescribed benzodiazepines, such as alprazolam and triazolam.

The widespread use of benzodiazepines comes from their effectiveness and relative safety compared with other sedative/hypnotic drugs. In general, benzodiazepines are well tolerated, but they do have side effects, the most common being drowsiness and sedation. The main drawbacks are associated with rebound and withdrawal syndromes that follow dosage decrease or drug discontinuation, causing unpleasant and potentially serious reactions. A step toward reducing side effects and the risks of psychological and physiological dependence in predisposed individuals is to increase our understanding of the pharmacology of benzodiazepines and the criteria for drug selection in different populations. Several thousand different benzodiazepines have been synthesized,[10] but only 50 are available worldwide for clinical use.[11] While there are many similarities among the various derivatives, there are also important differences. A careful discrimination among these differences will help in the selection of the most appropriate compound for a given clinical situation. This chapter provides an overview of benzodiazepines with respect to their chemical structure, pharmacological properties, and metabolism in neurobiological systems, as well as guidelines for their clinical use.

I. CHEMICAL STRUCTURE AND CLASSIFICATION

The chemical structure of benzodiazepine consists of a core benzene ring fused to a 7-member 1,4 diazepine ring. Almost all benzodiazepines also have a 5-aryl substituent ring. They differ in the chemical nature of the substituent groups at positions 1, 2, 3, 4 (of the diazepine ring), position 7 (of the benzene ring), and position 2' (of the 5-aryl substituent ring) (Figure 1).

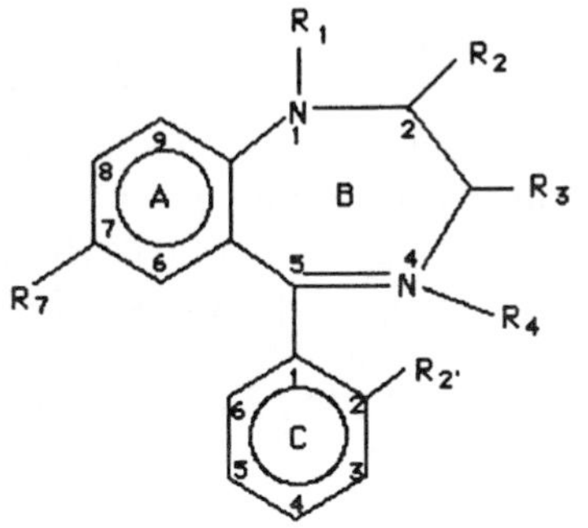

Figure 1
The structure of benzodiazepines.

R-groups represent the substituent groups, and their specific configuration makes each benzodiazepine derivative unique and chemically distinct from other derivatives. Based on R-group substituents, five pharmacologic subgroups have been defined: the alpha-keto benzodiazepines, the 3-OH benzodiazepines, the 7-nitro benzodiazepines, the triazolo benzodiazepines, and the imidazo benzodiazepine (midazolam). Derivatives in a given subgroup are metabolized in the liver by similar mechanisms and, therefore, have β elimination half-lives within the same ranges.[12,13] In clinical situations, a choice among benzodiazepines has usually been based on the differences in β elimination half-life of derivatives from various subgroups, namely, how long they take to be metabolized and excreted in the urine. β elimination half-life determines whether short-term intoxication or intoxication because of drug accumulation may be anticipated. Benzodiazepines with very similar chemical structures can differ greatly in their potency, rate of absorption, and on other important pharmacological parameters.[14,15] For instance, basic 1,4 benzodiazepines (such as diazepam and lorazepam), heterocyclic 1,4 benzodiazepines (such as triazolam), and 1,5 benzodiazepines (such as clobazam) possess similar properties (reduction of anxiety, sedation, and anticonvulsant action); however, they differ greatly in terms of their sedating properties.

II. GABA–BENZODIAZEPINE INTERACTION

Within the neuronal cell membrane are glycoproteins which act as high-affinity binding (recognition) sites for certain neurotransmitters. Neurotransmitters exert their chemical action on neural impulses by anchoring to neurons at these receptor sites and by regulating the flow of information into the brain, thereby ultimately influencing human behavior. Benzodiazepines produce their pharmacologic effects by interacting with benzodiazepine receptors. These binding sites are themselves part of the receptor for gamma-aminobutyric acid (GABA), the major inhibitory neurotransmitter of the central nervous system (CNS). When benzodiazepines bind to the benzodiazepine-GABA receptor complex, they augment the potential of GABA to depress neuronal excitability. Benzodiazepines are presumed to exert their anxiety-reducing effects by enhancing the inhibitory function of GABA, which blocks the neural impulses that stimulate the limbic system into triggering an anxiety response.[16] Receptor sites for GABA are especially numerous in the limbic system of the brain.[17] This system is composed of several brain regions (amygdala, thalamus, hypothalamus, and hippocampus) that facilitate diverse and adaptive behavioral and physiological responses, such as the emotional states of fear, anger, and pleasure.

GABA is the transmitter that is released at approximately 20 to 40% of all synapses in the mammalian brain. GABA is a simple amino acid molecule that is synthesized in GABAergic neurons through the decarboxylation of L-glutamic acid by the enzyme glutamic acid decarboxylase. GABA interacts with two distinct receptor types in target neurons: the postsynaptic $GABA_A$ receptor linked to a gated chloride ion channel, and the $GABA_B$ receptor linked to G protein–coupled Ca^{+2} and K^+ fluxes.[18] Benzodiazepines bind to a specific receptor located on the same macromolecular protein complex as the $GABA_A$ receptor[19] and act by increasing chloride ion conduction of endogenous GABA at this receptor.

The concentration of GABA varies in different regions of the brain, with the greatest concentrations found in the basal ganglia, hippocampus, cerebellum, and hypothalamus. Approximately 99% of the amount of GABA in mammals is found in the CNS,[20] although small amounts may be found in other areas, such as the autonomic ganglia. Plasma GABA seems to reflect cerebrospinal fluid (CSF) concentrations of GABA in normal humans.[21] Plasma GABA is relatively stable over time[22] and appears to be unaffected by gender, diet, exercise, menstrual cycle fluctuations,[23] and diurnal or circadian rhythm.[24] Few investigations have examined the effect of benzodiazepines on plasma GABA. Diazepam infusion was found by one study[25] to produce decreases in plasma GABA concentrations similar to levels found in patients with untreated panic disorder. Another study[21] showed that 5 mg of intravenous diazepam increased CSF GABA concentrations in 11 neurological patients. No study has yet to examine the effect of benzodiazepines on plasma GABA in normal humans.

Pharmacological evidence shows that GABA is involved in sleep induction, epilepsy, anxiety, memory, hypnosis, and sedation.[26,27] Certain drugs alter the synthesis and metabolism of GABA or act at postsynaptic GABA receptors. For instance, glutamic acid decarboxylase inhibitors lead to a reduction in brain concentrations of GABA which, in turn, produces convulsions.[26] Other drugs with anticonvulsant and sedative properties, such as diaminobutyric acid, block the reuptake of GABA from the presynaptic terminal which leads to increased concentrations of GABA.[28] Antagonists at the postsynaptic GABA receptor, such as bicuculline and picrotoxin, in large doses act as convulsants and in smaller doses alter memory.[29-31] The sedative and hypnotic effects of GABA are also found to have anesthetic applications. Numerous drugs, such as halothane, propofol, barbiturates, and neurosteroids, are now used to potentiate or to interact with the inhibitory action of GABA.[32-37] The antiepileptic action produced by the interaction of sodium valproate and GABA receptors is thought to result from the reduced metabolism of GABA.[26,38] More recently introduced drugs include GABA analogues, such as vigabatrin and systemic picrotoxin.[30,31,39] The latter acts antagonistically at the postsynaptic GABA receptor to enhance memory storage and retention.[30,31] In contrast, baclofen, inactive at the $GABA_A$ receptor but a potent GABAmimetic selective for the $GABA_B$ receptor, impairs memory, reduces spasticity, and prevents flexor spasms in a number of different neurological disorders.[27,40,41]

Cloning studies have since added to our knowledge of the heterogeneity of benzodiazepine-$GABA_A$ receptor complexes. Studies performed in the past decade indicate that the $GABA_A$ receptor may, in fact, be a group of different receptors rather than a single entity. Five types of subunits have been described: α, β, γ, δ and ε. The type of α subunit may determine the anxiolytic or sedative activity of a benzodiazepine.[17,32,42-44] Combinations of these subunits produce macromolecular complexes which include a chloride ion channel. The functional role of the β subtype of GABA receptor in the CNS remains the least understood.[45]

III. FUNCTIONAL CONSIDERATIONS: MECHANISMS OF ACTION

Highly specific receptors for benzodiazepines have been identified in the brain in the cerebral cortex (frontal cortex),[46-49] limbic system,[46-48] cerebellar cortex,[18] and brain stem.[49,50] While these binding sites are normally located in the neuronal membrane,[16] they are present in very different densities in most areas of the CNS and are highest in the areas cited above.[46-50]

The potency of a benzodiazepine is correlated with its affinity for its binding site, the benzodiazepine receptor.[13,47] Molecular biology studies originally proposed two subtypes of benzodiazepine receptor: (1) the benzodiazepine$_I$ (BZ_I) subtype (high affinity for triazolopyridazines) found throughout the brain with large concentrations in the cerebellum; and (2) the benzodiazepine$_{II}$ (BZ_{II}) subtype found principally in the cerebral cortex, spinal cord, and hippocampus.[17] Anxiolysis is ascribed to the BZ_I site. Benzodiazepine receptors are coupled to the $GABA_A$ receptor in a macromolecular complex which includes a chloride channel and binding sites for other modulators.[44,49,50]. This complex contains five subunits that are polymorphic and similar but not identical.[11,51] None can be definitively identified as a separate subtype because of the potential combinations of subunits.

Activation of the benzodiazepine–GABA receptor at a specific binding site in this macromolecular complex causes a conformational change in the receptor, enabling the specific neurotransmitter to bind more rapidly to its receptor.[52-54] Occupancy of the benzodiazepine receptor increases the affinity of the $GABA_A$ receptor for GABA.[55] This action increases the frequency of the opening of the chloride channel to allow negatively charged chloride ions to flow through the membrane into the neuron. The chloride influx makes the cell less responsive to excitatory synaptic inputs.[49] In the absence of GABA or an exogenous GABAmimetic (e.g., muscimol), stimulation of the benzodiazepine receptor does not increase the permeability of the chloride ion channel.[44] Benzodiazepines are not GABAmimetic per se.

Other types of sedative/hypnotic medications, such as barbiturates and ethanol, produce neuronal inhibition by acting at this same receptor complex to increase chloride conductance independently of GABA.[49] These agents are much more toxic than benzodiazepines which cannot increase neuronal inhibition above a maximum level since their action, even at high doses, is dependent on synaptic concentrations of GABA.[14] For this reason, despite concerns about drug addiction and abuse, benzodiazepines remain a drug of choice in the treatment of anxiety, insomnia, panic, psychotic agitation, and bipolar affective disorder.[56-63] Fatal overdoses are almost unknown,[64-67] and patients usually recover from large dosages of benzodiazepines (500 mg to 2000 mg of diazepam) within 24 to 48 h.[67] These drugs also have muscle relaxant properties usually without significant effect on coordination, and this action is probably mediated by the effects of GABA on spinal and supraspinal motor reflexes.[16] As the dose of benzodiazepine is increased, anxiolytic effects are first produced, followed by anticonvulsant effects and then a reduction in muscle tonus, followed by sedation and hypnosis.[27]

IV. PERIPHERAL BENZODIAZEPINE RECEPTORS

The peripheral benzodiazepine receptor (PBR) is a relatively small (17 to 18 kD) protein situated predominantly on the outer mitochondrial membrane in both peripheral and CNS tissues.[68-70] PBR densities are relatively high in the kidney, heart, lung, ovary, and testes and highest in the adrenal gland.[71-74] Peripheral benzodiazepine receptors in the CNS are linked to glial cells, and the highest densities are found in various regions of the brain, such as the choroid plexus, ependyma, pineal gland, olfactory bulb, and circumventricular organs.[75,76] PBR is thus chemically and anatomically distinct from the central benzodiazepine receptor (CBR), which is located on the postsynaptic GABA$_A$ receptor/chloride ion channel complex in the CNS.[77,78] Studies on molecular cloning, chromosomal localization, and DNA expression of peripheral-type benzodiazepine receptors confirm the molecular difference between PBR and CBR, suggesting that PBR is a distinct class of receptor.[79-81] Recent evidence suggests that PBR regulates steroidogenesis in both periphery and brain areas.[82,83] Various experiments have demonstrated the involvement of PBR ligands, including the endogenous peptide diazepam-binding inhibitor (DBI), in the stimulation of steroid biosynthesis in cultured adrenocortical, Leydig, granulosa, and glial cells.[82-86] This suggests that PBR serves a putative regulatory role in endocrine systems.[87] Finally, reports from several laboratories reveal that stress generally causes rapid and short-lived alterations in the density (Bmax) of PBR in rats.[88-90] This effect seems to be highly dependent on the type of stress involved and the tissues surveyed.

V. BENZODIAZEPINE RECEPTOR AGONISTS AND ANTAGONISTS

A diagram of the hypothetical GABA–benzodiazepine receptor/chloride ion channel complex has been constructed by Philip.[51] This complex, seen from the extracellular site and as a section of the cell membrane, is composed of four monomers that contain three major anatomical and functional domains: the chloride ion channel, the GABA receptor, and the benzodiazepine receptor. Each domain carries a ligand-binding site. Benzodiazepine receptor ligands are all those compounds, however different in structure, which are able to bind to the benzodiazepine receptor. Benzodiazepine receptor ligands include benzodiazepines and nonbenzodiazepines, such as β-carboline derivative abecarnil, imidazopyridine zolpidem, imidazopyrimidine divaplon, and cyclopyrrolone zopiclone. On the benzodiazepine receptor, there are sites for agonists, inverse agonists, and antagonists which represent three classes of active ligands. These ligands differ in their positive or negative modulatory action on GABA effects.

The multiplicity of actions of the different benzodiazepine receptor ligands has been well described by Haefely and associates.[44] Selective agonist ligands potentiate GABA effects, thereby maximizing the inhibition of synaptic function. Drugs which are benzodiazepine agonist ligands or indirect agonists of GABA$_A$ receptors, such as diazepam, midazolam, flurazepam, and clonazepam, make up the therapeutically useful benzodiazepines. In contrast, inverse benzodiazepine receptor agonists decrease the inhibitory efficacy of GABA on synaptic function. Drugs such as DMCM (a beta-carboline) or RO19-4603 (an imidazobenzodiazepinone), which act as negative modulators of GABA, are anxiogenic and proconvulsant. The third class of ligand, the antagonist drugs, have no modulatory activity at GABA$_A$ receptors. They compete with other ligands binding to the benzodiazepine receptor and block it, thus reducing the effect of both agonists and inverse agonists.[44] Flumazenil, a benzodiazepine receptor antagonist, stereospecifically and competitively blocks the therapeutic effects of agonists on the CNS as well as the opposite effects produced by inverse agonists. Flumazenil, structurally an imidazobenzodiazepine, is effective whether given before, simultaneously with, or after the agonist drug.[91,92] It will also block the few other drugs that are active at the benzodiazepine receptor; however, it is not effective against other CNS depressive drugs, such as opioids, which do not act through the benzodiazepine receptor, and barbiturates, which act at separate sites on the GABA receptor.[93] Reciprocal dose-dependent antagonism has been demonstrated between flumazenil and benzodiazepines such as midazolam.[94] Low doses of flumazenil are required to reverse high-dose midazolam effects such as hypnosis, whereas higher doses of flumazenil are required to reverse slight sedation. This action seems to be the result of concentration-related competition at receptor sites.[95] This agonist–antagonist relationship permits a variety of clinical applications for flumazenil, particularly in view of its low intrinsic toxicity[92] and rapid uptake.[96] It can be used to reverse some of the adverse effects of benzodiazepines, such as benzodiazepine-induced hypoventilation,[97] sedation,[98] memory loss, and amnesia,[99,100] as well as changes in electrical activity of the brain.[101] Flumazenil can also be used to reverse accidental or intentional benzodiazepine overdose,[102] but not ethanol intoxication.[103] In addition to its predominantly antagonist properties, flumazenil can produce agonist-like anticonvulsant activity when given in large oral doses. Flumazenil may also have weak, inverse agonist-like, anxiogenic effects.[92]

Partial agonists, such as alpidem and bretazenil, are the most promising new compounds since they share some of the anxiolytic and anticonvulsant properties of full agonists without inducing the same side effect profile. Partial agonists appear to stimulate only a small proportion of benzodiazepine receptors and, therefore, reduce anxiety without sedation and muscle relaxation.[44,45] They are less likely to induce tolerance and physical dependence. These agents were found to be effective in patients with generalized anxiety disorder (GAD) and panic disorder.[55]

The discovery of benzodiazepine receptors led to a search for naturally occurring ligands. Endogenous benzodiazepine-receptor ligands, which positively modulate GABA-activated Cl⁻ conduction, have recently been identified by extraction and purification from the human brain.[104] The extraction from the bovine brain of *n*-butyl-beta-carboline-3-carboxylate (beta-CCB), a compound that varies in concentration in the brain according to stress levels,[105] suggests that beta-CCB is an endogenous, inverse agonist active at benzodiazepine receptors and, perhaps, involved in the regulation of stress. An implication of this research is that the biological origins of anxiety disorders may include imbalances in endogenous ligands which have anxiogenic or anxiolytic effects.

VI. SELECTION OF BENZODIAZEPINES FOR CLINICAL USE IN DIFFERENT POPULATIONS

A. Tolerance, Dependence, and Discontinuation Syndromes

There are approximately 50 benzodiazepines marketed worldwide for clinical use which are considered to have anticonvulsant, anxiolytic, hypnotic, sedative, and muscle relaxative properties.[11] While exhibiting these therapeutic properties, they differ in their structure, pharmacological profiles, and clinical relevance. Benzodiazepines are usually classified according to elimination half-life (β half-life). Benzodiazepines that are more rapidly cleared from the body are defined as short to intermediate half-life drugs. Until recently, this pharmacokinetic attribute was a fundamental criterion in selecting the most appropriate benzodiazepine in a given clinical situation. Many authors[106] have questioned this approach to classifying benzodiazepines under the belief that the elimination half-life of a drug is no longer a reliable indicator of its duration of action, onset of effect, or clinical effects.

The concept of β elimination half-life should not be confused with duration of action. The onset and duration of action of a benzodiazepine depends upon its rate of absorption, uptake into the CNS, and binding at the benzodiazepine-GABA receptor complex, not on its elimination half-life. In the case of single-dose administration, rate of absorption and penetration of the blood-brain barrier are the criteria determining the onset of action of a compound. A drug such as diazepam, with a long elimination half-life, has a very rapid onset of action in comparison to a short half-life benzodiazepine such as oxazepam which is slowly absorbed and remains in the CNS longer than diazepam.

β elimination half-life has its greatest clinical significance when single-dose and multiple-dose kinetics are compared. When multiple doses are being prescribed over long periods of time, it takes longer to reach a steady state blood level with long half-life drugs than with short half-life compounds. Similarly, once steady state has been achieved, the rate of elimination of long half-life benzodiazepines is slower than that of short half-life drugs. β elimination half-life is important when a drug discontinuation regime is being calculated, even though the differences between short and long half-life compounds become less marked when these drugs are gradually tapered before termination.

An alternative to elimination half-life as a method to discriminate among the benzodiazepines is the therapeutic potency of each compound. Potency and half-life are not related. Potency, in pharmacological terms, is correlated with the affinity of a benzodiazepine compound or its active metabolites for benzodiazepine receptors *in vivo*. Benzodiazepines can be divided into two groups, depending on their biotransformation. Some benzodiazepines, such as lorazepam and oxazepam, do not produce active metabolites. Derivatives in the second group do produce active metabolites with kinetic properties which have therapeutic and toxic effects. For instance, diazepam produces the metabolites desmethyl diazepam, temazepam, and oxazepam. Until recently, there have been no data available on this more complex approach to benzodiazepine classification. Drug potency in a clinical sense refers to the inverse relationship between dosage and therapeutic efficacy: low dosage and therapeutic effect means higher potency.

Benzodiazepines present different risks of physiological dependence, even at therapeutic dosages, as well as risks of producing discontinuance symptoms. Physiological dependence and tolerance are related phenomena, but they are not synonymous. Physiological dependence implies a biological adaptation has occurred to the effects of the drug. Tolerance occurs when the efficacy of a drug diminishes with repeated use and higher doses are required to achieve the same effect. It has been difficult to establish a relationship between the development of tolerance to benzodiazepines and the development of dependence because of the differences between various substances and the fact that the development of tolerance does not proceed at a constant rate for all aspects of benzodiazepine clinical activity.[8] The development of tolerance to the sedative and psychomotor effects of benzodiazepines is well documented.[1] There is no consensus as yet on the development of tolerance to the anxiolytic effects of these compounds.[1] Tolerance is associated with the short-term effects of some benzodiazepines[107] and may follow long-term use. The need to use increasing doses of a drug in order to produce the same effect is receptor related; it may depend on altered receptor function or to a change in the number of receptors which may limit the potential of benzodiazepines to enhance the inhibitory function of GABA.

A variety of factors play a role in the alterations in CNS functioning which characterize benzodiazepine dependence: drug dose and duration, pharmacological differences among the derivatives, and the susceptibilities of different users. There is evidence that high-potency benzodiazepines that have short elimination half-lives lead to the neurophysiological problems of dependence.[107] Triazolam, alprazolam, and lorazepam, all potent benzodiazepines with relatively short half-lives, have been associated with dependence more than other compounds.[107] According to Tyrer,[107] these differences are partly pharmacokinetic in origin (rapid elimination of drugs) and partly pharmacodynamic, in that high-potency drugs may have a high affinity for benzodiazepine receptors. These mechanisms can also explain discontinuance symptoms.[15]

Benzodiazepines, like other sedative/hypnotic drugs, are commonly associated with characteristic discontinuance symptoms following abrupt discontinuation. Typically, the mirror image of the therapeutic effect of the drug is induced; for instance, discontinuance of a sedating medication produces arousal. The symptoms which follow abrupt drug termination are evidence of physiological adaptation. Results from clinical studies differ on the correlation between risk of withdrawal symptoms and dosage or prolonged use.[108] However, these symptoms have been seen even after short-term use at therapeutic doses.[109] The drug discontinuance syndrome consists of three types of symptoms: rebound, recurrence, and withdrawal. Rebound symptoms are like those for which the drug was originally prescribed, only more intense. When treatment is discontinued, they appear temporarily and worsen rapidly, as compared to baseline. When the same pattern and intensity of symptoms as those originally present return after drug termination and persist without worsening, these are known as recurrence or relapse symptoms. While often developing together with rebound phenomena, withdrawal symptoms are new symptoms which follow on drug discontinuation. Very common discontinuance symptoms are anxiety, insomnia, restlessness, agitation, irritability, and muscle tension.[1] Both rebound and withdrawal symptoms are more prominent after discontinuation of benzodiazepines with rapid elimination rates (shorter β half-life).

In choosing the most appropriate benzodiazepine for a clinical situation, it is important to weigh numerous pharmacological factors, such as the volume of absorption, distribution, retention, and elimination of drugs,[14] as well as their side effects on psychomotor performance, memory, learning, and cognition.[110-117] The appropriateness of acute, chronic, or maintenance treatment must also be considered.[14,37,106]

B. Anxious Patients with or without Generalized Anxiety Disorder

It is a normal part of life to experience anxiety in response to stressful situations. This situational anxiety can be very effectively treated on a short-term basis with benzodiazepines. There are also types and degrees of anxiety which are recognized as disorders. These fall into three clinical entities: generalized anxiety, panic disorder with or without agoraphobia, and obsessive-compulsive disorders (DSM-III-R). Anxiety can also be part of other psychiatric disorders and be considered secondary to a primary psychopathology, such as major depression, mania, or schizophrenia. Antianxiety or hypnotic drugs are given primarily for the treatment of minor (situational anxiety) or major GAD, with or without insomnia as the target symptom. Other antianxiety compounds are called antipanic or anti-obsessive-compulsive agents. An important development in understanding the pathophysiology of anxiety has come from new knowledge of the benzodiazepine–GABA macromolecular complex.[8] Clinical studies of benzodiazepines have shown how the brain, when subject to stress, uses the inhibitory neurotransmitter systems to prevent the breakdown of neural functions.

When anxiolytic benzodiazepine therapy is indicated, the most appropriate derivative must be selected since not all benzodiazepines are similar with respect to their antianxiety properties. In the case of patients with a pathological form of anxiety (GAD), the appropriate benzodiazepine will be one providing a smooth onset of action with minimal perceptible CNS effects after an acute dose (single dose) and a continuous pattern of anxiolytic action with multiple dosing.[118] Such a derivative will have relatively slow absorption and elimination rates so that plasma level fluctuations are minimized as accumulation occurs. The benzodiazepine of choice will be less likely to produce dependence or other disorders that are sometimes linked to anxiolytic treatment. Clonazepam and ketazolam, for instance, would be logical choices. Halazepam and prazepam also have these properties. Patients with chronic anxiety disorders benefit from long-term benzodiazepine therapy. Since these patients are not particularly prone to developing physiological dependence and to inappropriate drug-taking behavior, such as significant drug escalation,[119] most clinicians[66,120-123] agree that long-term therapy is appropriate for this population.

C. Insomniac Patients: Rebound Insomnia and Amnesia

Patients with sleep disturbances may suffer from primary insomnia or secondary insomnia; the appropriate benzodiazepine therapy will differ in each case. Although many instances of primary insomnia can be managed by treating the underlying cause and/or using a variety of nonpharmacological methods, judicious use of a benzodiazepine as part of the overall management of patients with this type of sleep disturbance is often beneficial. Flurazepam, nitrazepam, temazepam, and triazolam are four benzodiazepines marketed as hypnotics.[14] Other derivatives may also be used for nighttime sedation since they possess some sedative properties, although not all to the same degree.[37,66,124-127] Benzodiazepines are contraindicated when insomnia is associated with sleep apnea.

Several pharmacological factors must be considered when choosing a benzodiazepine for the treatment of primary insomnia since rate of absorption and elimination are associated with sedative action and side effect profile. During the first few days of use, both rapidly eliminated and slowly eliminated benzodiazepines are effective for inducing and maintaining sleep. Rapidly absorbed but slowly eliminated benzodiazepines, such as flurazepam, diazepam and clorazepate, are recommended for most patients with insomnia since they induce sleep, provide daytime anxiolysis, and are effective for a week or more. Slowly absorbed benzodiazepines such as oxazepam and ketazolam have been found more effective in sleep maintenance than in sleep induction.[14]

The development of pharmacodynamic tolerance to side effects varies with different derivatives. Slowly eliminated benzodiazepines, such as flurazepam, may be more frequently associated with impairment on performance tests the next day. On the other hand, rapidly absorbed and rapidly eliminated benzodiazepines, such as bromazepam, may be preferable (after the first dose) if optimal alertness and performance are important the next morning. After initiation of drug treatment, it is preferable that short-acting compounds not be used for more than a few nights since tolerance rapidly develops. In the case of flurazepam, tolerance to side effects usually develops after 1 week of nightly treatment; thus, while the drug no longer has an impact on performance tests,[128] hypnotic efficacy is maintained.[129] In contrast, rapidly eliminated benzodiazepines, such as triazolam, lorazepam, and alprazolam, which are initially effective, tend to lose their efficacy over the first few weeks of use[129-131] and may lead to early morning rebound insomnia[132] and next-day rebound anxiety.[133] Before completing treatment with a rapidly eliminated benzodiazepine, patients may also develop rebound symptoms despite gradual tapering of the dosage.[134] It is often helpful to switch to a slowly eliminated benzodiazepine before dose tapering since the risk of rebound anxiety is related in part to the elimination half-life of the benzodiazepine after short-term use (4 weeks).[135] Long-acting agents (diazepam) are associated with a lower risk of rebound reactions than that reported for short- and intermediate-acting benzodiazepines such as lorazepam and bromazepam.[136]

The effects of benzodiazepines on memory have been well documented.[112-115,137] Triazolam[138-141] and lorazepam[142-146] are the two derivatives that have been most often associated with cognitive impairment and anterograde amnesic effects (memory loss after the drug has been taken).[147,148] Memory impairment depends on the derivative, the dose, and route of administration.[1] Anterograde amnesia is more severe with increased doses, intravenous administration, faster absorption, and higher potency.[149,150] Tests of immediate and delayed (20 to 90 minutes) recall of word lists, after drug administration, found that immediate recall was unaffected and impairment was limited to delayed recall, an effect which was usually short lasting and reversible. Triazolam has also been reported to produce transient global amnesia or so-called traveler's amnesia.[151,152] Other benzodiazepines appear to have a similar but less pronounced effect on delayed recall. During chronic benzodiazepine administration, it appears that tolerance to this effect develops[153,154] and that the impairment is limited to a narrow window within 90 min after each dose.[155]

D. Drug Abusers and Alcohol Withdrawal

The usefulness of prescribing benzodiazepines to control anxiety and insomnia in abstinent alcoholics and drug abusers is at best controversial.[156,159] Patients with a history of drug abuse or alcoholism may treat a benzodiazepine as another substance to use inappropriately and without medical supervision.[119] Despite the fact that only very few drug abusers use benzodiazepines as their primary recreational drug,[66] chronic treatment regimens with benzodiazepines in drug-dependent individuals are not recommended since they will reinforce drug-seeking behavior. Benzodiazepines are usually given acutely to alleviate the undesirable effects of other drug abuse, such as anxiety provoked by amphetamines or heroin, or to ameliorate the withdrawal effects of cocaine and heroin. There are no prospective studies of rapid tranquilization of agitation in drug abusers; however, benzodiazepines are considered to be the drug of choice to treat patients intoxicated from cocaine or amphetamines. Although contraindicated in cases of alcohol intoxication, these compounds (diazepam in particular) are also the best treatment for alcohol withdrawal, when a gradual tapering of the dose is used, because of the cross-tolerance between alcohol and benzodiazepines.[158]

Benzodiazepines are at definite risk for abuse[158] in susceptible individuals, such as habitual polydrug users and alcoholics, but the derivatives differ in their attractiveness to the abuser. A rapidly absorbed drug such as diazepam will have a rapid onset of action and produce a more distinct peak effect than a slowly absorbed derivative such as oxazepam whose onset of effect is very gradual and often imperceptible.[12] The rapid peak effect is considered desirable by the drug abuser since it may be associated with a sensation of euphoria.[159] Benzodiazepines are not equally susceptible to abuse because of their different reinforcing properties. Clinical studies have shown that sedative drug abusers prefer lorazepam and diazepam to oxazepam, halazepam, and chlordiazepoxide.[160] Similarly, alcoholics are more likely to turn to diazepam and alprazolam[161] than to halazepam.[162]

E. Elderly Patients

Insomnia and anxiety are widespread psychological conditions in the general population.[163] However, in the elderly population, insomnia can be a "pathological" feature associated with age.[116,164] Most studies find that the rate of use of benzodiazepines is substantially higher in women than in men, and in older than in younger people.[165] Benzodiazepine hypnotic drugs have been reported to be prescribed with minimal physician follow-up among elderly patients,[166] despite the fact that benzodiazepine treatment in the elderly poses hazards which are not encountered in younger populations. Elderly patients are reported to be more sensitive than young patients to the effects of benzodiazepines on memory.[1] Benzodiazepine dependence in older people can cause memory impairment that persists into the early drug-free period.[164] A relationship has been proposed between benzodiazepine abuse, dementia in the elderly,[16,114] and the weak anticholinergic action (independent of GABA) responsible in part for the amnesic effect of benzodiazepines.[167-169] However, this has never been substantiated.

Studies of the effect of regular daily dosing with diazepam, chlordiazepoxide, nitrazepam, flurazepam, or temazepam have shown a greater degree of sedation and CNS depression in the elderly.[170,171] In the normal elderly, the hepatic function of the oxidase system becomes less efficient, whereas the efficiency of hepatic drug conjugation generally remains unchanged.[172] Thus, drugs such as diazepam, flurazepam, and chlordiazepoxide, when undergoing oxidative metabolism, have a decreased clearance and, consequently, a prolonged elimination rate which is reflected in an increase in β half-life and a greater accumulation with continued use.[171] However, when tolerance occurs, the degree of clinical sedation does not increase in parallel with the extent of drug accumulation.[14] Furthermore, in the absence of a correlation between drug accumulation and sedation, benzodiazepines that are not metabolized by oxidation are also thought to produce a sedative effect in the elderly population.[170,173] Therefore, the increased sensitivity to benzodiazepines which occurs with normal aging can also be attributed to pharmacodynamic changes.[15]

Benzodiazepine treatment of elderly patients is a cause of concern because of its sedative effects in this population. The incidence of hip fracture because of falls may be increased,[174-176] especially when benzodiazepines with a long half-life are used. However, there are fewer complications associated with benzodiazepines in the elderly than with tricyclic antidepressants and neuroleptics. For short-term hypnotic therapy, a rapidly absorbed and eliminated benzodiazepine derivative, such as bromazepam, is a safe treatment, whereas rapidly eliminated lorazepam and triazolam should be avoided on a long-term basis because of reports of cognitive impairment and anterograde amnesic effects[112-115,134-139,142-144] and rebound effects. In the clinical context, the relationship between benzodiazepine abuse and osteoporosis in the elderly must also be considered. Benzodiazepines may act as chelators, as do neuroleptics, when combined with iron,[177,178] to provoke a situation in which calcium intake by older people is insufficient to meet calcium losses and bone mineral reserves are depleted.[179] The use of psychotropic medication, including benzodiazepines, should be carefully considered in elderly patients who are at increased risk for dementia, visual impairment, postural hypotension, and neurological and musculoskeletal disability.[174]

VII. OTHER INDICATIONS FOR THE SELECTIVE USE OF BENZODIAZEPINES

A. Panic Disorder

Patients who take benzodiazepines for panic disorder with or without agoraphobic symptoms are a newly emerging group for whom long-term use may be appropriate. Most clinical studies report that the majority of these patients neither abuse the medication nor escalate the dosage.[1] Both alprazolam and clonazepam have demonstrated efficacy in panic disorder,[59-61,180-193] but they have significantly different pharmacokinetic properties. Clonazepam is much less lipophilic than alprazolam; that is, it has a lesser tendency to enter adipose tissue and to leave the central (plasma and brain) compartment.[14] This positive pharmacokinetic property of clonazepam allows a prolonged distribution phase (longer α half-life) and, therefore, a longer duration of action. Clonazepam is also more slowly metabolized and, hence, more slowly eliminated (longer β half-life). These properties allow twice-daily dosing which reduces fluctuations in plasma concentrations. Clonazepam represents the initial drug of choice for the treatment of panic disorder because it has a more rapid onset than tricyclic antidepressants or SSRIs, and fewer side effects than either tricyclics or alprazolam. Clonazepam therapy may be initiated, and then an SSRI may be started at low doses. When the patient is adjusted to SSRI therapy, clonazepam may be gradually reduced and then discontinued.

B. Depression

Most benzodiazepines, especially low-potency ones, appear relatively ineffective on the major symptoms of depression, including psychomotor retardation, guilt, suicidality, and diurnal variation.[194,195] The high-potency benzodiazepine, alprazolam, on the other hand, has been found to improve these symptoms[196-198] in mild to moderately depressed patients on a short-term basis. Alprazolam appears to be different from other benzodiazepines in this respect.[198]

The emergence of depressive symptoms has been shown to be a part of the side effect profile of several benzodiazepines when used to treat generalized anxiety. Depression is recorded as an emergent symptom if it was not initially present in the patient or if it intensified during treatment. Emergent depression has been found in patients with generalized anxiety when treated with lorazepam or bromazepam.[57] The emergence of depression in some patients with low initial anxiety scores may be related to the anxiolytic properties of the high-potency benzodiazepines with which they are treated. Thus, for less severely anxious patients, a lower dose might be more suitable and will avoid the risk of disinhibition leading to depressed mood. On the other hand, highly anxious patients with secondary depressive symptoms have shown a remission of depression as their anxiety decreased.[199] This is consistent with the findings of Rickels and colleagues[200] in a controlled study of anxious neurotic outpatients in which the benzodiazepine drug (bromazepam) provided better symptom relief for those who were initially more depressed. Shammas[201] also found bromazepam to be superior to amitriptyline and placebo in the treatment of moderate or severe anxiety-depressive neurosis. The development of depression in some patients treated with bromazepam or lorazepam contrasts to no reported cases in alprazolam-treated patients, another high-potency benzodiazepine.[57] These differences suggest that depression could be also induced by benzodiazepine drugs (except alprazolam) in susceptible individuals rather than being uncovered as the anxiety remits. Treatment-emergent depression has, however, been reported in patients with panic disorder treated with alprazolam[202] or tricyclics.[203] With respect to the depressive symptoms associated with panic disorder, clonazepam has been reported to be superior to placebo.[188,193,204] In a retrospective review of 177 patients with panic disorder treated with clonazepam, the incidence of emergent depression was reported to be 5.7%, but most cases were associated with a history of depression.[205]

It has been suggested that alprazolam be given for anxious depression with neurovegetative signs whereas endogenous melancholic forms of depression are more effectively treated by tricyclics[206] or SSRIs.[207-209] This remains controversial since the mechanism of antidepressant action of this benzodiazepine still needs to be elucidated. Animal studies suggest that alprazolam affects cortical β adrenoreceptors in a manner similar to tricyclic antidepressants and electroconvulsive therapy.[210,211] Studies with adinazolam (a triazolobenzodiazepine similar to alprazolam) and clonazepam suggest that they may also have antidepressant properties.[204,212] SSRIs are now the first-line treatment for major depression of the melancholic type and others.[207-209] The beneficial effects of benzodiazepines in depression may be because of their better side effect profile compared with tricyclic antidepressants.[213-218] A few studies have compared SSRIs with benzodiazepines.[219,220]

Paradoxically, patients undergoing long-term treatment of panic disorder with alprazolam and clonazepam have been reported to develop symptoms of major depression despite the improvement of their panic symptoms.[202] In light of these clinical data, it is recommended that caution be exercised especially during long-term treatment with alprazolam since depression may appear as an interdose rebound symptom. Alprazolam, although free of the anticholinergic side effects observed with tricyclics or other benzodiazepines such as triazolam,[168,169] can cause a variety of psychiatric problems such as the emergence of hostility,[221,222] manic behavior,[223-228] interdose anxiety during treatment,[229] and withdrawal symptoms after abrupt,[230] and even gradual,[153] discontinuation.

C. Mania, Psychotic Agitation, and Schizophrenia

Lithium monotherapy has been demonstrated to be efficacious in the treatment of acute mania,[231] but the delay in therapeutic onset that is estimated to be between 1 and 2 weeks[231,232] has required the use of adjunctive agents. Benzodiazepines or neuroleptics have been commonly used as adjuncts to lithium in the treatment of acute mania and bipolar affective disorder.[233-242] However, adjunctive therapy with neuroleptics often leads to the development of parkinsonian symptoms such as akathisia and may produce neurological complications.[243] In contrast, combination with benzodiazepines has proven advantages over neuroleptics, particularly in bipolar patients who have increased vulnerability to tardive dyskinesia and dystonia.[1] When used as adjunctive medication to lithium in mania, benzodiazepines avoid the risk of tardive dyskinesia and the possibility of rebound or supersensitivity psychosis associated with classic neuroleptics.[240] Tryptophan and benzodiazepines have been investigated as antimanic agents in an effort to reduce the dosage and duration of neuroleptic treatment.[243] For manic patients, there is evidence that clonazepam or lorazepam may be especially useful in controlling associated agitation in the period prior to the onset of the beneficial effects of lithium.[243] Manic patients appear to be less prone to dependency than schizophrenic patients. In the treatment of acute mania, clonazepam has been found to decrease neuroleptic requirements[56] and, therefore, to reduce the risk of drug-induced parkinsonism, neurotoxicity, and tardive dyskinesia.

The use of benzodiazepines and antipsychotic drugs for the tranquilization of severely agitated schizophrenic patients confirms the importance of adjunctive therapy with benzodiazepines.[244-252] Benzodiazepines are indicated only on a short-term basis when given in combination with neuroleptics for agitation in schizophrenic patients since there is a high risk of dependence in this group of patients. There is evidence to suggest that the GABA system interacts with the dopamine system in the pathophysiology of schizophrenia,[253-255] including the induction of negative symptoms.[254] Benzodiazepines are also known to enhance the GABAergic inhibition of dopamine neurons.[253-256] Recently, withdrawal from haloperidol maintenance treatment was found to be associated with a significant increase in CSF diazepam-binding inhibitors (DBI).[257] The DBI is a 9-kD neuropeptide and an endogenous ligand that interacts with the benzodiazepine-binding sites of the $GABA_A$ receptor and with the glial mitochondrial benzodiazepine receptor.[258] These results were found in paranoid schizophrenic and nonrelapsing patients,[257] suggesting that DBI is not a state-dependent marker. DBI acts as a negative allosteric modulator of $GABA_A$-receptor function.[259,260] Despite extensive investigation of full benzodiazepine agonists in schizophrenia,[243,251-260] their chronic use is not recommended because their beneficial effects have not been shown to outweigh the risk of dependency associated with their use. The results of full benzodiazepine agonists do not apply to partial agonists which can be shown to be beneficial in schizophrenia.

VIII. NEW DIRECTIONS FOR BENZODIAZEPINE USE

Recent studies[261,262] reported a role for dopamine in some forms of obsessive-compulsive disorder, which is considered a heterogeneous neurobiological disorder.[262] A dysregulation of dopamine function has also been associated with a variety of other conditions, such as high-dose stimulant abuse,[263] comorbid Tourette's and other chronic tic disorders,[264,265] and basal ganglia pathology,[16,262,266-268] among others. In these clinical situations, adjunctive therapy with benzodiazepines can be considered. Benzodiazepines may be combined with neuroleptic treatment in the most severe cases of obsessive-compulsive disorder with comorbid chronic tic disorders or other mixed syndromes in which dopamine function is implicated.

Certain derivatives appear to possess pharmacodynamic properties that are not shared by the entire benzodiazepine class; clinical trials have suggested the existence of antipanic properties for alprazolam and clonazepam, antidepressant properties for alprazolam, and antimanic properties for clonazepam and lorazepam. Benzodiazepine partial agonists will need to be further investigated in schizophrenia.

The side effect profile of benzodiazepines consists of sedation, psychomotor and cognitive impairment, memory loss, potentiation of other CNS depressants, and withdrawal and rebound symptoms. It has become increasingly clear that in some vulnerable individuals most benzodiazepines can produce dependence and tolerance. Although there are marked differences among individual derivatives in the nature, speed of onset, and extent of their sedative, anticonvulsant, and anxiolytic effects, tolerance is thought to occur with prolonged clinical exposure. There is no evidence of a correlation between the plasma half-life of a derivative and the development of tolerance or dependence. However, other pharmacokinetic variables, such as rate of absorption, lipophilicity, and receptor affinity, may produce effects that reinforce drug use and facilitate the onset of tolerance.[8] Choice of the appropriate benzodiazepine for a given indication must be based on the pharmacological differences among the derivatives, their side effects, and whether acute, short-term, or maintenance treatment is required.

Full benzodiazepine agonists are the first choice of treatment in primary anxiety disorders, especially GAD and insomnia. Benzodiazepines that are slowly absorbed and slowly eliminated are most appropriate for the anxious patient because these agents (clonazepam and ketazolam) produce a gradual and sustained anxiolytic effect. Rapidly absorbed and slowly eliminated benzodiazepines (flurazepam, diazepam, and clorazepate) are usually more appropriate for patients with sleep disturbances since the rapid absorption induces sleep and the slower elimination rate may induce fewer rebound and withdrawal symptoms. Although rapidly absorbed and rapidly eliminated benzodiazepines such as bromazepam are preferable, for instance, in primary insomnia when optimal alertness and performance are important the next morning, tolerance may render them less effective. In panic disorders, benzodiazepines are given to those patients who cannot tolerate SSRIs and to achieve rapid control of anxiety. In acutely agitated patients, they are most useful when given in combination with neuroleptics. Patients with psychiatric illnesses such as schizophrenia are given benzodiazepines acutely as adjunctive medication to treat severe agitation. Their chronic use in mania appears justified. Benzodiazepines are not recommended for individuals prone to drug or alcohol dependency.

The introduction of short-acting and intermediate-acting benzodiazepines has led to a greater risk of rebound reactions than is seen with long-acting agents. Clinicians must become acquainted with the different phenomena associated with benzodiazepine withdrawal and consider the β elimination half-life, the potency, and the unique pharmacological profile of each derivative. This is particularly the case when treating the elderly and the suspected drug abuser for whom choosing an optimal benzodiazepine is more problematic. Selection is made more difficult by the risk of physiological dependence, which is often related to tolerance that develops during treatment. Rebound and withdrawal syndromes are interrelated phenomena which appear after discontinuation of benzodiazepines with rapid elimination rates (short β half-lives). Withdrawal and rebound symptoms may also appear after sudden discontinuation of slowly eliminated derivatives but are less severe in this case. Withdrawal symptoms can be minimized by a stepwise procedure which consists of switching to a slowly eliminated benzodiazepine (preferably diazepam) and then gradually tapering the dose before discontinuation.

The future of benzodiazepines will depend on the development of new partial agonists. This will make available agents whose selective activity has therapeutic advantages, particularly for long-term use, since they may be less likely to be associated with dependence and discontinuance syndromes. Full benzodiazepine agonists, which include most of the compounds presently available, will continue to be used intramuscularly or intravenously, and on a short-term basis as monotherapy, or in combination with other psychotropic or anticonvulsant drugs.

REFERENCES

1. Task Force on Benzodiazepine Dependency. *Benzodiazepine Dependence, Toxicity, and Abuse. A Task Force Report of the American Psychiatric Association.* American Psychiatric Association, Washington, D.C., 1990.
2. Busto, U., Lanctot, K., Isaac, P., and Adrian, M., Benzodiazepine use and abuse in Canada. *Can. Med. Assoc. J.,* 141, 917, 1989.
3. Weissman, M. N., The epidemiology of anxiety disorders: rates, risks, and familial patterns. *J. Psychiatr. Res.,* 22, 99, 1988.
4. Labelle, A. and Lapierre, Y. D., Anxiety disorders. Part 2: pharmacotherapy with benzodiazepines. *Can. Fam. Physician,* 39, 2205, 1993.
5. Sadavoy, J., Lazarus, L. W., and Jarvik, L. F., Eds., *Comprehensive Review of Geriatric Psychiatry,* American Psychiatric Press, Washington, D.C., 1991.
6. Serfaty, M. and Masterton, G., Fatal poisonings attributed to benzodiazepines in Britain during the 1980s. *Br. J. Psychiatry,* 163, 386, 1993.
7. Dunbar, G. C., Perera, M. H., and Jenner, F. A., Patterns of benzodiazepine use in Great Britain as measured by a general population survey. *Br. J. Psychiatry,* 155, 836, 1989.
8. Hindmarch, I., Review and conclusion, in *Benzodiazepines: Current Concepts. Biological, Clinical and Social Perspectives,* Hindmarch, I., Beaumont, G., Brandon, S., and Leonard, B.E., Eds., John Wiley and Sons, Chichester, 1990, 273.
9. Bury, M. and Gabe, J., A sociological view of tranquillizer dependence: challenges and responses, in *Benzodiazepines: Current Concepts. Biological, Clinical and Social Perspectives,* Hindmarch, I., Beaumont, G., Brandon, S., and Leonard, B.E., Eds., John Wiley and Sons, Chichester, 1990, 211.
10. Hollister, L. E., Principles of therapeutic applications of benzodiazepines, in *The Benzodiazepines: Current Standards for Medical Practice,* Smith, D. E. and Wesson, D. R., Eds., MTP Press, Lancaster, England, 1985, 87.
11. Haefely, W., Kyburz, E., and Gerecke, M., Recent advances in the molecular pharmacology of benzodiazepine receptors and in the structure activity relationships of their agonists and antagonists. *Adv. Drug Res.,* 14, 165, 1985.
12. Greenblatt, D. J., Divoll, M., Abernethy, D. R., Ochs, H.R., and Shader, R.I., Benzodiazepine kinetics: implications for therapeutics and pharmacogeriatrics. *Drug Metab. Rev.,* 14, 251, 1983.
13. Harvey, S.C., Hypnotics and sedatives, in *Goodman and Gilman's The Pharmacological Basis of Therapeutics,* 7th ed., Goodman Gilman, A., Goodman, L. S., Dall, T. W., et al., Eds., Macmillan, New York, 1985, 339.
14. Teboul, E. and Chouinard, G., A guide to benzodiazepine selection. Part I: pharmacological aspects. *Can. J. Psychiatry,* 35, 700, 1990.
15. Teboul, E. and Chouinard, G., A guide to benzodiazepine selection. Part II: clinical aspects. *Can. J. Psychiatry,* 36, 62, 1991.
16. Kupfesmann, I., Hypothalamus and limbic system: peptidergic neurons, homeostasis, and emotional behavior, in *Principles of Neural Science,* 3rd ed., Kandel, E. R., Schwartz, J. H., and Jessell, T. M., Eds., Appleton & Lange, Norwalk, CT, 1991, 735.
17. Lippa, A. S., Critchett, D., Sano, M. C., Klepner, C. A., Greenblatt, E. N., Coupet, J., and Beer, B., Benzodiazepine receptors: cellular and behavioral characteristics. *Pharmacol. Biochem. Behav.,* 10, 831, 1979.
18. Matsumoto, R. R., GABA receptors: are cellular differences reflected in function? *Brain Res. Rev.,* 14, 203, 1989.
19. Tallman, J. and Gallagher, D. W., The GABA-ergic system a locus of benzodiazepine action. *Ann. Rev. Neurosci.,* 8, 21, 1985.
20. Zachmann, M., Tocci, P., and Nyhan, W.L., The occurrence of gamma-aminobutyric acid in human tissues other than brain. *J. Biol. Chem.,* 241, 1355, 1966.
21. Löscher, W. and Schmidt, D., Diazepam increases γ aminobutyric acid in human cerebrospinal fluid. *J. Neurochem.,* 49, 152, 1987.
22. Berrettini, W. H., Nurnberger, J. I., Jr., Hare, T. A., Simmons-Alling, S., Gershon, B. S., and Post, R. M., Plasma and CSF GABA in affective illness. *Br. J. Psychiatry,* 141, 483, 1982.
23. Petty, F., Kramer, G., and Feldman, M., Is plasma GABA of peripheral origin? *Biol. Psychiatry,* 22, 725, 1987.
24. Petty, F. and Kramer, G. L., Stability of plasma GABA with time in healthy controls. *Biol. Psychiatry,* 31, 743, 1992.
25. Roy-Byrne, P. P., Cowley, D. S., Hommer, D., Greenblatt, D. J., Kramer, G. L., and Petty, F., Effect of acute and chronic benzodiazepines on plasma GABA in anxious patients and controls. *Psychopharmacology,* 109, 153, 1992.

26. Roberts, E., GABA in nervous system function — an overview, in *The Nervous System, Vol. 1: The Basic Neurosciences*. Tower, D. B., Ed., Raven Press, New York, 1975, 541.

27. Goodchild, C.S., GABA receptors and benzodiazepines. *Br. J. Anaesth.*, 71, 127, 1993.

28. Davidoff, R. A., Gamma-aminobutyric acid antagonism and presynaptic inhibition in the frog spinal cord. *Science*, 175, 331, 1972.

29. Guo, Z., Gent, J. P., and Goodchild, C. S., Two GABA mechanisms for spinally mediated antinociception: evidence from antagonism of 5-hydroxytryptamine (5-HT) by bicuculine. *Br. J. Anaesth.*, 64, 589P, 1990.

30. Castellano, C. and McGaugh, J. L., Retention enhancement with post-training picrotoxin: lack of state dependency. *Behav. Neurol. Biol.*, 51, 165, 1989.

31. McGaugh, J. L., Castellano, C., and Brioni, J., Picrotoxin enhances latent extinction of conditioned fear. *Behav. Neurosci.*, 104, 264, 1990.

32. Galindo, A., Effects of procaine pentobarbital and halothane on synaptic transmission in the central nervous system. *J. Pharmacol. Exp. Ther.*, 169, 185, 1969.

33. Peduto, V. A., Concas, A., Santoro, G., Biggio, G., Gessa, G. L., Biochemical and electrophysiologic evidence that propofol enhances GABAergic transmission in the rat brain. *Anesthesiology*, 75, 1000, 1991.

34. Wesselman, J. P. M., Van Wilgenburg, H., and Long, S. K., The effects of pentobarbital and benzodiazepines on GABA-responses in the periphery and spinal cord *in vitro*. *Neurosci. Lett.*, 128, 261, 1991.

35. Ashton, D. and Willems, R., *In vitro* studies on the broad spectrum anticonvulsant loreclezole in the hippocampus. *Epilepsy Res.*, 11, 75, 1992.

36. Zaman, S. H., Shingai, R., Harvey, R. J., Darlison, M. G., and Barnard, E. A., Effects of subunit types of the recombinant GABA$_A$ receptor on the response to a neurosteroid. *Eur. J. Pharmacol.*, 225, 321, 1992.

37. Greenblatt, D. J., Pharmacology of benzodiazepine hypnotics. *J. Clin. Psychiatry*, 53, 7, 1992.

38. Cotariu, D. and Zaidman, J. L., Valproic acid and the liver. *Clin. Chem.*, 34, 890, 1988.

39. Halonen, T., Pitkänen, A., Saano, V., and Riekkinen, P. J., Effects of vigabatrin (gamma-vinyl GABA) on neurotransmission-related amino acids and on GABA and benzodiazepine receptor binding in rats. *Epilepsia*, 32, 242, 1991.

40. Castellano, C., Brioni, J. D., Nagahara, A. H., and McGaugh, J. L., Post-training systemic and intra-amygdala administration of the GABA-beta agonist baclofen impairs retention. *Behav. Neurol. Biol.*, 52, 170, 1989.

41. Waldmeier, P. C., Wicki, P., Feldtrauer, J. J., and Baumann, P. A., Potential involvement of a baclofen-sensitive autoreceptor in the modulation of release of endogenous GABA from rat brain slices *in vitro*. *N. S. Arch. Pharmacol.*, 337, 289, 1988.

42. Sieghart, W. and Karobath, M., Molecular heterogeneity of benzodiazepine receptors. *Nature*, 286, 285, 1980.

43. Barnard, E. A., Darlison, M. G., and Seeburg, P. H., Molecular biology of the GABA$_A$ receptor: the receptor/channel superfamily. *Trends Neurosci.*, 10, 502, 1987.

44. Haefely, W. E., Martin, J. R., Richards, J. G., and Schoch, P., The multiplicity of actions of benzodiazepine receptor ligands. *Can. J. Psychiatry*, 38, S102, 1993.

45. Price, G. W., Kelly, J. S., and Bowery, N. G., The location of GABA$_B$ receptor binding sites in mammalian spinal cord. *Synapse*, 1, 530, 1987.

46. Squires, R. F. and Braestrup, C., Benzodiazepine receptors in rat brain. *Nature*, 266, 732, 1977.

47. Möhler, H. and Okada, T., Benzodiazepine receptors: demonstration in the central nervous system. *Nature*, 198, 849, 1977.

48. Bossmann, H. B., Case, R., and Di Stephano, P., Diazepam receptor characterization: specific binding of a benzodiazepine to macromolecules in various areas of rat brain. *FEBS Lett*, 82, 368, 1977.

49. Hommer, D. W., Skolnick, P., and Paul, S. M., The benzodiazepine/GABA receptor complex and anxiety, in *Psychopharmacology: The Third Generation of Progress*, Meltzer, H. Y., Ed., Raven Press, New York, 1987, 977.

50. Haefely, W., The biological basis of benzodiazepine actions, in *The Benzodiazepines: Current Standards for Medical Practice*, Smith, D. E. and Wesson, D. R., Eds., MTP Press, Lancaster, England, 1985, 7.

51. Philip, B. K., Drug reversal: benzodiazepine receptors and antagonists. *J. Clin. Anesth.*, 5, 46S, 1993.

52. Haefely, W. E., Benzodiazepines. *Int. Anesthesiol. Clin.*, 26, 262, 1988.

53. Study, R. E. and Barket, J. L., Cellular mechanisms of benzodiazepine action. *J.A.M.A.*, 247, 2147, 1982.

54. Staley, K., Enhancement of the excitatory actions of GABA by barbiturates and benzodiazepines. *Neurosci. Lett.*, 146, 105, 1992.

55. Potokar, J. and Nutt, D. J., Anxiolytic potential of benzodiazepine receptor partial agonists. *CNS Drugs*, 1, 305, 1994.

56. Chouinard, G., Use of clonazepam in the maintenance treatment of manic depressive illness, in *Proceedings of the IVth World Congress of Biological Psychiatry 1985*. Elsevier, New York, 1986, 723.

57. Fontaine, R., Mercier, P., Beaudry, P., Annable, L., and Chouinard, G., Bromazepam and lorazepam in generalized anxiety: a placebo-controlled study with measurement of drug plasma concentrations. *Acta Psychiatr. Scand.*, 74, 451, 1986.

58. Chouinard, G., Clonazepam in acute and maintenance treatment of bipolar affective disorder. *J. Clin. Psychiatry*, 48, 29, 1987.

59. Davidson, J. R., Potts, N., Richichi, E., Krishnan, R., Ford, S. M., Smith, R., and Wilson, W. H., Treatment of social phobia with clonazepam and placebo. *J. Clin. Psychopharmacol.*, 13, 423, 1993.

60. Pollack, M. H., Otto, M. W., Tesar, G. E., Cohen, L. S., Meltzer-Brody, S., and Rosenbaum, J. F., Long-term outcome after acute treatment with alprazolam or clonazepam for panic disorder. *J. Clin. Psychopharmacol.*, 13, 257, 1993.

61. Sheehan, D. V., Raj, A. B., Harnett-Sheehan, K., Soto, S., and Knapp, E., The relative efficacy of high-dose buspirone and alprazolam in the treatment of panic disorder: a double-blind placebo-controlled study. *Acta Psychiatr. Scand.*, 88, 1, 1993.

62. Dinan, T.G. and Leonard, B. E., Triazolam as safe as other benzodiazepines. *Br. Med. J.*, 306, 1475, 1993.

63. Chouinard, G., Clonazepam in the treatment of psychiatric disorders, in *Use of Anticonvulsants in Psychiatry: Recent Advances*, McElroy, S. L. and Pope, H. G., Eds., Oxford Health Care, Inc., Clifton, NJ, 1988, 43.

64. Epstein, F. B. and Litovitz, T., Fatal benzodiazepine toxicity? (Letter). *Am. J. Emerg. Med.*, 5, 472, 1987.

65. Greenblatt, D. J., Shader, R. I., and Abernethy, D. R., Current status of benzodiazepines (second of two parts). *N. Engl. J. Med.*, 309, 410, 1983.

66. Marks, J., *The Benzodiazepines: Use, Overuse, Misuse, Abuse*, 2nd ed., MTP Press, Lancaster, England, 1985, 62.

67. Greenblatt, D. J., Woo, E., Allen, M. D., Orsulak, P. J., and Shader, R. I., Rapid recovery from massive diazepam overdose. *J.A.M.A.*, 240, 1872, 1978.

68. Anholt, R. R. H., Mitochondrial benzodiazepine receptors as potential modulators of intermediary metabolism. *Trends Pharmacol. Sci.*, 7, 506, 1986.

69. Antkiewicz-Michaluk, L., Guidotti, A., and Kreuger, K. E., Molecular characterization and mitochondrial density of a recognition site for peripheral-type benzodiazepine ligands. *Mol. Pharmacol.*, 34, 272, 1988.

70. Basile, A. S. and Skolnick, P., Subcellular localization of "peripheral-type" binding sites for benzodiazepines in rat brain. *J. Neurochem.*, 46, 305, 1986.

71. Benavides, J., Quarteronet, D., Imbault, F., Malgouris, C., Uzan, A., Renault, C., Dubroeucq, M. C., Gueremy, C., and Le Fur, G., Peripheral-type benzodiazepine binding sites in rat adrenals: binding studies with [^{3}H]PK11195 and autoradiographic localization. *Arch. Int. Pharmacodyn.*, 266, 38, 1983.

72. Davies, L. P. and Huston, V., Peripheral benzodiazepine binding sites in heart and their interaction with pyridamole. *Eur. J. Pharmacol.*, 73, 209, 1981.

73. DeSouza, E. B., Anholt, R. R. H., Murphy, K. M. M., Snyder, S. H., and Kuhar, M. J., Peripheral-type benzodiazepine receptors in endocrine organs: autoradiographic localization in rat pituitary, adrenal and testes. *Endocrinology*, 126, 567, 1985.

74. Fares, F., Bar-Ami, S., Brandes, J. M., and Gavish, M., Gonadotropin- and estrogen-induced increase of peripheral-type benzodiazepine binding sites in the hypophyseal-genital axis of rats. *Eur. J. Pharmacol.*, 133, 97, 1987.

75. Benavides, J., Malgouris, C., Imbault, F., Begassat, F., Uzan, A., Renault, C., Dubroeucq, M. C., Gueremy, C., and Le Fur, G., Labeling of "peripheral-type" benzodiazepine binding sites in the rat brain using [^{3}H]PK11195, an isoquinoline carboxamide derivative: kinetic studies and autoradiographic localization. *J. Neurochem.*, 41, 1744, 1983.

76. Doble, A., Malgouris, C., Daniel, M., Daniel, N., Imbault, T., Basbaum, A., Uzan, A., Gueremy, C., and Le Fur, G., Labeling of peripheral-type benzodiazepine binding sites in human with [^{3}H]PK11195: anatomical and subcellular distribution. *Brain Res. Bull.*, 18, 49, 1987.

77. McCabe, R. T. and Wamsley, J. K., Autoradiographic localization of subcomponents of the macromolecular GABA receptor complex. *Life Sci.*, 39, 1937, 1986.

78. Ticku, M. K., Benzodiazepine-GABA receptor-ionophore complex: current concepts. *Neuropharmacology*, 22, 1459, 1983.

79. Sprengel, R., Werner, P., Seeburg, P. H., Mukhin, A. G., Santi, M. R., Grayson, D. R., Guidotti, A., and Krueger, K. E., Molecular cloning and expression of cDNA encoding a peripheral-type benzodiazepine receptor. *J. Biol. Chem.*, 264, 20415, 1989.

80. Krueger, K. E., Mukhin, A. G., Antkiewicz-Michaluk, L., Santi, M. R., Grayson, D. R., Guidotti, A., Sprengel, R., Werner, P., and Seeburg, P. H., Purification, cloning, and expression of a peripheral-type benzodiazepine receptor, in *GABA and Benzodiazepine Receptor Subtypes*, Biggio, G. and Costa, E., Eds., Raven Press, New York, 1990, 1.

81. Riond, J., Mattei, M. G., Kaghad, M., Dumont, X., Guillemot, J. C., Le Fur, G., Caput, C., and Ferrara, P., Molecular cloning and chromosomal localization of a human peripheral-type benzodiazepine receptor. *Eur. J. Biochem.*, 195, 305, 1991.

82. Papadopoulos, V., Mukhin, A. G., Costa, E., and Krueger, K. E., The peripheral-type benzodiazepine receptor is functionally linked to Leydig cell steroidogenesis. *J. Biol. Chem.*, 265, 3772, 1990.

83. Papadopoulos, V., Berkovich, A., and Krueger, K. E., The role of diazepam binding inhibitor and its processing products at mitochondrial benzodiazepine receptors: regulation of steroid biosynthesis. *Neuropharmacology*, 30, 1417, 1991.

84. Amsterdam, A. and Suh, B.-S., An inducible functional peripheral benzodiazepine receptor in mitochondria of steroidogenic granulosa cells. *Endocrinology*, 129, 503, 1991.

85. Mukhin, A. G., Papadopoulos, V., Costa, E., and Krueger, K. E., Mitochondrial benzodiazepine receptors regulate steroid biosynthesis. *Proc. Natl. Acad. Sci. U.S.A.*, 86, 9813, 1989.

86. Yanagibashi, K., Ohno, Y., Nakamichi, N., Matsui, T., Hayashida, K., Takamura, M., Yamada, K., Tou, S., and Kawamura, M., Peripheral-type benzodiazepine receptors are involved in the regulation of cholesterol side chain cleavage in adrenocortical mitochondria. *J. Biochem.*, 106, 1026, 1989.

87. Holmes, P. V. and Drugan, R. C., Stress-induced regulation of the renal peripheral benzodiazepine receptor: possible role of the renin-angiotensin system. *Psychoneuroendocrinology*, 19, 43, 1994.

88. Ferrarese, C., Mennini, T., Pecora, N., Pierpaoli, C., Frigo, M., Marzorati, C., Gobbi, M., Bizzi, A., Codegoni, A., Garattini, S., and Frattoloa, L., Diazepam binding inhibitor (DBI) increases after acute stress in rat. *Neuropharmacology*, 30, 1445, 1991.

89. Holmes, P. V., Stringer, A. P., and Drugan, R. C., Impact of psychological dynamics of stress on the peripheral benzodiazepine receptor. *Pharmacol. Biochem. Behav.*, 42, 437, 1992.

90. Holmes, P. V. and Drugan, R. C., Amygdaloid central nucleus lesions and cholinergic blockade attenuate the response of the renal peripheral benzodiazepine receptor to stress. *Brain Res.*, 621, 1, 1993.

91. Philip, B. K., Flumazenil: the benzodiazepine antagonist, in *New Drugs in Anesthesia*, Ornstein, E., Ed., W. B. Saunders, Philadelphia, 1993, 303.

92. Klotz, U. and Kanto, J., Pharmacokinetics and clinical use of flumazenil. *Clin. Pharmacokinet.*, 14, 1, 1988.

93. Möhler, H. and Richards, J. G., The benzodiazepine receptor: a pharmacologic control element of brain function. *Eur. J. Anaesthesiol.*, Suppl. 2, 15, 1988.

94. Amrein, R., Hetzel, W., Bonetti, E. P., and Gerecke, M., Clinical pharmacology of Dormicum (midazolam) and Anexate (flumazenil). *Resuscitation*, 16, S5, 1988.

95. Amrein, R. and Hetzel, W., Pharmacology of Dormicum (midazolam) and Anexate (flumazenil). *Acta Anaesthesiol. Scand.*, 34, Suppl. 92, 6, 1990.

96. Persson, A., Ehrin, E., Eriksson, L., Farde, L., Hedstrom, C. G., Witton, J. E., Mindus, P., and Sedvall, G., Imaging of [^{11}C-] labeled Ro15-1788 binding to benzodiazepine receptors in the humain brain by positron emission tomography. *J. Psychiatr. Res.*, 19, 609, 1985.

97. Gross, J. B., Weller, R. S., and Conard, P., Flumazenil antagonism of midazolam-induced ventilatory depression. *Anesthesiology*, 75, 179, 1991.

98. Geller, E., Niv, D., Nevo, Y., Leykin, Y., Sorkin, P., and Rudick, V., Early clinical experience in reversing benzodiazepine sedation with flumazenil after short procedures. *Resuscitation*, 16, S49, 1988.

99. Pearson, R. C., McCloy, R. F., Morris, P., and Bardhan, K. D., Midazolam and flumazenil in gastroenterology. *Acta Anaesthesiol. Scand.*, 34, Suppl. 92, 21, 1990.

100. Hommer, D., Weingartner, H., and Breier, A., Dissociation of benzodiazepine-induced amnesia from sedation by flumazenil pre-treatment. *Psychopharmacology*, 112, 455, 1993.

101. Schulte, J. and Kochs, E., Midazolam and flumazenil in neuroanaesthesia. *Acta Anaesthesiol. Scand.*, 34, Suppl. 92, 96, 1990.

102. Höjer, J., Baehrendtz, S., Matell, G., and Gustafsson, L. L., Diagnostic utility of flumazenil in coma with suspected poisoning: a double-blind, randomized controlled study. *Br. Med. J.*, 301, 1308, 1990.

103. Clausen, T. G., Wolff, J., Carl, P., and Theilgaard, A., The effect of the benzodiazepine antagonist flumazenil on psychometric performance in acute ethanol intoxication in man. *Eur. J. Clin. Pharmacol.*, 38, 233, 1990.

104. Rothstein, J. D., Garland, W., Puia, G., Guidotti, A., Weber, R. J., and Costa, E., Purification and characterization of naturally occurring benzodiazepine receptor ligands in rat and human brain. *J. Neurochem.*, 58, 2102, 1992.

105. Pena, C., Medina, J. H., Novas, M. L., Paladini, A. C., De Robertis, E., Isolation and identification in bovine cerebral cortex of n-butyl beta carboline-3-carboxylate, a potent benzodiazepine binding inhibitor. *Proc. Natl. Acad. Sci. U.S.A.*, 83, 4952, 1986.

106. Bailey, L., Ward, M., and Musa, M. N., Clinical pharmacokinetics of benzodiazepines. *J. Clin. Pharmacol.*, 34, 804, 1994.

107. Tyrer, P., Pharmacological differences between benzodiazepines, in *Benzodiazepine Dependence*, Hallstrom, C., Ed., Oxford University Press, London, 1993, 124.

108. Fontaine, R., Chouinard, G., and Annable, L., Rebound anxiety in anxious patients after abrupt withdrawal of benzodiazepine treatment. *Am. J. Psychiatry*, 141, 848, 1984.

109. Noyes, R., Jr., Garvey, M. J., Cook, B. L., and Perry, P. J., Benzodiazepine withdrawal: a review of the evidence. *J. Clin. Psychiatry*, 49, 382, 1988.

110. Lader, M., Benzodiazepines: a risk-benefit profile. *CNS Drugs*, 1, 377, 1994.

111. Garvey, M. J., Panic disorder: guidelines to safe use of benzodiazepines. *Geriatrics*, 48, 49, 1993.

112. Ghoneim, M. M. and Mewaldt, S. P., Benzodiazepines and human memory: a review. *Anesthesiology*, 72, 926, 1990.

113. Weingartner, H. J., Joyce, E. M., Sirocco, K. Y., and Adams, C. M., Specific memory and sedative effects of the benzodiazepine triazolam. *J. Clin. Pharmacol.*, 7, 305, 1993.

114. Danion, J. M., Weingartner, H., File, S. E., and Jaffard, R., Pharmacology of human memory and cognition: illustrations from the effects of benzodiazepines and cholinergic drugs. *J. Clin. Pharmacol.*, 7, 371, 1993.

115. Barbee, J. G., Memory, benzodiazepines, and anxiety: integration of theoretical and clinical perspectives. *J. Clin. Psychiatry*, 54, 86, 1993.

116. Juergens, S. M., Problems with benzodiazepines in elderly patients. *Mayo Clin. Proc.*, 68, 818, 1993.

117. Marriott, S. and Tyrer, P., Benzodiazepine dependence. Avoidance and withdrawal. *Drug Safety*, 9, 93, 1993.

118. Weiershausen, U., Pharmacokinetic considerations in the treatment of chronic anxiety, in *The Benzodiazepines: Current Standards for Medical Practice*, Smith, D. E. and Wesson, D. R., Eds., MTP Press, Lancaster, England, 1985, 59.

119. Woods, J. H., Katz, J. L., and Winger, G., Use and abuse of benzodiazepines: issues relevant to prescribing. *J.A.M.A.*, 260, 3476, 1988.

120. Greenblatt, D. J., Shader, R. I., and Abernethy, D. R., Current status of benzodiazepines (second of two parts). *N. Engl. J. Med.*, 309, 410, 1983.

121. Busto, U., Isaac, P., and Adrian, M., Psychotropic drug utilization in Canada: 1978–1984 (abstract). *Acta Pharmacol. Toxicol. (Copenhagen)*, 5(Suppl.), 203, 1986.

122. Rosenbaum, J. and Gelenberg, A. J., Anxiety, in *The Practitioner's Guide to Psychoactive Drugs*, 3rd ed., Gelenberg, A. J., Bassuk, E. L., and Schoonover, S. C., Eds., Plenum Press, New York, 1991, 179.

123. Rickels, K., The clinical use of hypnotics: indications for use and the need for a variety of hypnotics. *Acta Psychiatr. Scand.*, 74 (Suppl.), 132, 1986.

124. Gillin, J. C. and Byerley, W. F., The diagnosis and management of insomnia. *N. Engl. J. Med.*, 322, 239, 1990.

125. Babbini, M., Gaiardi, M., and Bartoletti, M., Anxiolytic versus sedative properties in the benzodiazepine series: differences in structure-activity relationships. *Life sci.*, 5, 15, 1979.

126. Ongini, E. and Barnett, A., Hypnotic specificity of benzodiazepines. *Clin. Neuropharmacol.*, 8 (Suppl.), 517, 1985.

127. Kupfer, D. J., The use of benzodiazepine hypnotics: a scientific examination of a clinical controversy. *J. Clin. Psychiatry*, 53 (Suppl. 12), 84, 1992.

128. Roehrs, T., Kribbs, N., Zorick, R., and Roth, T., Hypnotic residual effects of benzodiazepines with repeated administration. *Sleep*, 9, 309, 1986.

129. Kales, A., Soldatos, C. R., and Vela-Bueno, A., Clinical comparison of benzodiazepine hypnotics with short and long elimination half-lives, in *The Benzodiazepines: Current Standards for Medical Practice*, Smith, D. E. and Wesson, D. R., Eds., MTP Press, Lancaster, England, 1985, 121.

130. Kales, A., Bixler, E. O., Vela-Bueno, A., Soldatos, C. R., and Manfredi, R. L., Alprazolam: effects on sleep and withdrawal phenomena. *J. Clin. Pharmacol.*, 27, 508, 1987.

131. Gillin, J. C., Spinweber, C. L., and Johnson, L. C., Rebound insomnia: a critical review. *J. Clin. Psychopharmacol.*, 9, 161, 1989.

132. Kales, A., Soldatos, C. R., Bixler, E. O., and Kales, J. D., Early morning insomnia with rapidly eliminated benzodiazepines. *Science*, 220, 95, 1983.

133. Morgan, K. and Oswald, L., Anxiety caused by a short half-life hypnotic. *Br. Med. J.*, 284, 942, 1982.

134. Mellman, T. A. and Uhde, T. W., Withdrawal syndrome with gradual tapering of alprazolam. *Am. J. Psychiatry*, 143, 1464, 1986.

135. Chouinard, G., Rebound anxiety: incidence and relationship to subjective cognitive impairment. *J. Clin. Psychiatry Monogr.*, 4, 12, 1986.

136. Chouinard, G., Additional comments on benzodiazepine withdrawal. *C.M.A.J.*, 139, 119, 1988.

137. Woods, J. H., Katz, J. L., and Winger, G., Benzodiazepines: use, abuse and consequences. IV. Adverse behavioral consequences of benzodiazepine use. *Pharmacol. Rev.*, 44, 207, 1992.

138. Spinweber, C. L. and Johnson, L. C., Effects of triazolam (0.5 mg) on sleep, performance, memory, and arousal threshold. *Psychopharmacology*, 76, 5, 1982.

139. Spinweber, C. L., Johnson, L. C., and Webb, S. C., Triazolam (0.25 and 0.5 mg): effects on memory, performance, and subjective mood. *Sleep Res.,* 14, 60, 1985.

140. Bixler, E. O., Kales, A., Manfredi, R. L., Vgontzas, A. N., Tyson, K. L., and Kales, J. D., Next-day memory impairment with triazolam use (letter). *Lancet,* 337, 827, 1991.

141. Weingartner, H. J., Eckardt, M. J., Hommer, D. W., Mendelson, W., and Wolkowitz, O. M., Specificity of memory impairments with triazolam use. *Lancet,* 338, 883, 1991.

142. Scharf, M. B. and Jacoby, I. A., Lorazepam — efficacy, side effects, and rebound phenomena. *Clin. Pharmacol. Ther.,* 31, 175, 1982.

143. Healey, M., Pickens, R., Meisch, R., and McKenna, T., Effects of clorazepate, diazepam, lorazepam and placebo on human memory. *J. Clin. Psychiatry,* 44, 436, 1983.

144. Mac, D. S., Kumar, R., and Goodwin, D. W., Anterograde amnesia with oral lorazepam. *J. Clin. Psychiatry,* 46, 137, 1985.

145. Scharf, M. B., Khosla, N., Lysaght, R., and Scharf, S., Anterograde amnesia with oral lorazepam. *J. Clin. Psychiatry,* 44, 362, 1983.

146. Lister, R. G. and File, S. E., The nature of lorazepam-induced amnesia. *Psychopharmacology,* 83, 183, 1984.

147. Scharf, M. B., Khosla, N., Brocker, N., and Goff, P., Differential amnestic properties of short- and long-acting benzodiazepines. *J. Clin. Psychiatry,* 45, 51, 1984.

148. Scharf, M. B., Hirschowitz, J., Woods, M., and Scharf, S., Lack of amnestic effects of clorazepate on geriatric recall. *J. Clin. Psychiatry,* 46, 518, 1985.

149. Kirk, T., Roache, J. D., and Griffiths, R. R., Dose-response evaluation of the amnestic effects of triazolam and pentobarbital in normal subjects. *J. Clin. Psychopharmacol.,* 10, 160, 1990.

150. Scharf, M. B., Fletcher, K., and Graham, J. P., Comparative amnestic effects of benzodiazepine hypnotic agents. *J. Clin. Psychiatry,* 49, 134, 1988.

151. Brown, J. and Lewis, V., A comparison between transient amnesias induced by two drugs (diazepam or lorazepam) and amnesia of organic origin. *Neuropsychologia,* 20, 55, 1982.

152. Morris, H. H. and Estes, M. L., Traveler's amnesia: transient global amnesia secondary to triazolam. *J.A.M.A.,* 258, 945, 1987.

153. Peedicayil, J., Abraham, A., and Thomas, M., The effect of diazepam on memory in a group of patients with anxiety neurosis. *Curr. Ther. Res. Clin. Exp.,* 44, 385, 1988.

154. Golombok, S., Moodley, P., and Lader, M., Cognitive impairment in long-term benzodiazepine users. *Psychol. Med.,* 18, 365, 1988.

155. Lucki, I. and Rickels, K., The effect of anxiolytic drugs on memory in anxious subjects. *Psychopharmacol. Bull.,* 6, 128, 1988.

156. Annitto, W. J., Alcoholics use of benzodiazepines (letter). *Am. J. Psychiatry,* 145, 683, 1989.

157. Ciraulo, D. A., Sands, B. F., and Shader, R. I., Dr Ciraulo and associates reply (letter). *Am. J. Psychiatry,* 145, 684, 1989.

158. Sellers, E. M., Ciraulo, D. A., DuPont, R. L., Griffiths, R. R., Kosten, T. R., Romach, M. K., and Woody, G. E., Alprazolam and benzodiazepine dependence. *J. Clin. Psychiatry,* 54, (Suppl.), 64, 1993.

159. Busto, A. and Sellers, E. M., Pharmacokinetic determinants of drug abuse and dependence: A conceptual perspective. *Clin. Pharmacokinet.,* 11, 144, 1986.

160. Griffiths, R. R. and Roache, J. D., Abuse liability of benzodiazepines: a review of human studies evaluating subjective and/or reinforcing effects, in *The Benzodiazepines: Current Standards for Medical Practice,* Smith, D. E. and Wesson, D. R., Eds., MTP Press, Lancaster, England 1985, 209.

161. Ciraulo, D. A., Barnhill, J. G., Greenblatt, D. J., Shader, R. I., and Ciraulo, A. M., Abuse liability and clinical pharmacokinetics of alprazolam in alcoholic men. *J. Clin. Psychiatry,* 49, 333, 1988.

162. Jaffe, J. H., Ciraulo, D. A., Nies, A., Dixon, R. B., and Monroe, L. L., Abuse potential of halazepam and of diazepam in patients recently treated for alcohol withdrawal. *Clin. Pharmacol. Ther.,* 34, 623, 1983.

163. Saletu, B., Anderer, P., Brandstätter, N., Frey, R., Grünberger, J., Klosch, G., Mandl, M., Wetter, T., and Zeitlhofer, J., Insomnia in generalized anxiety disorder: polysomnographic, psychometric and clinical investigations before, during and after therapy with a long versus a short half-life benzodiazepine (quazepam vs. triazolam). *Neuropsychobiology,* 29, 69, 1994.

164. Rummans, T. A., Davis, L. J., Morse, R. M., and Ivnik, R. J., Learning and memory impairment in older, detoxified, benzodiazepine-dependent patients. *Mayo Clin. Proc.,* 68, 731, 1993.

165. Balter, M. B., Manheimer, D. I., Mellinger, G. D., and Uhlenhuth, E. H., A cross-national comparison of anti-anxiety/sedative drug use. *Curr. Med. Res. Opin.,* 8, Suppl. 4, 5, 1984.

166. Shorr, R. I., Bauwens, S. F., and Landefeld, C. S., Failure to limit quantities of benzodiazepine hypnotic drugs for outpatients: placing the elderly at risk. *Am. J. Med.,* 89, 725, 1990.

167. Cooper, J. R., Bloom, F. E., and Roth, R. H., *The Biochemical Basis of Neuropharmacology,* 6th ed., Oxford University Press, New York, 1991, 190.

168. Rektor, I., Anticholinergic action of benzodiazepines. *Eur. Bull. Cogn. Psychol.,* 12, 547, 1992.

169. Suhara, T., Inoue, O., Kobayashi, K., Satoh, T., and Tateno, Y., An acute effect of triazolam on muscarinic cholinergic receptor binding in the human brain measured by positron emission tomography. *Psychopharmacology*, 113, 311, 1994.

170. Cooks, P. J., Benzodiazepines hypnotics in the elderly. *Acta Psychiatr. Scand.*, 74, Suppl. 332, 149, 1986.

171. Meyer, B. R., Benzodiazepines in the elderly. *Med. Clin. North Am.*, 66, 1017, 1982.

172. Roth, J. A., Drug metabolism, in *Textbook of Pharmacology*, Smith, C. M. and Reynard, A. M., Eds., W. B. Saunders, Philadelphia, 1992, 42.

173. Salzman, C., Shader, R. I., Greenblatt, D. J., and Harmatz, J. S., Long versus short half-life benzodiazepines in the elderly. *Arch. Gen. Psychiatry*, 40, 293, 1983.

174. Tinetti, M. E., Speechley, M., and Ginter, S. F., Risk factors for falls among elderly persons living in the community. *N. Engl. J. Med.*, 319, 1701, 1988.

175. Ray, W. A., Griffin, M. R., Schaffner, W., Baugh, D. K., and Melton, L. J., Psychotropic drug use and the risk of hip fracture. *N. Engl. J. Med.*, 316, 363, 1987.

176. Ray, W. A., Griffin, M. R., Downey, W., Benzodiazepines of long and short elimination half-life and the risk of hip fracture. *J.A.M.A.*, 262, 3303, 1989.

177. Ben-Shachar, D., Finberg, J. P. M., and Youdim, M. B. H., Effect of iron chelators on dopamine D2 receptors. *J. Neurochem.*, 45, 999, 1985.

178. Ben-Shachar, D. and Youdim, M. B. H., Neuroleptic induced dopamine receptor supersensitivity and tardive dyskinesia may involve altered brain iron metabolism. *Br. J. Pharmacol.*, 90 (Suppl.), 95, 1987.

179. Harrison, J. E. and McNeill, K. G., The skeletal system, in *Nutrition and Metabolism in Patient Care*, Kinney, J. M., Jeejeebhoy, K. N., Hill, G. L., and Owen, O. E., Eds., W. B. Saunders, Philadelphia, 1988, 701.

180. Chouinard, G., Annable, L., Fontaine, R., and Solyom, L., Alprazolam in the treatment of generalized anxiety and panic disorders: a double-blind, placebo-controlled study. *Psychopharmacology*, 77, 229, 1982.

181. Sheehan, D. V., Current views on the treatment of panic and phobia disorders. *Drug Ther. Hosp.*, 7, 74, 1982.

182. Liebowitz, M. R., Fyer, A. J., Gorman, J. M., Campeas, R., Levin, A., Davies, S. R., Goetz, D., and Klein, D. F., Alprazolam in the treatment of panic disorders. *J. Clin. Psychopharmacol.*, 6, 13, 1986.

183. Ballenger, J. C., Burrows, G. D., DuPont, R. L., Jr., Lesser, I. M., Noyes, R., Jr., Pecknold, J. C., Rifkin, A., and Swinson, R. P., Alprazolam in panic disorder and agoraphobia: results from a multicenter trial. I. Efficacy in short-term treatment. *Arch. Gen. Psychiatry*, 45, 413, 1988.

184. Chouinard, G., Labonté, A., Fontaine, R., and Annable, L., New concepts in benzodiazepine therapy: rebound anxiety and new indications for more potent benzodiazepines. *Prog. Neuropsychopharmacol. Biol. Psychiatry*, 7, 669, 1983.

185. Fontaine, R. and Chouinard, G., Antipanic effect of clonazepam (letter). *Am. J. Psychiatry*, 141, 149, 1984.

186. Beaudry, P., Fontaine, R., Chouinard, G., and Annable, L., Clonazepam in the treatment of patients with recurrent panic attacks. *J. Clin. Psychiatry*, 47, 83, 1986.

187. Pollack, M. H., Tesar, G. E., Rosenbaum, J. F., and Spier, S. A., Clonazepam in the treatment of panic disorder and agoraphobia: a one year follow-up. *J. Clin. Psychopharmacol.*, 6, 302, 1986.

188. Tesar, G. E. and Rosenbaum, J. F., Successful use of clonazepam in patients with treatment-resistant panic disorder. *J. Nerv. Ment. Dis.*, 174, 473, 1986.

189. Tesar, G. E., Rosenbaum, J. F., Pollack, M. H., Herman, J. B., Sachs, G. S., Mahoney, E. M., Cohen, L. S., McNamara, M., and Goldstein, S., Clonazepam versus alprazolam in the treatment of panic disorder: interim analysis of data from a prospective, double-blind, placebo-controlled trial. *J. Clin. Psychiatry*, 48, 16S, 1987.

190. Annable, L., Beauclair, L., Fontaine, R., Holobow, N., and Chouinard, G., Clonazepam in panic disorder. *VIII World Congress of Psychiatry, Book of Abstracts*, 422, 116, 1989.

191. Herman, J. B., Rosenbaum, J. F., and Brotman, A. W., The alprazolam to clonazepam switch for the treatment of panic disorder. *J. Clin. Psychopharmacol.*, 7, 175, 1987.

192. Chouinard, G. and Landry, P., New drugs for the treatment of generalized anxiety, panic and obsessive-compulsive disorders, in *Handbook of Anxiety, Vol. 4: The Treatment of Anxiety*, Noyes, R., Jr., Roth, M., and Burrows, G. D., Eds., Elsevier, New York, 1990, 271.

193. Beauclair, L., Fontaine, R., Annable, L., Holobow, N., and Chouinard, G., Clonazepam in the treatment of panic disorder: a double-blind, placebo-controlled trial investigating the correlation between clonazepam concentrations in plasma and clinical response. *J. Clin. Psychopharmacol.*, 14, 111, 1994.

194. Schatzberg, A. F. and Cole, J. O., Benzodiazepines in depressive disorders. *Arch. Gen. Psychiatry*, 35, 1359, 1978.

195. Cassano, G. B., Castrogiovanni, P., and Conti, L., Drug responses in different anxiety states under benzodiazepine treatment. Some multivariate analyses for the evaluation of "rating scale for depression" scores, in *The Benzodiazepines*, Garattini, S., Mussini, E., and Randall, L. O., Eds., Raven Press, New York, 1973, 379.

196. Fabre, L. F., Pilot open-label study with alprazolam (U-31.889) in outpatients with neurotic depression. *Curr. Ther. Res.*, 19, 661, 1976.

197. Rickels, K., Feighner, J. P., and Smith, W. T., Alprazolam, amitriptyline, doxepin and placebo in the treatment of depression. *Arch. Gen. Psychiatry*, 42, 134, 1985.

198. Rickels, K., Chung, H. R., Csanalosi, I. B., Hurowitz, A. M., London, J., Wiseman, K., Kaplan, M., and Amsterdam, J. D., Alprazolam, diazepam, imipramine and placebo in outpatients with major depression. *Arch. Gen. Psychiatry*, 44, 862, 1987.

199. Fontaine, R., Annable, L., Chouinard, G., and Ogilvie, R., Bromazepam and diazepam in generalized anxiety: a placebo-controlled study with measurement of drug plasma concentrations. *J. Clin. Psychopharmacol.*, 3, 80, 1983.

200. Rickels, K., Pereira-Ogan, J. A., Chung, H. R., Gordon, P. E., and Landis, W. B., Bromazepam and phenobarbital in anxiety: a controlled study. *Curr. Ther. Res.*, 15, 679, 1973.

201. Shammas, E., Controlled comparison of bromazepam, amitriptyline, and placebo in anxiety-depressive neurosis. *Dis. Nerv. Syst.*, 38, 201, 1977.

202. Lydiard, R. B., Laraia, M. T., Ballenger, J. C., and Howell, E. F., Emergence of depressive symptoms in patients receiving alprazolam for panic disorder. *Am. J. Psychiatry*, 44, 664, 1987.

203. Aronson, T. A., Treatment-emergent depression with antidepressants in panic disorder. *Compr. Psychiatry*, 30, 267, 1989.

204. Kishimoto, A., Kamata, K., Sugihara, T., Ishiguro, S., Hazama, H., Mizukawa, R., and Kunimoto, N., Treatment of depression with clonazepam. *Acta Psychiatr. Scand.*, 77, 81, 1988.

205. Cohen, L. S. and Rosenbaum, J. F., Clonazepam: new uses and potential problems. *J. Clin. Psychiatry*, 48, S50, 1987.

206. Goldberg, S. C., Ettigi, P., Schulz, P. M., Hamer, R. M., Hayes, P. E., and Friedel, R. O., Alprazolam versus imipramine in depressed out-patients with neurovegetative signs. *J. Affective Disord.*, 11, 139, 1986.

207. Nemeroff, C. B., Paroxetine: an overview of the efficacy and safety of a new selective serotonin reuptake inhibitor in the treatment of depression. *J. Clin. Psychopharmacol.*, 13, 10S, 1993.

208. Preskorn, S., Targeted pharmacotherapy in depression management: comparative pharmacokinetics of fluoxetine, paroxetine and sertraline. *Int. Clin. Psychopharmacol.*, 9, S13, 1994.

209. Song, F., Freemantle, N., Sheldon, T. A., House, A., Watson, P., Long, A., and Mason, J., Selective serotonin reuptake inhibitors: meta-analysis of efficacy and acceptability. *Br. Med. J.*, 306, 683, 1993.

210. Sethy, V. H. and Hodges, D. H., Role of β-adrenergic receptors in the antidepressant activity of alprazolam. *Res. Commun. Chem. Pathol. Pharmacol.*, 36, 329, 1982.

211. Sethy, V. H. and Hodges, D. H., Antidepressant activity of alprazolam in a reserpine-induced model of depression. *Drug Dev. Res.*, 5, 179, 1985.

212. Dunner, D., Myers, J., Khan, A., Avery, D., Iskiki, D., and Pyke, R., Adinazolam — a new antidepressant: findings of a placebo-controlled, double-blind study in outpatients with major depression. *J. Clin. Psychopharmacol.*, 7, 170, 1987.

213. Feighner, J. P., Aden, G. C., Fabre, L. F., Rickels, K., and Smith, W. T., Comparison of alprazolam, imipramine and placebo in the treatment of depression. *J.A.M.A.*, 249, 3057, 1983.

214. Overall, J. E., Donachie, N. D., and Faillace, L. A., Implications of restrictive diagnosis for compliance to antidepressant drug therapy: alprazolam versus imipramine. *J. Clin. Psychiatry*, 48, 51, 1987.

215. Overall, J. E., Biggs, J., Jacobs, M., and Holden, K., Comparison of alprazolam and imipramine for treatment of outpatient depression. *J. Clin. Psychiatry*, 48, 15, 1987.

216. Rush, A. J., Erman, M. K., Schlesser, M. A., Roffwarg, H. P., Vasavada, N., Khatami, M., Fairchild, C., and Giles, D. E., Alprazolam versus amitriptyline in depressions with reduced REM latencies. *Arch. Gen. Psychiatry*, 42, 1154, 1985.

217. Ansseau, M., Ansoms, C., Beckers, G., Bogaerts, M., Botte, L., De Buck, R., Diricq, S., Dumortier, A., Jansgers, E., and Owieczka, J., Double-blind clinical study comparing alprazolam and doxepin in primary unipolar depression. *J. Affective Disord.*, 7, 287, 1984.

218. Fawcett, J., Edwards, J. H., Kravitz, H. M., and Jeffriess, H., Alprazolam: an antidepressant? Alprazolam, desipramine, and an alprazolam-desipramine combination in the treatment of adult depressed outpatients. *J. Clin. Psychopharmacol.*, 7, 95, 1987.

219. Otero, F. J., Hernandez-Herrero, C., Martinez-Arevalo, M. J., Garrido, J., Armenteros, S., and Velasco, J., Fluoxetine/bentazepam combination in the treatment of dysthymic disorders. *Curr. Ther. Res.*, 55, 519, 1994.

220. Fux, M., Taub, M., and Zohar, J., Emergence of depressive symptoms during treatment for panic disorder with specific 5-hydroxytryptophan reuptake inhibitors. *Acta Psychiatr. Scand.*, 88, 235, 1993.

221. Rapoport, M. and Braff, D. L., Alprazolam and hostility (letter). *Am. J. Psychiatry,* 141, 146, 1985.

222. Rosenbaum, J. F., Wood, S. W., Groves, J. E., and Klerman, J. L., Emergence of hostility during alprazolam treatment. *Am. J. Psychiatry,* 141, 792, 1984.

223. Strahan, A., Rosenthal, J., Kaswan, M., and Winston, A., Three case reports of acute paraoxysmal excitement associated with alprazolam treatment. *Am. J. Psychiatry,* 142, 859, 1985.

224. Aranam G. W., Pearlmanm, C., and Shader, R. I., Alprazolam-induced mania: two clinical cases. *Am. J. Psychiatry,* 142, 368, 1985.

225. France, R. D. and Krishnan, K. R. R., Alprazolam-induced manic reaction (letter). *Am. J. Psychiatry,* 141, 1127, 1984.

226. Pecknold, J. C. and Fleury, D., Alprazolam-induced manic episode in two patients with panic disorder. *Am. J. Psychiatry,* 143, 652, 1986.

227. Mayerhoff, D., Vital-Herne, J., Lesser, M., and Brenner, R., Alprazolam-induced manic reaction. *N.Y. State J. Med.,* 86, 320, 1986.

228. Goodman, W. K. and Charney, D. J., A case of alprazolam, but not lorazepam, inducing manic symptoms. *J. Clin. Psychiatry,* 48, 117, 1987.

229. Herman, J. B., Brotman, A. W., and Rosenbaum, J. F., Rebound anxiety in panic disorder patients treated with short-acting benzodiazepines. *J. Clin. Psychiatry,* 48, S22, 1987.

230. Noyes, R., Jr., Clancy, J., Coryell, W. H., Crowe, R. R., Chaudhry, D. R., and Domingo, D. V., A withdrawal syndrome after abrupt discontinuation of alprazolam. *Am. J. Psychiatry,* 142, 114, 1985.

231. Prien, R. F., Caffey, E. M., and Klett, C. J., Comparison of lithium carbonate and chlorpromazine in the treatment of mania. *Arch. Gen. Psychiatry,* 26, 146, 1972.

232. Takahashi, R., Sakuma, A., Itoh, K., Itoh, H., and Kurihara, M., Comparison of efficacy of lithium carbonate and chlorpromazine in mania. *Arch. Gen. Psychiatry,* 32, 1310, 1975.

233. Dunner, D. L. and Clayton, P. J., Drug treatment of bipolar disorder, in *Psychopharmacology: The Third Generation of Progress,* Meltzer, H. Y., Ed., Raven Press, New York, 1987, 1077.

234. Dubin, W. R., Rapid tranquilization: antipsychotics or benzodiazepines? *J. Clin. Psychiatry,* 49, S5, 1988.

235. Goldney, R. D. and Spence, N. D., Safety of the combination of lithium and neuroleptic drugs. *Am. J. Psychiatry,* 143, 882, 1986.

236. Lenox, R. H., Newhouse, P. A., Creelman, W. L., and Whitaker, T. M., Adjunctive treatment of manic agitation with lorazepam versus haloperidol: a double-blind study. *J. Clin. Psychiatry,* 53, 47, 1992.

237. Dilsaver, S. C., Swann, A. C., Shoaib, A. M., Bowers, T. C., and Halle, M. I., Depressive mania associated with nonresponse to antimanic agents. *Am. J. Psychiatry,* 150, 1548, 1993.

238. Sernyak, M. J., Griffin, R. A., Johnson, R. M., Pearsall, H. R., Wexler, B. E., and Woods, S. W., Neuroleptic exposure following inpatient treatment of acute mania with lithium and neuroleptic. *Am. J. Psychiatry,* 151, 133, 1994.

239. Chouinard, G., The use of benzodiazepines in the treatment of manic-depressive illness. *J. Clin. Psychiatry,* 49, 15, 1988.

240. Chouinard, G., Clonazepam in the treatment of bipolar psychotic patients after discontinuation of neuroleptics (letter). *Am. J. Psychiatry,* 146, 1642, 1989.

241. Bodkin, J. A., Emerging uses for high-potency benzodiazepines in psychotic disorders. *J. Clin. Psychiatry,* 51, S41, 1990.

242. Sernyak, M. J. and Woods, S. W., Chronic neuroleptic use in manic-depressive illness. *Psychopharmacol. Bull.,* 29, 375, 1993.

243. Chouinard, G., Annable, L., Turnier, L., Holobow, N., and Szkrumelak, N., A double-blind randomized clinical trial of rapid tranquilization with I.M. clonazepam and I.M. haloperidol in agitated psychotic patients with manic symptoms. *Can. J. Psychiatry,* 38, S114, 1993.

244. Wolkowitz, O. M., Turetsky, N., Reus, V. I., and Hargreaves, W. A., Benzodiazepine augmentation of neuroleptics in treatment-resistant schizophrenia. *Psychopharmacol. Bull.,* 28, 291, 1992.

245. Jibiki, I. and Yamaguchi, N., Beneficial effect of the high-dose benzodiazepine derivative clotiazepam on intractable auditory hallucinations in chronic schizophrenic patients. *Acta Therapeutica,* 19, 383, 1993.

246. Ungvari, G. S., Leung, C. M., Wong, M. K., and Lau, J., Benzodiazepines in the treatment of catatonic syndrome. *Acta Psychiatr. Scand.,* 89, 285, 1994.

247. Jimerson, D. C., van Kammen, D. P., Post, R. M., Docherty, J. P., and Bunney, W. E., Jr., Diazepam in schizophrenia: a preliminary double-blind trial. *Am. J. Psychiatry,* 139, 489, 1982.

248. Arana, G. W., Ornsteen, M. L., Kanter, F., Friedman, H. L., Greenblatt, D. J., and Shader, R. I., The use of benzodiazepines for psychotic disorders — a literature review and preliminary clinical findings. *Psychopharmacol. Bull.,* 22, 77, 1986.

249. Altamura, A. C., Mauri, M. C., Mantero, M., and Brunetti, M., Clonazepam/haloperidol combination therapy in schizophrenia: a double-blind study. *Acta Psychiatr. Scand.,* 76, 702, 1987.

250. Csernansky, J. G., Riney, S. J., Lombrozo, L., Overall, J. E., and Hollister, L. S., Double-blind comparison of alprazolam, diazepam, and placebo for the treatment of negative schizophrenic symptoms. *Arch. Gen. Psychiatry*, 45, 655, 1988.

251. Wolkowitz, O. M., Breier, A., Doran, A., Kelsoe, J., Lucas, P., Paul, S. M., and Pickar, D., Alprazolam augmentation of the antipsychotic effects of fluphenazine in schizophrenic patients. *Arch. Gen. Psychiatry*, 45, 664, 1988.

252. Wolkowitz, O. M. and Pickar, D., Benzodiazepines in the treatment of schizophrenia: a review and reappraisal. *Am. J. Psychiatry*, 148, 714, 1991.

253. Enna, S. J., γ-Aminobutyric acid (GABA), pharmacology and neuropsychiatric illness, in *Psychiatry Update, Annual Review*, Vol. 4, Hales, R. E. and Frances, A. J., Eds., American Psychiatric Press, Washington, D.C., 1985, 67.

254. van Kammen, D. P., Sternberg, D. E., Hare, T. A., Waters, R. N., and Bunney, W. E., Jr., CSF levels of γ-aminobutyric acid in schizophrenia: low values in recently ill patients. *Arch. Gen. Psychiatry*, 39, 91, 1982.

255. van Kammen, D. P. and Gelertner, J., Biochemical instability in schizophrenia. II. The serotonin and γ-aminobutyric acid systems, in *Psychopharmacology: The Third Generation of Progress*, Meltzer, H. Y., Ed., Raven Press, New York, 1987, 753.

256. van Kammen, D. P., γ-Aminobutyric acid (GABA) and the dopamine hypothesis of schizophrenia. *Am. J. Psychiatry*, 134, 138, 1977.

257. van Kammen, D. P., Guidotti, A., Kelley, M. E., Gurklis, J., Guarneri, P., Gilbertson, M. W., Yao, J. K., Peters, J., and Costa, E., CSF diazepam binding inhibitor and schizophrenia: clinical and biochemical relationships. *Biol. Psychiatry*, 34, 515, 1993.

258. Costa, E. and Guidotti, A., Minireview: Diazepam binding inhibitor (DBI): a peptide with multiple biological actions. *Life Sci.*, 49, 325, 1991.

259. Alho, H., Costa, E., Ferrero, P., Fujimoto, M., Cosenza-Murphy, D., and Guidotti, A., Diazepam-binding inhibitor: a neuropeptide located in selected neuronal populations of rat brain. *Science*, 229, 179, 1985.

260. Ferrarese, C., Appollonio, I., Frigo, M., Piolti, R., Tamma, F., and Frattola, L., Distribution of a putative endogenous modulator of the GABAergic system in human brain. *Neurology*, 39, 443, 1989.

261. Goodman, W. K., McDougle, C. J., Price, L. H., Riddle, M. A., Pauls, D. L., and Leckman, J. F., Beyond the serotonin hypothesis: a role for dopamine in some forms of obsessive-compulsive disorder? *J. Clin. Psychiatry*, 51, 36, 1990.

262. McDougle, C. J., Goodman, W. K., and Price, L. H., Dopamine antagonists in tic-related and psychotic spectrum obsessive-compulsive disorder. *J. Clin. Psychiatry*, 55, S24, 1994.

263. Gawin, F. H. and Ellinwood, E. H., Jr., Cocaine and other stimulants: actions, abuse, and treatment. *N. Engl. J. Med.*, 318, 1173, 1988.

264. Sandyk, R., Dopaminergic supersensitivity factors in Tourette's syndrome: a hypothesis. *Int. J. Neurosci.*, 44, 169, 1989.

265. Golden, G. S., Tourette syndrome: recent advances. *Neurol. Clin.*, 8, 705, 1990.

266. Modell, J. G., Mountz, J. M., Curtis, G. C., and Greden, J. F., Neurophysiologic dysfunction in basal ganglia/limbic striatal and thalamocortical circuits as a pathogenetic mechanism of obsessive-compulsive disorder. *J. Neuropsychiatry*, 1, 27, 1989.

267. Elkashef, A. M., Egan, M. F., Frank, J. A., Hyde, T. M., Lewis, B. K., and Wyatt, R. J., Basal ganglia iron in tardive dyskinesia: an MRI study. *Biol. Psychiatry*, 35, 16, 1994.

268. Elkashef, A. M., Buchanan, R. W., Gellad, F., Munson, R. C., and Breier, A., Basal ganglia pathology in schizophrenia and tardive dyskinesia: an MRI quantitative study. *Am. J. Psychiatry*, 151, 752, 1994.

Chapter **13**

COGNITION-ENHANCING (NOOTROPIC) DRUGS

Manfred Windisch

CONTENTS

0-8493-8386-0/96/$0.00+$.50
© 1996 by CRC Press, Inc.

I. THE PROBLEM OF DEMENTIA AND
COGNITIVE IMPAIRMENT

A major challenge for medicine of the future is the steadily increasing life expectancy in industrialized countries. The fastest growing segment in the total population is persons aged above 65 years. Between 1990 and 2040 the U.S. elderly population is expected to grow from 31.6 to 68.1 million.[1] It can be expected that the number of individuals at risk for age-associated disorders such as dementia and Parkinson's disease will significantly increase.

A. Prevalence of Dementia and Its Economic Impact

Data from several population studies show an overall prevalence rate of 5 to 6% for moderate and severe dementia at the age of 65 and over.[2] This rate rises exponentially with age and doubles approximately every 5 years.[3] For example, the prevalence rates range from 0.7% in the age group from 60 to 65 years to approximately 38.6% in the age group of 90 to 95 years. The most frequent causes of dementia are Alzheimer's disease (AD) and vascular dementia. Independently of the age group, the contribution of AD is approximately 55 to 65%, vascular dementia 20 to 25%, and mixed forms of dementia 15 to 20%. AD seems to be predominant in most countries, with the exceptions of Japan and Russia reporting higher rates of vascular dementia, which could reflect different approaches to diagnosis in these countries.[4] Together with increasing age, a positive family history of dementia is one of the few established risk factors for developing AD.[5] There is much evidence for reduced life expectancy of patients with AD, compared with the general population. An increase in neurodegenerative disease mortality by 119 to 231% is expected in the U.S. alone, exceeding mortality increases from lung cancer, liver cirrhosis, and other diseases.[6]

From recent epidemiological studies it appears that 3.75 million individuals suffer from AD in the U.S., and this number is expected to increase up to 9 million by the year 2040.[7] Dementia imposes a substantial medical, social, and psychological burden on patients and their relatives and considerable financial burden on the community as well. The total costs for caring for a patient with AD is calculated to be approximately $47,000 U.S. per year regardless of whether the patient lives at home or in a nursing home.[8] It can be estimated that overall annual costs of caring for dementia patients can be approximately $58 billion.[8-9] These facts show the importance of developing efficient strategies for treatment of vascular dementia and particularly AD.

B. Diagnosis and Differential Diagnosis of Dementing Illnesses

Dementia is a clinical syndrome of acquired, global cognitive impairment. This syndrome includes disturbances of attention, perception, concentration, memory, thinking, and orientation, as well as changes in personality. The most widely used criteria for diagnosis of dementia are those developed by the American Psychiatric Association for the *Diagnostic and Statistical Manual of Mental Disorders* (DSM-IIIR and DSM-IV).[10] In addition, there have been fully operationalized questionnaires developed; they are known as Agecat,[11] Camdex,[12] and Sidam.[13]

Estimation of the severity of dementia is important for evaluation of treatment effects. Available scales include the Alzheimer Disease Assessment Scale (ADAS),[14] the Global Deterioration Scale (GDS),[15] and the Gottfries-Brane-Steen Geriatric-Scale (SCAG).[17] Testing procedures include the Mini Mental State Examination (MMSE),[18] the Selective Reminding Test (SRT),[19] and the Syndrome Short Test (SKT).[20]

At the present time there are no specific markers available for AD, and the diagnosis of this disease is a diagnosis of exclusion with histopathological confirmation sometimes possible post-mortem. With the help of standardized diagnostic criteria, clinical diagnosis of AD can achieve a reasonable degree of accuracy (85 to 100%) as compared with the neuropathological diagnosis.[21,22] Widely used are the criteria of the National Institute of Neurological and Communicative Disorders and Stroke–Alzheimer's Disease and Related Disorders Association (NINCDS-ADRDA) working group. For diagnosis of vascular dementia, e.g., multiinfarct dementia, scores developed by Hachinsky[23] and their modifications[24,25] have been used. Diagnostic accuracy of these instruments is only 55 to 60% in cases of multiinfarct dementia and lower in cases of mixed forms of vascular and neurodegenerative dementia. Imaging techniques, like CT, SPECT, and MRI, are valuable tools for the diagnosis of vascular lesions in the brain. However, the relationship between brain vascular lesions and cognitive deficits remains unclear.

II. PHARMACOLOGICAL APPROACH TO TREATMENT OF DEMENTIA

In very general terms, nootropic drugs are compounds that act on the central nervous system (CNS) enhancing the cognitive performance of the individual. In particular, these drugs are expected to improve memory, perception, attention, judgment, and orientation, as well as social activities of daily living (ADL). Synonyms for nootropic drugs are "cognition enhancers," "cerebral metabolic activators," "cerebroactive drugs," "antidementia drugs," "intelligence boosters," "neurodynamics," etc. The term *nootropic drug* was created by Giurgea[26] in 1972, who also developed a pharmacological characterization of this heterogeneous group of substances.

Research for developing new nootropic drugs is mostly focused on their efficacy in AD and vascular dementia. Other indications for nootropic drugs include cerebral ischemia (acute stage and rehabilitation), cerebrocranial trauma (acute stage and rehabilitation), dementia caused by alcohol and drug abuse, coma, delirium, and cognitive disturbances in children suffering from mild to moderate mental retardation or learning disabilities. It must be mentioned here that the average efficacy of nootropic drugs is usually low (10 to 20% compared with placebo). However, in a particular individual sometimes spectacular improvements can be achieved.

A. Pharmacological Profile of Nootropic Drugs

It has been thought that nootropic drugs work by "enhancing resistance of the central nervous system against damage," improving learning and memory, and increasing cortical–subcortical control function, as well as improving the information transfer in the telencephalon.[27] Improvement of cerebral microcirculation can result from metabolic activities or from effects on blood flow properties, such as increases of erythrocyte flexibility or prevention of platelet aggregation.[28] Other authors consider the ability of cognition enhancers to protect neuronal cells and to facilitate the regeneration of brain tissue following damage the most important mechanisms of action of effective nootropics.[29]

Unfortunately, most data about the mode of action of nootropics are acquired from animal experiments. Although several animal experimental models for testing cognition enhancers have been established, there has been no single animal model accurately reflecting pathological changes occurring in humans with dementia. Moreover, it is not entirely clear which types of animal behavior reflect higher cognitive functions in humans. Therefore, one must expect that the predictive validity of preclinical data on nootropic drugs is rather poor.[30]

In summary, efficacy of a given nootropic drug can be established in randomized, prospective, controlled, double-blind clinical studies. A successful cognition enhancer must significantly improve the psychopathological condition of the patient, enhance cognitive performance, and improve ADL.

B. Chemical Heterogeneity and Classification of Nootropic Drugs

One common property of nootropic drugs is their activity on higher integrative functions of the brain. From the chemical point of view, this is a highly heterogeneous group, and even substances with similar pharmacological activity are derived from different chemical substance classes. What follows is an attempt to give an overview of currently available substances.

III. PHARMACOLOGY OF NOOTROPIC DRUGS

A. Tacrine and Related Substances

There is evidence indicating that cognitive disturbances occurring in patients with AD can be due to a deficiency in cholinergic neurotransmission.[31] Biochemical findings from brains of AD patients show decreased activity of choline acetyltransferase and cholinesterase, and a decline in the number of muscarinic and nicotinic acetyl-choline receptors.[32] These changes are accompanied by a significant loss of cholinergic neurons in the basal forebrain nuclei.[33] There have been several different strategies developed aiming to increase cholinergic neurotransmission. The main approaches included acetylcholine precursor loading, direct cholinergic receptor stimulation, and cholinesterase inhibition. The most promising clinical results so far have been obtained by using anticholinesterases like physostigmine or tetrahydroaminoacridines (i.e., tacrine and velnacrine).

Tacrine was first synthesized in 1945 and was approved by the U.S. Food and Drug Administration (FDA) for treatment of AD in 1993 following encouraging results from clinical trials in AD patients that were reported in 1986.[34]

1. Pharmacological Profile

Tacrine is a competitive reversible inhibitor of tissue and plasma cholinesterases.[35] Tacrine produces an allosteric inhibition by binding to a hydrophobic region in the active zone of the enzyme. In addition, tacrine interacts with muscarinic and nicotinic receptors.[36] This interaction modulates acetylcholine release and may increase the therapeutic value of the drug. Tacrine inhibits monoamine oxidases (MAO), thus directly interfering with catecholaminergic neurotransmission.[37] It has been reported that tacrine stimulates release of the neurotransmitters noradrenalin, dopamine, and serotonin and blocks the reuptake of these substances. Effects on monoaminergic neurotransmission could be responsible for amelioration of emotional and motor disturbances found in AD patients. Effects on muscarinic receptors and the increasing cytoskeletal protein–protein interactions caused by tacrine could significantly alter neuronal structure and function in AD.[40] Pharmacokinetic investigations show rapid absorption and a significant first-path metabolism of tacrine, reducing bioavailability after oral dosage[41] and rapid uptake into the brain, where the concentration of the substance is approximately tenfold that found in the plasma.[42]

2. Clinical Effects

Most clinical trials have been performed on patients suffering from mild to moderate forms of AD. Early trials with tacrine were performed as crossover studies with comparatively small numbers of subjects and short treatment durations. The results of these trials with dosages ranging from 30 to 200 mg of tacrine per day were controversial.[43] With dosages ranging from 150 to 200 mg of tacrine per day, dramatic effects were found in 10 of 17 patients, and in 16 of 17 patients a global improvement could be demonstrated.[44] Several large-scale clinical trials, conducted at multiple sites, further confirmed the efficacy of tacrine within the observation period, which ranged from 6 to 30 weeks, and the dosage, which ranged from 20 to 160 mg of tacrine per day. In these studies large numbers of patients dropped out during the initial dose-titration phases because of intolerable side effects. The remaining patients showed moderate but significant and consistent improvement of cognition. Significant differences could be shown using the ADAS, a clinician's interview-based impression of change and information provided by caregivers.[45-46] There seems to be a close correlation between plasma concentrations of tacrine and the extent of cognitive improvement, with best responses shown by patients with plasma levels above 10 ng/ml. Based on tacrine clinical trials data, a statistical model was developed to describe the effects on the ADAS cognitive subscale.[47] It has been concluded that at a dose of tacrine of 80 mg/day, the progression of the disease is delayed by 2.99 ADAS cognitive subscale points per year, which means that the treatment can slow the progression of AD by approximately 50%. However, long-term studies of tacrine treatment have not been done so far.

3. Side Effects

Nausea or vomiting have been reported to occur in approximately 30% of the patients treated with tacrine, and the nausea could not be tolerated by approximately 3% of patients. Diarrhea and abdominal pain occur in a significant number of patients. Hepatotoxicity is the most common clinically significant side effect associated with tacrine.[46] As many as 42% of patients exposed to the drug show elevated alanine aminotransferase (AST) levels, whereas approximately 21% have elevation in these levels to at least three times the upper limit of normal. A rise in ALT level is approximately twice as common in women as compared with men. The rise in aminotransferase levels usually occurs after the first 7 weeks of exposure to tacrine; maximal effects are found after 10 weeks. Liver enzymes usually return to normal levels after a mean of 5 weeks of drug withdrawal. The hepatotoxicity of tacrine appears to be dose dependent.

B. Co-Dergocrine Mesylate

Co-dergocrine mesylate is an ergot alkaloid, consisting of dihydroergocornin, dihydroergrocristin, dihydro-alpha-ergocryptin, and dihydro-beta-ergocryptin in the ratio of 3:3:2:1. This drug was approved by the FDA in 1981 for the treatment of cerebrovascular insufficiency.

1. Mode of Action

The components of co-dergocrine interact with alpha-adreno receptors without any significant effect on beta receptors.[49] The alkaloids also act as central serotonin receptor agonists.[50] There is evidence of interaction with dopamine receptors, particularly with D_2 receptors and postsynaptic D_1 receptors.[51] There are no direct actions on cholinergic receptors, but long-term treatment with co-dergocrine can enhance activity of choline acetyltransferase and increase the density of cholinergic receptors. This is probably related to co-dergocrine's effects on high-energy phosphates in the brain and on cerebral glucose utilization, especially in old animals.[55]

2. Clinical Effects

There have been more than 40 placebo-controlled, double-blind trials performed on patients with dementia receiving co-dergocrine mesylate. Dosages ranging from 3 to 6 mg/day were used, and the duration of treatment varied from 6 to 52 weeks.[56-61] In most of these clinical trials, the SCAG scale has been used to measure the outcome. The data show a consistent but moderate improvement of perception, cognition, motivation, and ADL in patients receiving co-dergocrine for 7 to 9 weeks. EEG evaluation in demented patients, treated with 3 to 4.5 mg co-dergocrine shows increased alpha activity and a decrease of slow EEG waves suggesting a direct influence on brain function.

Co-dergocrine mesylate is rapidly but incompletely (25%) absorbed after oral administration. Approximately 50% of the absorbed dose is degraded after first-path metabolism in the liver. The excretion of the drug is biliary. In older patients, plasma concentration of co-dergocrine mesylate can increase because of changes in absorption and diminished first path metabolism.

3. Side Effects

Only a very low incidence of side effects has been reported. The most common complaints are nausea, vomiting, and gastrointestinal discomfort. Long-term studies conducted for more than a 5-year period did not reveal any significant long-term adverse effects.[62]

C. Nicergoline

Nicergoline is a hydrated semisynthetic ergoline derivative with alpha-receptor antagonist properties. It has a property to cause vasodilatation and has been used for treatment of vascular or mixed forms of dementia.

1. Pharmacological Properties

Nicergoline acts as an antagonist on dopaminergic presynaptic receptors increasing dopaminergic neurotransmission.[63] In addition, nicergoline is thought to regulate glucose uptake and utilization and protein biosynthesis.[64] It reportedly blocks platelet aggregation and increases erythrocyte flexibility.[65,66]

2. Clinical Effects

Different studies of patients receiving nicergoline demonstrated an increase of alpha and beta EEG activities accompanied by a decrease in theta and delta waves which correlated with an improvement of attention.[67]

Approximately 20 double-blind trials investigating effects of nicergoline have been reported, but only a few of them fulfill modern criteria of antidementia drug testing. The degree of improvement in patients receiving nicergoline seems to be significant albeit moderate and was similar to that observed with co-dergocrine.[67-70]

Nicergoline shows a rapid and almost complete resorption from the gastrointestinal tract. In contrast to co-dergocrine, it is eliminated in urine. The improvement of symptoms in patients with dementia responding to nicergoline is slow and can be observed after approximately 12 weeks of treatment.

3. Side Effects

Side effects because of nicergoline treatment are rare and tend to decrease with prolonged treatment.[71] Erythema, heat sensation, tiredness, disturbances of sleep, and gastrointestinal discomfort were reported most frequently. Because of the alpha antagonistic effect of nicergoline, orthostatic hypotension is common.

D. Piracetam and Related Compounds

Piracetam (2-oxo-pyrrolidine-1-acetamid) is a derivative of cyclic gamma-aminobutyric acid. This drug has been used for treatment of patients suffering from AD, multiinfarct, or mixed forms of dementia. Other indications include stroke, cerebrocranial trauma, neurosurgery, alcohol withdrawal, and learning disabilities in children.

1. Mode of Action

Piracetam can influence cerebral glucose metabolism, enhance glucose oxidation, and diminish production of lactic acid, probably by stimulation of pentose-phosphate shunt.[72-75] Under hypoxic or ischemic conditions, piracetam reduces decreases in cerebral ATP levels and can stimulate the resynthesis of ATP.[76] It increases synthesis of neuronal membrane lipids such as phosphatidylinositol, phosphatidylcholine, and phosphatidylethanol amine.[77] Other reports suggest that piracetam can modulate neuronal activity by increasing cholinergic, noradrenergic, dopaminergic, and serotonergic synaptic transmission.[78-80] Piracetam increases erythrocyte flexibility and at the same time reduces aggregation and diminishes adhesion on vascular endothelium.[82] The effect of piracetam on learning and memory, as well as its neuroprotective properties, has been demonstrated in animal models.[83-84]

2. Clinical Effects

Several findings from animal experiments could be reproduced in clinical pharmacological studies showing that piracetam in patients with AD improves brain perfusion, increases glucose metabolism, and improves attention.[85-87] There are more than 25 reports showing clinically significant effects of piracetam in patients with dementia.[88] However, other reports show no changes in patients' condition following a prolonged treatment with the drug.[89] Apparently, piracetam is more effective in treating neurological and psychiatric consequences of stroke.[90] Efficacy of piracetam has been reported in treatment of postanoxic myoclonus syndrome.[91] The combined treatment of piracetam with neuroleptic drugs has been shown to shorten the duration of delirium and reduce alcohol withdrawal symptoms in patients with alcohol abuse.[92]

Piracetam shows an almost complete resorption after oral treatment, with a biological half-life of 5 h. There is a high passage through the blood-brain barrier with an accumulation of the substance in the cytoplasm of the neurons.[93] Piracetam is not metabolized and is eliminated in urine.

3. Side Effects

Piracetam is usually well tolerated. The most commonly reported side effect is agitation. There are rare reports of insomnia, depression, anxiety, weight gain, and increased libido.

4. Other 2-Oxo-Pyrrolidines

Several new cognition enhancers have been recently synthesized. Such substances as oxiracetam, aniracetam, nefiracetam, pramiracetam, and others are already in clinical phase III trials, and some of them (aniracetam) have already been registered in some countries. The clinical profile and mechanisms of action of these drugs are similar to those of piracetam. Aniracitam has been shown to reduce the rate of AMPA receptor desensitization[81] (see Chapter 7) suggesting that this class of drugs may have important interactions with glutamate-mediated neurotransmission.

E. Calcium Channel Blocker Nimodipine

Enhanced calcium influx and excessive intracellular accumulation of this ion seem to be involved in mechanisms of nerve cell death (see also Chapter 7 of this volume). The relationship of glutamate excitotoxicity and calcium is a well-established fact in the mechanism of hypoxic ischemic cell injury.[95] There is increasing evidence that alterations in calcium homeostasis also play a critical role in neuronal degeneration, associated with AD.[96-98] These findings provide a good rationale for the use of calcium channel antagonists in the treatment of ischemic as well as neurodegenerative disorders.

Nimodipine is a calcium channel blocker which belongs to the 1,4-dihydropyridine group. It has been registered in the U.S. since 1989 for treatment of neurological deficits associated with subarachnoidal hemorrhage and prevention of cerebral vasospasms. It has been approved in some other countries for treatment of impaired cognitive functions because of organic conditions, AD, and vascular dementias.

1. Pharmacological Properties

Nimodipine blocks L-type calcium channels relaxing smooth muscles of blood vessels, including cerebral arterioles,[99,100] significantly increasing cerebral blood flow.[101,102] It does not directly affect brain metabolism but appears to antagonize metabolic disturbances following ischemia or hypoxia.[103] It reduces the volume of ischemic lesions, reduces neurological deficits, and decreases mortality.[104] There are data indicating that nimodipine can prevent occurrence of behavioral and learning deficits following brain injury.[105,106] Nimodipine reportedly slowed age-dependent changes in behavior and antagonized performance decline.[108,109] It decreased blood viscosity by increasing red blood cell flexibility and deformability, probably by antagonizing erythrocyte calcium overload without affecting platelet aggregation.[110,111] However, some reports do not support these findings.[107]

2. Clinical Effects

Several open and placebo-controlled trials of nimodipine were performed on patients with subarachnoidal hemorrhage. There has been an improvement of neurological deficits and even a reduction in patient mortality.[112,113] There are reports of improved glucose and oxygen metabolism in stroke patients accompanied by significant improvement of their neurological status and reduced mortality.[114,115] On the other hand, several trials found no difference between placebo and nimodipine treatment. In these investigations, positive effects could only be demonstrated after analyses of subgroups, where major differences depending on severity of clinical symptoms, onset of treatment, and dosage could be shown.[116-117]

More recent publications have focused on the efficacy of nimodipine in treatment in AD and vascular dementia. Approximately 15 clinical trials have been done, some of them following the most recent guidelines of antidementia drug testing. Major outcome variables were clinical symptomatology, memory, concentration, objective cognitive performance, and clinical global impression.[118-120] All these trials show significant albeit minimal to moderate improvement in these patients.[121] Interestingly, it was found that nimodipine can be more effective in AD patients than in vascular dementia patients, suggesting that altered calcium homeostasis in these patients can be influenced by the drug.[122] Despite numerous investigations of nimodipine in treating dementia patients, its clinical effects are still unclear and need further investigation.[123]

Nimodipine is almost completely absorbed after oral administration. It shows a considerable first-path metabolism, and its bioavailability is about 13%. The plasma protein binding is over 95%. The drug is eliminated by the kidneys and biliary tract.

3. Side Effects

Among side effects the most common are hypotension, dizziness, and lightheadedness. Among very rare side effects are peripheral edema, skin rashes, and tachycardia. Other infrequent side effects include headache, nausea, sleeplessness, hyperactivity, irritability, and depression.

F.   Pentoxifylline

Pentoxifylline is a dimethylxantine derivative that has been approved in the U.S. for treatment of peripheral vascular disease and in European countries for treatment of disturbances of cerebral blood flow.[124,125]

1.   Mode of Action

Pentoxifylline increases blood element flexibility,[126-128,132] apparently by elevating ATP concentration in the cells.[129,130] It prevents platelet aggregation and adhesion on blood vessel walls[128] probably by stimulating prostocycline synthesis and secretion and by blocking synthesis of thromboxanes.[131] In animal models of cerebral ischemia, the drug reduces edema formation and decreases mortality.[133-134]

2.   Clinical Effects

There have been clinical trials investigating pentoxifylline's effects in patients suffering from vascular forms of dementia. Doses of 1200 mg/day were used over comparatively long periods of time (up to 1 year). Most studies report improvement of cognition and reduction of such symptoms as dizziness or headache.[135-137] The effectiveness of pentoxifylline in acute stroke patients remains unproven.[138] Pentoxifylline is absorbed almost completely after oral administration showing some first-path metabolism. Its metabolites are eliminated renally.

3.   Side Effects

Side effects are comparatively rare and tend to be mild. There have been reports of nausea, vomiting and diarrhea, headache, and dizziness. In rare cases flushing, tachycardia, hypotension, or angina pectoris occurred, particularly in patients receiving high doses of the drug.

G.   *Gingko biloba* Leaf Extracts

Extracts from the leaves of the *Gingko biloba* tree have been used therapeutically for centuries in China and western European countries. *Gingko* extracts are among the most widely prescribed drugs in Germany and France. The main indications for these drugs are peripheral vascular disease and symptoms usually occurring in the elderly population, such as difficulties of concentration and memory, absentmindedness, confusion, tiredness, depressive mood, dizziness. In many cases these symptoms may be related to the early stages of dementia.[139] The clinically used preparations of *Gingko* contain approximately 25% flavone glycosides and 6% terpenoids.[140]

1.   Pharmacological Properties

These extracts are able to influence erythrocyte flexibility and block red blood cell aggregation.[141] One of the components, gingkolide B, is an antagonist for platelet-aggregating factor (PAF).[142] PAF production is enhanced during episodes of cerebral ischemia and contributes to platelet aggregation, neutrophil degranulation, and production of toxic oxygen radicals.[143] The described PAF antagonist properties of *Gingko*, together with free oxygen radical scavenger function,[144,145] may be potentially useful against hypoxic and ischemic brain damage.[147,148] In addition, long-term treatment with *Gingko* preparations seems to increase muscarinic acetylcholine receptor density.[146]

2. Clinical Effects

There have been more than 40 clinical trials reported on the clinical efficacy of standardized *Gingko* extracts. Eight of these clinical trials followed modern guidelines of antidementia drug testing. Treatment periods varied from 6 weeks to 1 year. All eight trials concluded that improvements of cognitive function have been noted in patients receiving *Gingko* extracts. Specifically, clinically relevant improvement of concentration and memory and reduction of anxiety, dizziness, headache, and tinnitus were noted. The degree of improvement varied among the trials but was always reproducible.[149-152] So far, there has been only one clinical trial of *Gingko* extracts reporting negative results in patients suffering from vascular dementia.[153]

The bioavailability of gingkolides and bilobalides ranges between 70 and 80% after oral dosage.[140] The elimination seems to occur by exhalation and excretion in urine and feces.[154]

3. Side Effects

There have been no serious side effects reported in the literature. The most common side effects are mild gastrointestinal disturbances, headaches, and allergic skin reactions. Decrease in blood pressure and appearance of dizziness have been reported following parenteral administration of the drug.

H. Cerebrolysin

Cerebrolysin was developed in the 1950s, by using proteolytic degradation of lipid-free protein components from porcine brain. The aqueous solution ready for injection contains free amino acids and low molecular weight, biologically active peptides (MW < 10 kDa). It is free of pyrogenes and allergenes. The drug is registered in many European countries for treatment of AD and vascular dementia, acute stroke, head injuries, and disturbances of brain function following neurosurgery. Initially, it was believed that cerebrolysin effects were due to nutritive effects of the amino acids present in the solution. It turned out that biological activity is likely mediated by the peptides which possess neurotrophic activity similar to that of naturally occurring growth factors, such as nerve growth factor (NGF) and others.[155] The role of neurotrophic factors in neurodegenerative disorders has been discussed in the literature.[156-158]

1. Pharmacological Properties

Cerebrolysin enhances glucose breakdown, increases aerobic energy production, decreases lactic acid levels, and influences brain protein synthesis.[159,160] It can influence passive avoidance behavior in developing animals.[161] One important finding was the neurotrophic activity of cerebrolysin.[155] This finding has been subsequently reproduced by several other groups in experiments on tissue cultures and *in vivo* models.[162,163] Cerebrolysin can induce differentiation of nerve cells, increase long-term survival of neurons in primary tissue culture, and shows neuroprotective activity in different lesion models. In animal experiments it protects against ischemic brain damage, decreasing mortality by 50%, and shows significant activity in animal models of neurodegenerative diseases (e.g., fimbria fornix transection).[162,164] Most impressive is the ability of the drug to antagonize all lesion-dependent behavioral deficits almost completely.[165,166]

2. Clinical Efficacy

There have been more than 50 clinical trials reporting that cerebrolysin improved cognition, memory, concentration, and ADL. Seven of these trials meet the guidelines of modern nootropic drug testing in neurodegenerative as well as in vascular dementia.[167-169] A very recent trial in a selected population of patients suffering from senile dementia of Alzheimer's type showed a dramatic improvement in all investigated parameters, with an impressively fast onset of clinical efficacy, reaching statistical significance after 2 weeks of intravenous infusion treatment.[170] Continuation of treatment for an additional 4 weeks further improved patients' condition. A persistent improvement in cerebrolysin-treated subjects remained following the cessation of treatment for 6 months.

A major disadvantage of cerebrolysin is the fact that its peptide components have not been characterized yet, and virtually no information is available about their pharmacokinetics.[171]

3. Side Effects

Clinical trial data and experience in millions of patients treated so far indicate that cerebrolysin is a rather safe drug. The drug seems to have no potential to induce hypersensitivity reactions. In extremely rare cases, fever and headache were reported, and there were single cases of confusion, dizziness, and hot flushes. There is one report of a grand mal seizure in a patient suffering from epilepsy following cerebrolysin treatment.

I. Other Cognition Enhancers

1. Pyritinol

Pyritinol is a pyridoxine derivative which affects glucose and phospholipid metabolism and interacts with cholinergic neurotransmission. There have been several clinical trials investigating pyritinol in patients with dementia. The reported degree of improvement in cognitive functions seems to be mild to moderate but consistent. More recent clinical trials in patients with AD have shown that the drug may slow the progress of neurodegeneration.

2. MAO-B Inhibitors

Deprenyl is an MAO-B inhibitor synthesized 25 years ago. It has a beneficial therapeutic effect in patients with Parkinson's disease. The recent finding that MAO-B may be involved in neuronal degeneration by mediating oxygen free radical–induced cell damage generated a significant interest in a possible neuroprotective action of deprenyl. Several clinical trials in AD patients showed an improvement in the overall performance of L-deprenyl–treated subjects. The main effect seems to be a specific stimulation of motivation, attention, and increased performance in memory tasks and cognitive tasks,[172-173] although these data are inconclusive so far. An interesting possibility is the combination of deprenyl and cholinergic drugs.[173,174]

IV. FUTURE PERSPECTIVES

One conclusion that could be drawn from the data on cognition-enhancing drugs is that not a single substance is available at present that could cure a demented patient. However, given the dramatic impact of AD and other dementias on patients and society, even the smallest improvements, the most moderate reductions of symptoms, the shortest delays in disease progression must be seen as meaningful changes. It is evident that all drugs available today are most effective in patients with mild to moderate cognitive impairment. It is necessary, therefore, to develop new diagnostic procedures for early stages of dementia.

REFERENCES

1. Lilienfeld, D. E., Berl, D. P., Projected neurodegenerative disease mortality in the United States, 1990–2040, *Neuroepidemiology,* 12, 219, 1993.
2. Hafner, H., Epidemiology of Alzheimer's disease, in *Alzheimer's Disease. Epidemiology, Neuropathology, Neurochemistry and Clinics,* Maurer, K., Riederer, P., and Beckmann, H., Eds., Springer Verlag, New York, 1990, 23.
3. Jorm, A. F., Corten, A. E., Henderson, A. S., The prevalence of dementia: a quantitative integration of the literature, *Acta Psychiatr. Scand.,* 76, 465, 1987.
4. Hofman, A., Rocca, W. A., Brayne, C., et al., The prevalence of dementia in Europe: a collaborative study of 1980–1990 findings, *Int. J. Epidemiol.,* 20, 736, 1991.
5. Amaducci, L., Falcini, M., Lippi, A., Descriptive epidemiology and risk factors for Alzheimer's disease, *Acta Neurol. Scand.,* 139, 21, 1992.
6. Breteler, M. M. B., Claus, J. J., Van Duijn, C. M., Launer L. J., Hofman, A., Epidemiology of Alzheimer's disease, *Epidemiol. Rev.,* 14, 59, 1992.
7. Evans, D., Estimated prevalence of Alzheimer's disease in the United States, *Milbank Q.,* 68, 267, 1990.
8. Max, W., The economic impact of Alzheimer's disease, *Neurology,* 43, S6, 1993.
9. Rise, D. B., Fox, P. J., Max, W., Webber, P. A., Lindemann, D. A., Hauck, W. W., Segura, E., The economic burden of Alzheimer's disease care, *Data Watch-Health Aff.,* 1, 164, 1993.
10. American Psychiatric Association, *Diagnostic and Statistical Manual of Mental Disorders,* 4th Ed., American Psychiatric Association, Washington, D.C., 1994.
11. Copeland, J. R. M., McWilliam, C., Dewey, M. E., Forshaw, D., Shiwach, R., Abed, R., Muthu, M. F., Wood N., The early recognition of dementia in the elderly: a preliminary communication about a longidutinal study using the GMS-AGE-CAT package (community version), *Int. J. Geriatr. Psychol.,* 1, 63, 1986.
12. Roth, M., Tym, E., Mountjoy, C. Q., Huppert, F. A., Hendrie, H., Verma, S., Goddard, R., Camdex-A standardised instrument for the diagnosis of mental disorder in the elderly with special reference to the early detection of dementia, *Br. J. Psychiatry,* 149, 698, 1986.
13. Zaudig, M., Mittelhammer, J., Hiller, W., Bauls, A., Thora, C., Morvinigo, A., Mombour, W., Sidam — a structured interview for the diagnosis of dementia of the Alzheimer type, multiinfarct dementia and dementias of other etiology according to ICD 10 and DSM-III-R, *Psychol. Med.,* 21, 225, 1991.
14. Mohs, R. C., Rosan, W. G., Davis, K. L., The Alzheimer Disease Assessment Scale: an instrument of treatment efficacy, *Psychopharmacol. Bull.,* 19, 448, 1983.
15. Reisberg, P., Ferris, S. H., De Leon, M. J., Crook, T., The Global Deterioration Scale (GDS): an instrument for the assessment of primary degenerative dementia (PDD), *Am. J. Psychiatry,* 139, 1136, 1982.
16. Gottfries, C. G., Braine, G., Steen, G., A new rating scale for dementia syndromes, *Gerontology,* 28, 20, 1982.
17. Shader, R. I., Harmatz, J. S., Saltzman, C., A new scale for clinical assessment in geriatric populations: Sandoz-Clinical-Assessment-Geriatric (SCAG), *J. Am. Geriatr. Soc.,* 22, 107, 1974.
18. Folstein, M. F., Folstein, S. E., McHugh, P. R., Minimental State. A practical method for grading the cognitive state of patients for the clinician, *J. Psychiatr. Res.,* 12, 189, 1975.
19. Buschke, H., Selective Reminding for analyses of memory and learning, *J. Verbal Learning Verbal Behav.,* 12, 543, 1973.
20. Erzigkeit, H., *Syndromkurztest,* 2nd rev. ed., Vless Verlag, Ebersberg, 1986.
21. Jellinger, K., Danielczyk, W., Fischer, P., Gabriel, E., Clinico-pathological analysis of dementia disorders in the elderly, *J. Neurol. Sci.,* 95, 239, 1993.
22. McKhann, G., Drachman, D., Folstein, M., Katzmann, R., Price, D., Stadlan, E. N., Clinical diagnosis of Alzheimer's disease: report of the NINCDS-ADADA work group under the auspices of the Department of Health and Human Services Task Force on Alzheimer's Disease, *Neurology,* 34, 939, 1984.

23. Hachinsky, V. D., Iliff, L. D., Zihka, E., Cerebral blood flow in dementia, *Arch. Neurol.*, 32, 632, 1975.

24. Rosen, W. G., Terry, R. D., Fuld, B. A., Datzman, R., Peck, A., Pathological verification of ischemic score in differentiation of dementias, *Ann. Neurol.*, 7, 486, 1980.

25. Loeb, C., Gandolfo, C., Diagnostic evaluation of degenerative and vascular dementia, *Stroke*, 14, 399, 1983.

26. Giurgea, C. E., Vers une pharmacologie de l'activite integrative au servo. Dendative du concept nootrope n psychopharmacologie, *Actual Pharmacol.*, 25, 115, 1972.

27. Giurgea, C. E., The nootropic concept and its prospective implications, *Drug Def. Res.*, 2, 441, 1982.

28. Skondia, V., Criteria for the clinical development and classification of nootropic drugs, *Clin. Ther.*, 2, 316, 1979.

29. Schmidt, J., About pharmacology of nootropic drugs, *Beitr. Wirkstofforsch., Akad. Ind. Komplex Arzneimittelforsch.*, (Berlin), 17, 1984.

30. Sarter, M., Taking stock of cognition enhancers, *TIPS Rev.*, 12, 456, 1991.

31. Coyle, J. D., Price, D. L., De Long, M. R., Alzheimer's disease: a disorder of cortical cholinergic innervation, *Science*, 219, 1184, 1983.

32. Kopeman, M. D., The cholinergic neurotransmitter system in human memory and dementia, a review, *Q. J. Exp. Psychol.*, 38A, 535, 1986.

33. Katzman, R., Saitoh, D., Advances in Alzheimer's disease, *FASEB J.*, 5, 278, 1991.

34. Summers, W. K., Majorsky, L. V., Marsh, G. M., Tachiki, K., Kling, A., Oral tetrahydroaminoacridine in long-term treatment of senile dementia, Alzheimer type, *N. Engl. J. Med.*, 315, 1241, 1986.

35. Sherman, K. A., Messamore, E., Blood cholineesterase inhibition as a guide to the efficacy of putative therapies for Alzheimer's dementia: comparison of tacrine and physostigmine, In *Current Research in Alzheimer Therapy*, Giacobini, E., Becker, R. E., Eds., Taylor and Francis, New York, 1988, 73.

36. Adem, A., Mohammed, A., Nordberg, A., Winblad, B., Tetrahydroaminoacridine and some of its analogues: effects on the cholinergic system, In *Basic, Clinical and Therapeutic Aspects of Alzheimer's and Parkinson's Diseases*, Nagazu, D., Fisher, A., Yoshida, M., Eds., Plenum Press, New York, 1990, 2, 387.

37. Adem, A., Jossan, S. S., Oreland, L., Tetrahydroaminoacridine inhibits human and rat brain monoaminoxidase, *Neurosci. Lett.*, 107, 113, 1989.

38. Drukarch, B., Kits, K. S., Van der Meer, E. G., Lodder, J. C., Stoof, J. C., 9-Amino-1,2,3,4-tetrahydroacridine (THA), an alleged drug for the treatment of Alzheimer's disease, inhibits acetylcholinesterase activity and slow outward K+ current, *Eur. J. Pharmacol.*, 141, 153, 1987.

39. Rogawski, M. A., Tetrahydroaminoacridine blocks voltage-dependent ion channels in hippocampal neurons, *Eur. J. Pharmacol.*, 142, 169, 1987.

40. Butterfield, D. A., Hensley, K., Hall, N., Umhauer, S., Carney, J., Interaction of tacrine and velnacrine with neocortical synaptosomal membranes: relevant to Alzheimer's disease, *Neurochem. Res.*, 18, 989, 1993.

41. McMelley, W., Roth, M., Yang, R., Quantitative whole body autoradiographic determination of tacrine tissue distribution in rats, following intravenous or oral dose, *Pharm. Res.*, 6, 924, 1989.

42. Neilsen, J. A., Mena, E. E., Williams, I. H., Correlation of brain levels of 9-amino-1,2,3,4-detrohydroaminoacridine (DMA) as neurochemical and behavioral changes, *Eur. J. Pharmacol.*, 173, 53, 1989.

43. Schneider, L. S., Clinical pharmacology of aminoacridines in Alzheimer's disease, *Neurology*, 43, S64, 1993.

44. Eagger, S. A., Levy, R., Sahakian, N. B. J., Tacrine in Alzheimer's disease, *Lancet*, 337, 989, 1991.

45. Davis, K. L., Thal, L. J., Gamzu, E. R., Davis, C. S., et al., A double-blind placebo controlled multicenter study of tacrine for Alzheimer's disease, *N. Engl. J. Med.*, 327, 1253, 1992.

46. Knapp, M. J., Knopman, D. S., Solomon, P. R., Bendlebury, W. W., et al., A thirty-week randomized controlled trial of high-dose tacrine in patients with Alzheimer's disease, *J. Am. Med. Assoc.*, 271, 985, 1994.

47. Holford, N. H. G., Peace, K., Results and validation of a population pharmacodynamic model for cognitive effects in Alzheimer patients treated with tacrine, *Proc. Nat. Acad. Sci., U.S.A.*, 89, 11471, 1992.

48. Rothlin, E., Bruegger, J., Quantitative investigations about sympaticolytical effects of ergot alkaloids and their dehydro-derivatives on isolated uterus of rabbits, *Helv. Physiol. Acta*, 3, 519, 1945.

49. Goldstein, M., Ergot alkaloids and central monominergic receptors, *J. Pharmacol.*, 16, 19, 1985.

50. Markstein, R., Hydergine: interaction with neurotransmitter systems in the central nervous system, *J. Pharmacol.*, 16, 1, 1985.

51. Horowski, R., McDonald, R. J., Experimental and clinical aspects of ergoderivatives used in the treatment of age-related disorders, in *Aging Brain and Ergot Alkaloids*, Agnoli, A., Crepaldi, G., Spano, P. F., Trabucchi, M., Eds. Raven Press, New York, 1983, 283.

52. Dravid, A. R., Hiesdand, B., Deficit des activites enzymatiques cholinergiques dans les zones septo-temporales de l'hippocampe de rat age et dans le cerveau anterieur de souris agee. Action dutrait-ement par l'hydergine, *J. Pharmacol.* 16, 29, 1985.

53. Le Poncin-Lafitte, M., Rapin, J. R., Duterte, D., Learning and cholinergic neurotransmission in old animals: the effects of Hydergine, *J. Pharmacol.*, 16, 57, 1985.

54. Imperato, A., Obinu, M. C., Dazzi, L., Crata, G., Mascia, M. S., Casu, M. A., Gessa, G. L., Co-dergocrine (Hydergine) regulates striatal and hippocampal acetylcholine release through D2 re-ceptors, *Neuroreport*, 5, 674, 1994.

55. Maier-Ruge, W., Effects of prolonged co-dergocrine mesylate (Hydergine) treatment on local cere-bral glucose uptake in aged Fisher 344 rats. *Arch. Gerontol. Geriatr.*, 5, 65, 1986.

56. Matejcek, M., EEG of the aging brain. Contribution of quantitative electroencephalography to geriatric research: application with ergot-mesylate, in *Aging Brain and Ergot Alkaloids*, Agnoli, A., Crepaldi, G., Spano, P. F., Trabucchi, M., Eds., Raven Press, New York, 1983, 97.

57. Oswald, W. D., Mateijec, M., Lukaschek, K., About the relevance of psychometrically operationa-lised therapy effects in treatment of age-inventory, *Drug Res.*, 32, 584, 1982.

58. Fanchamps, A., Dihydroergotoxin in senile cerebral insufficiency, in *Aging Brain and Ergot Alkaloids*, Agnoli, A., Crepaldi, G., Spano, P. F., Trabucchi, M., Eds. Raven Press, New York, 1983, 311.

59. Chierchetti, S., Cucinotta, D., Santini, V., in *Aging 2000: Our Health Care Destiny*, Vol. I, Gaitz, C. M., Samorajski, D., Eds., Springer, New York, 1985, 377.

60. Thienhaus, O. J., Wheeler, B. G., Simon, S., A controlled double-blind study of high-dose dihydro-ergotoxin mesylate (Hydergine) in mild dementia, *J. Am. Geriatr. Soc.*, 35, 219, 1987.

61. Van Loveren-Huyben, D. M. B., Engelaar, H. F. W. J., Hermans, M. B. M., Van der Bom, J. A., Leer-ing, C., Munnichs, J. M. A., Double-blind clinical and psychological study of ergoloid mesylates (Hydergine) in subjects with senile mental deterioration, *J. Am. Geriatr. Soc.*, 32, 584, 1984.

62. Huber, F., Koberle, S., Prestele, H., Spiegel, R., Effects of long-term ergoloid mesylates (Hydergine) administration in healthy pensioners; 5 year results, *Curr. Med. Res. Opin.*, 10, 256, 1986.

63. Moretti, A., Carfagna, N., Caccia, C., Carpentieri, M., Effects of ergolines on neurotransmitter systems in the rat brain, *Arch. Int. Pharmacodyn.*, 294, 33, 1988.

64. Le Poncin-Lafitte, N., Grosdemouge, C., Billon, C., Rapin, J. R., Nicergoline: the oxyglucose uptake and cerebral blood flow under hypoxia, in *Pharmacologie Experimentale et Clinique*, Les Alpha-Bloquants, Masson, Paris, 1981, 321.

65. Bolli, R., Ware, J. A., Brandon, B. A., Weilbaecher, D. G., Maze, N. L., Platelet-mediated thromboses in stenosed canine coronary arteries: inhibition by nicergoline, a platelet-active alpha-adrenergic antagonist, *J. Am. Coll. Cardiol.*, 3, 1417, 1984.

66. Kuzuya, F., Hayakawa, M., Einfluss von nicergolin auf die thrombozytenaggregation und eryth-rozytenverformbarkeit, in *Therapeutische wirksamkeitsnachweise beinootropen und vasoakitiven substan-zen: Fortschritte in der klinischen und experimentellen nicergolin-forschung*, Heidrich, H., Ed. Springer, New York, 1985, 193.

67. Saletu, B., Anderer, P., Grunberger, J., Dopographic brain mapping of EEG after acute application of ergot alkaloids in the elderly, *Arch. Gerontol. Geriatr.*, 11, 1, 1990.

68. Nicergoline Cooperative Study Group, A double-blind randomised study of two ergot derivatives in mild to moderate dementia, *Curr. Ther. Res.*, 48, 597, 1990.

69. Ladurner, G., Erhart, P., Erhart, C., Scheiber, V., Therapy of the organic brain syndrome by nicer-goline administration in a single daily dose regimen, *Wien. Klin. Wochenschr.*, 103, 8, 1991.

70. Borromei, A., Adjuvant therapy for Parkinson's disease: preliminary data of a multicenter study with nicergoline, *Funct. Neurol.*, 4, 63, 1989.

71. Saletu, P., Codergocrine type drugs in the pharmacotherapy of mental and cognitive disorders in aging, in Proceedings of the 5th World Congress of Biological Psychiatry, Florence, June, *Exc. Med. Int. Cong. Ser.*, 1991.

72. Ostrovskaya, R. U., Hoffmann, W., Molodwakin, G. M., Untersuchungen zur antihypoxischen wirk-samkeit von disochromid (16-244) im vergleich zu piracetam, *Pharmacie*, 38, 251, 1983.

73. Burnotte, R. E., Gobert, J. G., Temmermann, J. J., Piracetam (2-pyrrolidinonacetamid) induced mod-ifications of the brain polyribosome pattern in aging rats, *Biochem. Pharmacol.*, 22, 811, 1973.

74. Domanska-Janik, K., Zaleska, M., The action of piracetam on 14C glucose metabolism in normal and posthypoxic cerebral cortex slices, *Pol. J. Pharmacol. Pharm.*, 29, 111, 1977.

75. Bonifaci, J. F., Ferran, C., Laborit, H., Actions du piracetam sur certeaines etaps du metabolism du servo de rat, *Agressologie*, 23, 127, 1982.

76. Benzi, G., Passdoris, O., Villa, R. R., Guffrida, A. M., Influences of aging and exogenous substances on cerebral energy metabolism in postglycemic recovery, *Biochem. Pharmacol.*, 34, 1477, 1985.

77. Pellegata, R., Cyclic GABA-GABOB analogues. III. Synthesis and biochemical activity of new alkyl and acyl-derivatives of 4-hydroxy-2-pyrrolidinone, *June Farmaco (ed. sci.)*, 36, 845, 1981.

78. Wurtmann, R. J., Magil, B. C., Reinstein, D. K., Piracetam diminishes hippocampal acetylcholine levels in rats, *Life Sci.*, 28, 1091, 1981.

79. Apud, J. A., Masotto, C., Racagni, G., New perspectives in the mechanisms of action of nootropic drugs, in 11th Int. Congress on Nootropic Drugs and Organic Brain Syndrome, Rome, 48, 14, 10, 1983.

80. Peckov, V. D., Grahovska, D., Peckov, V. V., Constantinova, E., Stancheva, S., Changes in the brain bioorganic monoamines of rats induced by piracetam and aniracetam, *Acta Physiol. Pharmacol. Bulgarica*, 10, 6, 1984.

81. Behring, B., Muller, W. E., The interaction of piracetam with several neurotransmitter receptors in the central nervous system. The relative specificity for H-glutamate sites, *Drug Res.*, 35,1350, 1985.

82. Fleischmann, J. A., Piracetam effect on sickle and diabetic erythrocytes as demonstrated in enucleated human eye trabecular meshwork filterability studies: clinical implications, 10th Int. Symposium on Nootropic Agents, Paris, October 1982, 61, 1982.

83. Bryand, R. C., Patty, F., Byrne, W. L., Effects of piracetam (SKF 38462) on acquisition, retention and activity in the goldfish, *Psychopharmacologica*, 29, 121, 1973.

84. Buresova, O., Bures, J., Piracetam induced facilitation of interhemispheric transfer of visual information in rats, *Psychopharmacologia*, 46, 93, 1976.

85. Herrschaft, H., Die wirkung von piracetam auf die gehirndurchblutung des menschen. Quantitative regionale hirndurchblutungsmessungen bei der akuten cerebralen ischamie, *Med. Klin.*, 73, 1995, 1978.

86. Heiss, W.-D., Hebold, D., Klinkhammer, P., Ziffling, P., Szelies, B., Pawlik, G., Herholz, K., Effect of piracetam on cerebral glucose metabolism in Alzheimer's disease as measured by positron emission tomography, *J. Cereb. Blood Flow Metab.*, 8, 613, 1988.

87. Herrschaft, H., Die wirksamkeit von piracetam bei der akuten cerebralen ischamie des menschen. Klinisch kontrollierte doppelblindstudie piracetam/10% dextran 40 versus placebo/10% dextran 40, *Med. Klin.*, 83, 667, 1988.

88. Fleischmann, U. M., Wirkeffekte nootroper substanzen bei alterspatienten — eine sekundaranalyse am beispiel von piracetam, *Z. Gerontol. Psychiatrie*, 4, 285, 1990.

89. Croisile, B., Trillet, M., Vondarai, J., Laurenz, B., Mauguiere, F., Billarton, M., Long-term and high-dose piracetam treatment of Alzheimer's disease, *Neurology*, 43, 301, 1993.

90. Platt, D., Horn, J., Summa, J.-D., Schmitt-Ruth, R., Kauntz, J., Kronert, E., On the efficacy of piracetam in geriatric patients with acute cerebral ischemia: a clinically controlled double-blind study, *Arch. Gerontol. Geriatr.*, 16, 149, 1993.

91. Peuvot, J., Piracetam in the treatment of myoclonus, in *Piracetam, 5 Year Progress in Pharmacology and Clinics*, Int. Symposium, Athens, 1990, Ciencia Y Medicina, Madrid, 101, 1990.

92. Snel, H., Lehman, E., Velikomja, M., Piracetam in der Behandlung des alkoholbedingten Delirs, *Münch. Med. Wochenschr.*, 42, 947, 1983.

93. Ostrowski, J., Keil, M., Autoradiographische untersuchungen zur verteilung von 14-C-piracetam im affengehirn, *Drug Res.*, 28, 29, 1978.

94. Reifenrath, M., Kaltenbach, T., Buck, W., Vorteile einer parenteralen initialtherapie mit piracetam, *Psycho.*, 16, 158, 1990.

95. Ginsberg, M. D., Lim, B., Morikawa, E., Dietrich, W. D., Busto, R., Globus, M. Y.-T., Calcium antagonists in the treatment of experimental cerebral ischemia, *Drug Res.*, 41, 334, 1991.

96. Landfield, B.W., Calcium homeostasis in brain aging and Alzheimer's Disease, in *Diagnosis and Treatment of Senile Dementia*, Bergener, M., Reisberg, P., Eds., Springer-Verlag, Berlin, 1989, 276.

97. Peal, D. P., Gajdusek, D. C., Garruto, R. M., Yanagihara, R. T., Aluminum accumulation in amyotrophic lateral sclerosis in parkinsonism dementia of Guam, *Science*, 217, 1053, 1982.

98. Iacopino, A. M., Christakos, S., Specific reduction of calcium-binding (28-kD Calbindin-D) gene expression in aging and neurodegenerative diseases, *Proc. Nat. Acad. Sci., U.S.A.*, 87, 4078, 1990.

99. Van Zwieten, B. A., Differentiation of calcium entry blockers into calcium channel blockers and calcium overload blockers, *Eur. Neurol.*, 25, 57, 1986.

100. Schmidli, J., Santilann, G. G., Saeed, M., Palmieri, D., Bing, R. J., The effects of nimodipine, a calcium antagonist, on intracortical arterioles in the cat brain, *Curr. Ther. Res.*, 38, 94, 1985.

101. Auer, L. M., Oberhauser, R. W., Schalk, H. V., Human pial vascular reaction to intravenous nimodipine infusion during EC-IC bypass surgery, *Stroke*, 14, 210, 1983.

102. Gaab, N. R., Haubitz, I., Brawanski, A., Korn, A., Czech, Th., Acute effects of nimodipine on the cerebral blood flow and intracranial pressure, *Neurochirurgia*, 28, 93, 1985.

103. Bielenberg, G. W., Beck, D., Saver, D., Burniol, M., Krieglstein, J., Effects of cerebroprotective agents on cerebral blood flow and on postischemic energy metabolism in the rat brain, *J. Cereb. Blood Flow Metab.*, 7, 480, 1987.

104. Mohammed, A. A., Gotoh, O., Graham, D. I., Osborne, H. A., McCullough, J., Effect of pretreatment with a calcium antagonist nimodipine on local cerebral blood flow and histopathology after middle cerebral artery occlusion, *Ann. Neurol.*, 18, 705, 1985.

105. LeVere, T. E., Recovery of function after brain damage: the effects of nimodipine on the chronic behavioral deficit, *Psychobiology*, 21, 125, 1993.

106. Bannon, A. W., McMonagle-Strucko, K., Fanelli, R. J., Nimodipine prevents medial septal lesion induced performance deficits in the Morris water maze, *Psychobiology*, 21, 209, 1993.

107. Barneett, G. H., Bose, B., Little, J. R., Jones, S. C., Friel, H. T., Effects of nimodipine on acute focal cerebral ischemia, *Stroke*, 17, 886, 1986.

108. Schuurmann, T., Raaba, J., Old rats as an animal model of senile dementia: behavioral effects of nimodipine, in *Diagnosis and Treatment of Senile Dementia*, Bergener, N., Reisberg, B., Eds., Springer, New York, 1989, 295.

109. Nomura, M., Effect of nimodipine on brightness discrimination learning tests in Wistar rats and continuously hypertensive rats, *Drug Res.*, 38, 1282, 1988.

110. Forconi, S., Guerrini, N., Cappelli, N., Bicchi, M., Von Tani, R., Trabalzini, L., Vattimo, A., Bertelli, P., Burroni, L., Effects of treatment with a calcium entry-blocker nimodipine on the cerebral blood flow (SPECT) and blood viscosity of patients effected by cerebral vascular insufficiency, *Clin. Hemorrheology*, 12, 697, 1992.

111. Feinberg, W. M., Bruck, D. C., Effect of oral nimodipine on platelet function, *Stroke*, 24, 10, 1993.

112. Ulrich, G., Zur wirkung von nimodipine auf die topische verteilung der absoluten alphaleistung im EEG, sowie die aktuelle befindlichkeit gesunder probanden, *Drug Res.*, 37, 541, 1987.

113. Pickard, J. D., Murray, G. D., Illingworth, R., Shaw, M. D. M., Teasdale, G. M., Effect of oral nimodipine on cerebral infarction and outcome after subarachnoid hemorrhage: British aneurysm nimodipine trial, *Br. Med. J.*, 298, 636, 1989.

114. Holthoff, V. A., Heiss, W.-D., Pawlik, G., Neveling, M., Positron emission tomography in nimodipine treated patients with acute ischemic stroke, in *Nimodipine*, Scriabine, A., Teasdale, G. M., Tetenborn, D., Young, W., Eds. Springer, New York, 1991, 133.

115. Gelmers, H. J., The effects of nimodipine on the clinical course of patients with acute ischemic stroke, *Acta Neurol. Scand.*, 69, 232, 1984.

116. Kramer, G., Tetenborn, D., Rodacher, G., Hacke, W., Busse, O., Hornig, C. R., Aichner, F., Ladurner, G., Nimodipine German-Austrian Stroke Trial, *Neurology*, 40, 415, 1990.

117. The American Nimodipine Study Group, Clinical trial of nimodipine in acute ischemic stroke, *Stroke*, 23, 3, 1992.

118. Kanowski, S., Fischlof, P., Hiersemenzel, R., Rohmel, J., Kern, U., Wirksamkeitsnachweis von nootropika am beispiel von nimodipin — ein beitrag zur entwicklung geeigneter klinischer prufmodelle, *Z. Gerontopsychol. Gerontopsychiatrie*, 1, 35, 1988.

119. Tollefson, G. D., Short term effects of the calcium channel blocker nimodipine (Bay-e-9736) in the management of primary degenerative dementia, *Biol. Psychiatry*, 27, 1132, 1990.

120. Schmage, N., Boehme, K., Dycka, J., Schmitz, H., Nimodipine for psychogeriatric use: methods, strategies, and considerations based on the experience with clinical trials, in *Diagnosis and Treatment of Senile Dementia*, Bergener, M., Reisberg, B., Eds., Springer, New York, 1989, 374.

121. Grobe-Einsler, R., Clinical aspects of nimodipine, *Clin. Neuropharmacol.*, 16, 39, 1993.

122. Fischhof, P. D., Divergent neuroprotective effects of nimodipine in PDD and MID provide indirect evidence of disturbances in calcium homeostasis in dementia, *Med. Find. Exp. Clin. Pharmacol.*, 15, 549, 1993.

123. Jarvik, L. F., Calcium channel blocker nimodipine for primary degenerative dementia, *Biol. Psychiatry*, 30, 1171, 1991.

124. Grotta, J., Ackerman, R., Correia, J., Fallick, G., Zhang, J., Whole blood viscosity parameters and cerebral blood flow, *Stroke*, 13, 296, 1982.

125. Tohgi, H., Uchimaya, S., Ogawa, M., Tabuchi, M., Nagura, H., Yamanouchi, H., Matsuda, T., The role of blood constituents in the pathogenesis of cerebral infarction, *Acta Neurol. Scand.*, 72, 616, 1979.

126. Seiffge, D., Kiesewetter, H., Filterability investigations with red blood cell (RBC) suspensions: effects of different blood components and pentoxifylline on RBC flow rate, *Ric. Clin. Lab.*, 11, 117, 1981.

127. Angelkort, B., Thrombozytenfunktion, plasmatische blutgerinnung und fibrinolyse bei chronish arterieller verschlusskrankheit, *Med. Welt.*, 30, 1239, 1979.

128. Ott, E., Fazekas, F., Lechner, H., Hemorrheological effects of pentoxifylline in disturbed blood flow behavior in patients with cerebrovascular disease, *Eur. Neurology*, 22, 105, 1983.

129. Stefanovich, V., John, J. B., Influence of pentoxifylline on brain metabolism of normal and anoxic rats, *Drug Res.*, 28, 2097, 1978.

130. Le Comte, M. C., Boivin, P., Effects of methylxanthine derivatives on red cell phosphorylation, *Scand. J. Clin. Lab. Invest.*, 41, 291, 1981.

131. Pohanka, E., Sinzinger, H., Effects of a single pentoxifylline administration on platelet sensitivity, plasma factor activity, plasma 6 oxo-PGF1 and thromboxan B2 in healthy volunteers, *Prostagland. Leukot. Med.*, 22, 191, 1986.

132. Del Zoppo, G. J., Schmidt-Schonbein, G. W., Mori, E., Copeland, B. R., Chang, C. H. M., Polymorphonuclear leukocytes occlude capillaries following middle cerebral artery occlusion and reperfusion in baboons, *Stroke*, 22, 1276, 1991.

133. Hartmann, J. F., Becker, R. A., Cohen, M. N., Effects of pentoxifylline on cerebral ultrastructure of normal and ischemic gerbils, *Neurology*, 27, 77, 1977.

134. Bluhm, R. E., Molnar, J., Cohen, M. N., The effect of pentoxifylline in the energy metabolism of ischemic gerbil brain, *Clin. Neuropharmacol.*, 8, 180, 1985.

135. Harwart, D., The treatment of chronic cerebrovascular insufficiency. A double-blind study with pentoxifylline ("Trental 400"), *Curr. Med. Res. Opin.*, 6, 73, 1979.

136. Parnetti, L., Ciufetti, G., Mercuri, M., et al., The role of hemorrheological factors in the aging brain. Long-term therapy with pentoxifylline ("Trental 400") in elderly patients with initial mental deterioration, *Pharmatherapeutica*, 4, 617, 1986.

137. Black, R. S., Barclay, L. L., Nolan, K. A., Thaler, H. D., Hardiman, S. D., Blass, J. P., Pentoxifylline in cerebral vascular dementia, *J. Am. Geriatr. Soc.*, 40, 237, 1992.

138. Hsu, C. Y., Morris, J. W., Hogan, E. L., Bladin, P., Dinsdale, H. B., Yatsu, F. M., Ernest, M. P., et al., Pentoxifylline in acute nonhemorrhagic stroke, a randomized, placebo-controlled double-blind trial, *Stroke*, 19, 716, 1988.

139. Warburton, D. M., *Gingko biloba* extracts and cognitive decline, *Br. J. Clin. Pharmacol.*, 36, 137, 1993.

140. Kleijnen, J., Knipschild, P., *Gingko biloba*, *Lancet*, 340, 1136, 1992.

141. Ernst, E., Matrai, A., Hamorheologische *in vitro* effekte von *Gingko biloba*, *Herz/Kreislauf*, 18, 258, 1986.

142. Klein, P., Untersuchung uber die Hemmwirkung von *Gingko biloba*-Extrakt, PAF-induzierte thrombozytenaggregation *in vitro* und *in vivo* und spontanaaggregation — untersuchung mit tebonin, *Therapiewoche*, 38, 2379, 1988.

143. Braquet, P., Hosford, D., Ethnopharmacology and the developement of natural PAF antagonists as therapeutic agents, *J. Ethnopharmacol.*, 32, 135, 1991.

144. Oyama, J., Ueha, P., Hayashi, A., Chikahisa, L., Noda, K., Flow cytometric estimation of the effects of *Gingko biloba* extracts on the content of hydrogen peroxides in dissociated mammalian brain neurons, *Jpn. J. Pharmacol.*, 60, 385, 1992.

145. Barth, S. A., Inselmann, G., Engemann, R., Heidemann, H. P., Influences of *Gingko biloba* on cyclosporine A induced lipid peroxidation in human liver microsomes in comparison to vitamine E, glutathione and nacetylcystein, *Biochem. Pharmacol.*, 41, 1521, 1991.

146. Taylor, J. E., Liaison des neuronmediateurs a leurs recepteurs dans le cerveau des rats, *Presse Med.*, 15, 1491, 1986.

147. Winter, E., Effects of an extract of *Gingko biloba* on learning and memory in mice, *Pharmacol. Biochem. Behav.*, 38, 109, 1991.

148. Krieglstein, J., Heuer, H., Uber die brauchbarkeit eines modells der akuten hypoxie zur testung cerebroprotektiver pharmaka, *Drug Res.*, 36, 1568, 1986.

149. Eckmann, F. Hirnleistungsstorungen — Behandlung mit *Gingko biloba*-Extrakt, *Fortschr. Med.*, 108, 557, 1990.

150. Wesnes, K., Simons, D., Rook, M., Simpson, B., A double-blind placebo controlled trial of tanakan in the treatment of ideopathic cognitive impairment in the elderly, *Hum. Psychopharmacol.*, 2, 159, 1987.

151. Vorberg, G., Schenk, N., Schmidt, U., Wirksamkeit eines neuen *Gingko biloba*-Extrakts bei 100 patienten mit cerebraler insuffizienz, *Herz. Gefabe*, 9, 936, 1989.

152. Bruchert, E., Heinrich, S. E., Ruf-Kohler, P., Wirksamkeit von LI 1370 bei alteren patienten mit hirnleistungsschwache. Multizentrische doppelblindstudie des fachverbands deutscher allgemeinarzte, *Münch. Med. Wochenschr.*, 133, S9, 1991.

153. Hartmann, A., Frick, M., Wirkung eines *Gingko*-spezialextraktes auf psychometrische parameter bei patienten mit vaskularbedingter demenz, *Münch. Med. Wochenschr.*, 133, S23, 1991.

154. Moreau, J. P., Eck, J., McCabe, J., Skinner, S., Absorption, distribution et elimination de l'extrait marque de feuilles de *Gingko biloba* chez le rat, *Presse Med.*, 15, 1458, 1986.

155. Shimazu, S., Iwamoto, N., Itoh, P., Akasako, A., Seki, H., Fujimoto, M., Neurotrophic activity of cerebrolysin, 2nd Int. Springfield Symposiums on Advantage in Alzheimer Therapy, Springfield, 51, 1991.

156. Pepeu, G., Ballerini, L, Pugliese A. M., Neurotrophic factors in the aging brain: a review, *Arch. Gerontol. Geriatr.*, 2, 151, 1991.

157. Vantini, G., Skaper, S. D., Neurotrophic factors: from physiology to pharmacology, *Pharmacol. Res.*, 26, 1, 1992.

158. Saffran, N. B., Should intracerebroventricular nerve growth factor be used to treat Alzheimer's disease? *Perspect. Biol. Med.,* 35, 471, 1992.

159. Windisch, M., Piswanger, A., Influence of 7 days treatment with cerebrolysin on respiratory activity of brain mitochondria, *Neuropsychiatrie,* 1, 83, 1987.

160. Piswanger, A., Paier, B., Windisch, M., Modulation of protein synthesis in a cell-free system from rat brain by cerebrolysin during development and aging, in *Aminoacids: Chemistry, Biology and Medicine,* Lubec, G., Rosenthal, G. A., Eds., Escom Science Publishers, Amsterdam, 1990, 651.

161. Paier, B., Eggenreich, U., Windisch, M., Postnatal administration of 2 peptides solutions affects passive avoidance behavior of young rats, *Behav. Brain Res.,* 51, 23, 1992.

162. Akai, F., Hiruma, S., Satoh, P., Iwamoto, M., Fujimoto, M., Ioku, M., Hashimoto, S., Neurotrophic factor like effect of FPF 1070 on septal cholinergic neurons after transections of fimbria fornix, *Histol. Histopathol.,* 7, 213, 1992.

163. Albrecht, E., Hingel, S., Crailsheim, K., Windisch, M., The effects of cerebrolysin on the vitality and sprouting of neurons from hemispheres and from the brain stem of chicken embryos *in vitro,* in *Alzheimer's Disease and Related Disorders,* Nicolini, M., Zatta, B. F., Coraine, B., Eds., Pergamon Press, Oxford, 1993, 341.

164. Windisch, M., Gschanes, A., Valouskova, V., The effects of different growth factors in rats after lesions of the somatosensoric cortex, *Eur. J. Neurosci.,* Suppl. 6, 542, 1993.

165. Francis-Turner, L., Zinjuk, L., Valouskova, V., The effect of NGF and cerebrolysin on spatial memory after unilateral fimbria fornix lesion, abstracts of the 17th Annual Meeting of the European Neuroscience Association, *Eur. J. Neurosci.,* Suppl. 7, 24.08, 1994.

166. Valouskova, V., Gschanes, A., Windisch, M., The influence of NGF, I-FGF and cerebrolysin on spatial navigation after bilateral lesions of the sensori-motor cortex, abstract of the 17th Annual Meeting of the European Neuroscience Association, *Eur. J. Neurosci.,* Suppl. 7, 55.14, 1994.

167. Vereschagin, M. V., Lebedeva, N. V., Suslina, Z. A., Mild forms of multiinfarct dementia: effectiveness of cerebrolysin, *Sowjetskaja Med.,* 11, 6, 1991.

168. Suchanek-Frohlich, H., Wunderlich, E., Randomised double-blind placebo controlled study with an amino acid peptide extract, *Prakt. Arzt,* 11, 1027, 1987.

169. Kofler, B., Erhart, C., Erhart, P., Harrer, G., A multidimensional approach testing nootropic drug effects (cerebrolysin), *Arch. Gerontol. Geriatr.,* 10, 129, 1990.

170. Ruther, E., Ritter, R., Apecechea, M., Freytag, S., Windisch, M., Efficacy of the peptidergic nootropic drug cerebrolysin in patients with senile dementia of the Alzheimer type (SDAT), *Pharmacopsychiatry,* 27, 32, 1994.

171. Ono, T., Takahashi, M., Nakamura, Y., Report on phase-I-clinical trial of FPF 1070 (Cerebrolysin) injection — 1 time administration and repetitive administration test, *Jpn. Pharmacol. Ther.,* 20, 4, 1992.

172. Agnoli, A., Martucci, N., Fabbrini, G., Buckley, A. E., Fioravanti, M., Monoamine oxidase and dementia: treatment with an inhibitor of MAO-B activity, *Dementia,* 1, 31, 1992.

173. Schneider, L. S., Olin, J. T., Pawluczyk, S., A double-blind cross over study of l-deprenyl (selegiline) combined with cholinesterase inhibitors in Alzheimer's disease, *Am. J. Psychiatry,* 150, 321, 1993.

174. Sunderland, T., Molchan S., Lawlor, B., A strategy of "combination chemo-therapy" in Alzheimer's disease: rational and preliminary results with physostigmine plus deprenyl, *Int. Psychogeriatry,* 4, Suppl. 2, 291, 1992.

INDEX

INDEX

O

P